Meat Animals
Growth and Productivity

NATO ADVANCED STUDY INSTITUTES SERIES

A series of edited volumes comprising multifaceted studies of contemporary scientific issues by some of the best scientific minds in the world, assembled in cooperation with NATO Scientific Affairs Division.

The series is published by an international board of publishers in conjunction with NATO Scientific Affairs Division

A	Life Sciences	Plenum Publishing Corporation
B	Physics	New York and London
C	Mathematical and Physical Sciences	D. Reidel Publishing Company Dordrecht and Boston
D	Behavioral and Social Sciences	Sijthoff International Publishing Company Leiden
E	Applied Sciences	Noordhoff International Publishing Leiden

Meat Animals
Growth and Productivity

Edited by

D. Lister and D. N. Rhodes

Meat Research Institute
Langford, Bristol, England

and

V. R. Fowler and M. F. Fuller

Rowett Research Institute
Bucksburn, Aberdeen, Scotland

PLENUM PRESS • NEW YORK AND LONDON
Published in cooperation with NATO Scientific Affairs Division

Library of Congress Cataloging in Publication Data

Symposium on Growth and Productivity of Meat Animals, Prestbury, Eng., 1974.
  Meat animals: growth and productivity.

  (NATO advanced study institutes series: Series A, Life sciences; v. 8)
  Includes index.
  1. Stock and stock-breeding—Congresses. 2. Feed utilization efficiency—Congresses.
3. Meat—Congresses. I. Lister, D. II. Title. III. Series.
SF5.S88 1974   636.08'83   76-985

Proceedings of a Symposium on Growth and Productivity of Meat Animals
held at Mottram Hall, Prestbury, Cheshire, England, December 2-6, 1974

©1976 Plenum Press, New York
Softcover reprint of the hardcover 1st edition 1976

A Division of Plenum Publishing Corporation
227 West 17th Street, New York, N.Y. 10011

United Kingdom edition published by Plenum Press, London
A Division of Plenum Publishing Company, Ltd.
Davis House (4th Floor), 8 Scrubs Lane, Harlesden, London, NW10 6SE, England

ISBN 978-1-4615-8905-1   ISBN 978-1-4615-8903-7 (eBook)
DOI 10.1007/978-1-4615-8903-7

Acknowledgments

This Meeting was sponsored by the North Atlantic Treaty Organisation, Scientific Affairs Division, as one of their Advanced Study Institutes. Additional support, for which we offer our grateful thanks, came from The Frank Parkinson Agricultural Trust, the Meat and Livestock Commission and the Underwood Fund of the Agricultural Research Council.

A number of colleagues and associates gave freely of their time and expertise to assist in the organisation of the Meeting and the preparation of this volume and it would be impossible to name them all. However, a few must be singled out for special thanks. The Chairmen of the Sessions: Professor J. Kielanowski, Dr. R. W. Pomeroy, Professor J. A. F. Rook, Professor G. E. Lamming, Professor I. A. M. Lucas, Professor Alan Robertson, Professor F. W. H. Elsley, Dr. J. W. B. King, Professor C. R. W. Spedding and Professor M. Ingram, exercised admirable control, tact and judgement in the face of, at times, overwhelming odds. We should like to thank Drs. B. N. Perry and J. D. Wood for their part in the organisation and in the making of the permanent record of the proceedings, and Mrs. Eileen Rhodes for taking excellent care of the domestic arrangements and registration. We should also like to record our appreciation for the hospitality accorded us by Sir Bernard Lovell and his staff at the University of Manchester's Radio Telescope complex at Jodrell Bank.

Mrs. A. J. Smith and Miss Davina J. Bicknell were responsible for typing the camera-ready manuscripts. We offer them our warmest thanks for their careful and willing contribution. Mrs. G. A. Garton assisted with the preparation of the index.

Finally we extend our appreciation to the Directors of the Meat Research Institute and the Rowett Research Institute for their interest and support.

We should like to thank the following for permission to reproduce certain of the illustrations which appear in this book:

The Institute of Dialect and Folk Life Studies, The University of Leeds, for Fig. 1, p. 277; The Athlone Press, for Fig. 9, p. 314; Human Biology, for Figs. 1, 2 and 7 on pp. 329, 330, 337; The Editor and Archives of Disease in Childhood, for Fig. 5, p. 335; Science, for Figs. 9 and 10, pp. 345, 346 (Copyright 1974 by the American Association for the Advancement of Science); Social Biology, for Fig. 11, p. 349; and John Wiley & Sons, for Table 2, p. 332.

Preface

Dramatic shortfalls in crop production in various regions of the world have led some people to question the relatively inefficient use of cereal grains for feeding meat animals instead of their direct use for human food. There is no doubt, however, that meat offers a nutritionally valuable, attractive and widely accepted food, the world demand for which increases daily. Thus it is not enough simply to condemn the consumption of meat as an irresponsible extravagance; rather it is preferable to examine how the demand for meat can be met most efficiently and effectively, which requires a fundamental enquiry into how meat is 'grown'. The importance of fat, for instance, both to the growing animal and to the consumer, needs to be established in view of the 'expense' involved in its deposition by the animal and the extent to which it is discarded at many points in the chain from the slaughterhouse to the consumer.

We were aware that there existed a wealth of information on the physiology of growth which, because of its having been collected as part of investigations in many other disciplines and the inevitable communication gap, had not been incorporated into the science of animal production. Similarly there were principles and techniques of animal husbandry which, if known in other disciplines, might enable more pertinent questions to be asked. The biochemical and physiological pathways by which animals utilise feed to produce body protein, fat and other components are intriguing problems which are receiving considerable attention. We were concerned, however, about our ability to identify and use to advantage any potentially important physiological capabilities which might exist within and between animal species.

Our meeting, which was held from 2nd-6th December 1974, at Mottram Hall, Prestbury, Cheshire, England, brought together expertise from a wide range of disciplines to promote the interchange and pooling of ideas about the relevant topics which could offer novel foci for fundamental research and identify bases for the development of new technologies in animal production.

Contents

Welcome

WELCOME

Professor J. R. Norris

Agricultural Research Council
Meat Research Institute
Langford, Bristol

It is a pleasure for me to welcome you all to this meeting
and I would like to take the opportunity of making a few brief
remarks by way of introduction.

Firstly, I would like to thank all of you for agreeing to
participate in the meeting and, in many cases, for travelling very
large distances to do so. I know that a great deal of work has
gone into the organisation of the meeting and I sincerely hope that
you will find the next three days both enjoyable and profitable.

Certainly the topic is important.

I came into the meat world quite recently from a very
different background and I have enjoyed a fascinating year and a
half exploring the highways and by-ways of the meat industry.
Impressions have come thick and fast and comparisons with my
earlier experience in industry are inevitable. The main impression
is that, in fact, there is no such thing as a United Kingdom meat
industry. There are farmers, abattoir operators, distributors,
butchers, retailers and so forth, but little evidence that they
work together as an industry to convert feedstuffs into meat to
the satisfaction of the consumer.

The organisation with which I was previously associated produced natural gas in Brunei and shipped it in liquid form to Japan
where it was supplied to Tokyo Gas and Tokyo Electric Companies.
This operation, which was incredibly complex, involved the passage
of the gas between some twenty different companies. Transfer
prices were negotiated in different currencies at many stages,

and these negotiations required the assistance of a computer
simulation exercise in London. The whole exercise finished up as
a profitable venture with a well satisfied consumer. The essence
of this operation is that everyone involved is fully committed to
the objective of taking gas out of the ground and selling it under
satisfactory conditions to the consumer. It is an exercise in
linear programming and that, it seems to me, is what we badly need
in the UK industry today.

During the past few months I have stood on Herefordshire
hillsides and discussed with obviously successful breeders the
important criteria that they use in selecting breeding animals to
improve their stock. Not unfrequently have I heard comments like
'the white mark on the back shall not exceed one-third of the
length of the animal'. Questions about the eating quality of meat
from animals of different conformation have received some strange
answers, often with an assurance that the animal in question will
be 'a good little eater', whatever that may mean.

At the other end of the line, I have listened to the criticisms
of housewives who complain of the variability of beef and the
impossibility of judging from its appearance whether a piece of
meat will be tough or tender. I have traced the history of fat
from its origins in the animal through the cooking process, where
it is stated by some of the trade to have a mystical and quite
undemonstrable effect, until it is discarded to the tune of many
millions of pounds'worth per annum on the sides of plates, and I
have looked in vain for a pricing structure which reflects the
real cost of producing this fat. I have heard farmers talk about
feed efficiency in terms of the weight of an animal without
apparently realising that weight is made up of fat and lean and
that two animals weighing the same may have very different compositions and very different eating qualities. And so it goes on,
and I marvel at the survival of an industry which, in a competitive
environment, disregards some of the basic rules of business and
makes scant provision for that all-important feedback from
consumer to producer.

We must know more about the control of growth and productive
efficiency; about that subtle interplay of physiology, genetics
and environment during growth which determines the nature and
quality of the ultimate product, if we are to satisfy the consumer
and if our industry is to prosper.

There is another reason why this meeting is timely and
important. It was conceived against a very different background
of world economy than prevails today. The forecasts of the Club
of Rome, far-fetched a year or so ago, seem all too real now.
What does this mean for the producer of meat animals in the future?
Should he not be abandoning intensive rearing schemes and should

we not be re-examining the energy balance of our industry? Is it now appropriate to reconsider the larger, more economic carcasses of the big breeds? Are such breeds really economic? Do they eat too much?

Surely today, more than ever before, we need an effort of integration in the industry, an identification of a common purpose and a reappraisal of the factors involved in growth and productive efficiency.

I hope that we shall be able to bear these questions in mind during our discussions this week.

Introduction

WHAT DO WE WANT FROM THE CARCASS?

D. N. Rhodes

Meat Research Institute
Bristol BS18 7DY

The production of meat is undoubtedly the most complex
operation in the food industry; it involves contributions from a
series of sub-industries starting from the breeders, through
rearers, raisers, abattoir operators and wholesale and retail
distributors. At many of these levels adventitious factors exert
powerful constraints which do not always contribute to output or
quality: for example breeders are concerned in cattle with the
dairy herd and in sheep with wool as well as meat; the farmer must
integrate his animal production with the remainder of his agricultural activities; the meat trade is concerned with changing
consumer demand which may influence, for example, desirable
carcase size; and overall are the potent influences of the
economics of buying animals and selling meat which are, apparently,
both independent and unpredictable.

This symposium is intended to consider the scientific bases
of manipulations in breeding and growing which can influence production of a meat carcass; principles which should be applicable
independently of any constraints and capable of being used by the
practical man to achieve the best possible results within any
given circumstances. This introductory paper will try to establish the criteria by which these results should be judged and to
define the sort of carcass which should be produced.

The product moves from conception to consumption but the
cash flow moves in the opposite direction. The consumer in the
end pays everybody's bills hence it is consumer satisfaction which
should be the governing consideration in all production decisions.
Such a realistic approach receives little consideration in the
meat industry: the turnover time in beef may be as long as three

years and at least one year in lamb and pork and, as the consumer
is so many hands away from the producers, how can her approval or
dissatisfaction ever be reflected back through the chain? Rather,
the producing industry grows what it can or what it estimates will
be most profitable and it regards the distribution side as a means
of sorting the output and matching it to the appropriate geograph-
ical, social and economic outlets. To improve this state of
affairs it is necessary to identify those production and handling
variables which significantly affect eating quality and thus
provide a meaningful basis for rational decisions on how to breed
and grow. Then, if eating quality could be more reliably predicted
before retail sale, the better product will come to command a
premium price and an economic incentive to produce more of the
desired product will develop.

Unfortunately the application of scientific method to meat
production has evolved in quite the reverse direction. The domestication of wild species and improvement by rule of thumb
selection has a history as long as civilisation and its formal-
ization into animal science followed naturally in the eighteenth
and nineteenth centuries. Only recently have the harvesting,
storage and preservation of foodstuffs received scientific study
and, of these, meat has remained until the last. It is not
surprising, therefore, to find that the vast majority of early
papers in animal science, dealing with every facet of producing
meat animals, stops abruptly at the last weighing of the animal on
the farm. Two major reasons for this are obvious: firstly, the
farmer is paid for weight on the hoof and experimentation directed
at efficiency of production has used this as the only real measure
of success; secondly, the costs of carcass analysis and eating
quality assessments are high, the facilities required are considerable and they cannot conveniently be located at the same place as
the animal experimentation. The latter factor has, at least, been
alleviated somewhat by coordination between research organisations
especially in the USA. In recent years some measurements of
carcass dimensions, carcass grades, visual assessment of fatness
levels, gross analysis of tissue composition or some form of
cutting value have been made, and a number of workers have
carried through their observations to include estimation of eating
quality. Of 141 papers in the Journal of Animal Science during the
last four years, about one quarter made quality assessments; of
those in Animal Production only 10%; but at least over 90% of
workers in USA and UK deemed it desirable to make compositional
assessments of the carcasses they had so laboriously grown
(Table 1).

Various end points can be used in definitions of productive
efficiency, such as gross weight on the hoof, gross carcass weight,
fat, lean and bone, or yields of saleable meat; this paper seeks
to go further and to look at the validity of the current beliefs

Table 1

Criteria used in papers studying effects of management, heredity and feed of meat animals published in
J. Anim. Sci. or in Anim. Prod. 1970-74

		Growth	Data collected on carcass	Eating quality
J.A.S.	Pigs	47	41	7
	Cattle	70	64	23
	Sheep	24	20	6
	Total	141	125	36
A.P.	Pigs	34	34	5
	Cattle	22	18	2
	Sheep	11	14	0
	Total	67	66	7
	Overall %	100	92	21

about what constitutes, in terms of the consumer, a useful piece
of meat.

THE PLACE OF MEAT IN THE DIET

Nutrition

The first consideration of a staple food must be its contribution to the nutritive value of the diet; a question loaded
with pitfalls of a statistical nature. If we take the official
statistics of the food supplies available in 1972 (1), meat,
offal, bacon and ham can be calculated (2) to contribute 26g
protein per person per day; this compares with 20g from cereals,
12 from milk and 19 from fish, poultry, cheese and eggs. The
recommended daily intake of protein is 40-50g/day (3), hence
meat gives us about one third of our intake and one half of that
desirable. These figures are average values and it is certain
that some sections of our population, such as the old age
pensioners, eat a diet inadequate in protein and that other sections
are grossly malnourished with a surfeit.

Amongst the vitamins and trace elements, meat contributes
significantly to the daily requirements of nicotinic acid, ribo-
flavin, thiamin, B6, B12 and iron in varying amounts up to 30%.
The iron is particularly valuable in that it is easily assimilated.
The same reservations apply to these figures as to protein,

in addition to further uncertainties about vitamin requirements.
Meat also supplies varying amounts of expendable calories within
the limits of the eaters' preferences for fat; but there is an
impressive body of evidence accepted by many authorities that the
consumption of saturated fatty acids in meat glycerides (especially of ruminants) is a factor predisposing the human to circulatory disease (4). Thus reduction of the intake of meat fat is
indicated on health grounds, certainly for the fats of beef and
lamb though less so for pork.

The reasons for eating meat clearly cannot rest on nutritional
grounds, and the protein which would be released if the growing of
meat animals were entirely abandoned would, if that released were
consumable by humans, exceed by many times the minimum amount
needed by our population.

A Food of Choice

Meat is eaten in preference to all other foodstuffs because
it gives a unique eating satisfaction and people have continued
to purchase meat in almost unchanging quantities despite the
enormous increase in its price relative to other foods. This is
not only a matter of a traditional taste established because of
local indigenous food supplies; when meat is introduced into
countries with other staple animal protein foods, it rapidly
becomes popular and as soon as a meat industry is established,
demand leads supply. The prime objective of meat producers and
of research workers supporting them should be, therefore, to
maintain or improve quality since this is the major component of
the popularity of the product. So far, as the analysis of the
literature shows, little serious thinking has been done in this
direction.

Eating Quality

Colour and odour are the first components of eating quality
appreciated by the consumer; in the mouth, texture is immediately
judged and, if satisfactory, juiciness and flavour are considered.
Experimental assessment of these components can be made with
instruments for colour and texture but odour, flavour and overall
acceptability requires the use of taste panels. During the last
six years at the Meat Research Institute we have been trying to
define eating quality in objective terms to study the effects of
the more important variables in production on the eating quality
of the final product. We have to date examined samples from about
400 animals all of known genetic history and controlled growth
from Experimental Husbandry Farms, sister agricultural institutes
and various trials in University farms or run by the Meat and
Livestock Commission. These 400 beasts cannot in any way form a
valid sample for estimating statistics to describe British Beef,

because of the rather specialised nature of the original experimental conditions, but it does contain some adequately replicated
groups and interesting conclusions can be drawn about the beliefs
of the industry which are, at present, being used to describe the
desirable characteristics of a carcass.

FACTORS AFFECTING EATING QUALITY

Age and Weight at Slaughter

Age and weight at slaughter depend upon rate of growth, hence
commercial animals, raised on various systems, show a wide range
of interaction. It is commonly believed that eating quality is
highly dependent upon age; flavour is alleged to improve or
intensify and tenderness to decrease. In beef there are real
differences between veal and mature beef, but after 10 months the
eating quality characteristics do not show any great changes until
well above the normal limit of age for beef animals, that is over
3 years. The literature contains many studies which are summarised
in Table 2. When the experiments covered a wide range of age,
that is including veal, a decrease in tenderness was sometimes
observed; in other cases when the normal 10-30 month range was
examined no effect was recorded. In the large studies by Berry
et al (15) and Cross et al (16), significant decreases in tenderness were obtained only when animals above 4 years of age were
included.

We do not need to consider this effect in lamb or pork where
the animals are physiologically very young. Even up to 100 kg
carcass weight, or slaughter age of 390 days, pig meat retains its
excellent eating quality (17) and we have looked at the occasional
4-5 year old boar or sow and they turn out not to be inedible as
might be expected but acceptable, though dark coloured, meat.
It may certainly be concluded that there is no evidence to justify
any heavy emphasis on slaughter age in beef animals on grounds of
eating quality of the meat. Despite this, the USDA quality grades
use age or "maturity" as a major segregating factor within the
range below 24 months.

Heritability of Eating Quality

Very few serious attempts have been made to investigate this
question. Thornton et al (17) were able to show differences in
tenderness between Yorkshire and Duroc pigmeat but this probably
arose from the variation in fatness; similar results were given
by Hiner et al (18).

In cattle the evidence is conflicting; we have examined the
tenderness of the progeny of 10 Charolais bulls on random Friesian

Table 2

The effect of age and weight at slaughter
on the tenderness of beef

Age range (mo)	Wt range (kg)	Breed/ sex	Finding	Ref.
Wide range	–	–	Decrease	(5)
"	–	–	Decrease	(6)
"	–	–	Decrease	(7)
Narrow range			No effect	(8)
"			"	(9)
"			"	(10)
"			"	(7)
–	390–460	AA, st	"	(11)
18	400–500	F, st	"	(12)
10–20	400	F, crypt	"	(12)
6–9	–	various	"	(13)
–	340–430	H, st h	Increase	(14)
USDA A to D	–	–	No effect	(15)
" A to E	–	–	Decrease	(15)
12–48	–	H, f	No effect	(16)
12–160	–	"	Decrease	(16)

dams and found no effect; Epley et al (13) found no effect
among 200 progeny of 12 Aberdeen Angus and Hereford bulls on
Angus and Hereford dams, and McBee et al (8) found no better
meat from sires selected by index than from random sires on two
comparable herds of dams. On the other hand some papers
describe significant differences between sires (20,21,22,23,
24) and Alsmeyer et al (25) were able to assign variability in
tenderness in the progeny of Brahman and Shorthorn bulls
slaughtered between 8 and 87 months as follows; breed 14%,
sire 14%, carcass grade 12%, age 8%, marbling 7%. Bryce-Jones
et al (26) found differences between progeny grown at two farms
sired by five Hereford bulls attributable both to sire and to
farm and closely linked with fatness level; the mean differences
were, however, less than 1 unit in a ten unit scale, an insufficient difference in taste panel scores to warrant prediction of an effect of real importance to the consumer eating
at home.

Estimates of the heritability index for tenderness as
measured by shear value or by panel assessment vary between
28% (Palmer et al (27)), 40 to 70% (Alsmeyer et al (25);
Cartwright et al (20); Zinn, (28); Gregory (29); Palmer et al

(27)) and between 71 and 92% (DuBose et al (30); Kieffer et al (23),
Alsmeyer et al (25); Cover et al (22)). These figures would
indicate a high probability of improving beef tenderness by selection, but the difficulties in collecting and examining meat of
progeny in a meaningful way would be enormous.

Breed

The beef and dairy breeds are well established entities in
most countries and the belief in the superiority of eating quality
of meat from the former is widely accepted. Books on meat written
30 years ago treated the dairy breeds with scant respect as meat
producers, but today more than half of our beef comes from
Friesian dams and serious attention should be paid to comparisons
of eating quality between breeds and crosses, especially as the new
sires from the Continent and elsewhere are introduced. We have
examined the meat from a variety of pure breeds and crosses by
objective methods and part of the study is shown in Table 3.
However, there is no clear differentiation between the beef and
dairy sires on Friesian dams, nor do the pure Devons and Aberdeen
Angus (AA) stand clearly as the most tender beef, (the pure Friesian
being about the same level). Among the crosses, the Friesian and
Simmental bulls on Ayrshire cows produced outstandingly tender
meat. These comparisons must be subject to major reservations
because the trials cover a period of five years during which
membership of the taste panels continually changed. This was con-
founded too with variations in age and fatness, and the pre - and
post-slaughter treatment of the animals was uncontrolled; for
example, the two Friesian x Ayrshire groups (6 and 10) originated
from different farms and abattoirs. In a similar broad study
Weniger et al (31) found only 2.5% of the variability in tenderness
in 173 carcasses to be attributable to breed (German Schwarzbunte
and Fleckvieh).

More acute analysis is possible where direct comparisons were
made on carcasses from the same abattoir. In this way we found no
differences in acceptability between HxF beef raised in Somerset
and either Devon or AA (Scotch beef) except in tenderness where
the pure breeds were slightly better (1, 7, 9). Similarly the
three European breeds (2, 3 and 4) were compared directly as were
the Friesian and Simmental crosses on Ayrshire (10 and 11). Bryce-
Jones et al (32) tested Charolais x Friesian against Hereford,
Devon and Friesian x Friesian and also Charolais x Ayrshire against
Hereford and Devon x Ayrshire; the Charolais beef was found to be
less tender than any of the other comparable crosses, but no
estimate of the size of the difference was made.

Table 3

Tenderness of roast eye muscle of beef
measured by taste panel. Scale -7 extremely tough
to +7 extremely tender   (Rhodes, unpublished observations)

	Cross		No	Mean	Variance
1.	H	x F	35	0.2	6.3
2.	Sw Sim	x F	8	1.8	1.8
3.	Ger Sim	x F	6	1.9	0.7
4.	Lim	x F	8	2.2	1.6
5.	Lin Red	x HF	20	2.2	3.0
6.	F	x Ayr	19	2.5	4.0
7.	Dev	x Dev	11	2.9	3.9
8.	F	x F	31	2.9	2.2
9.	AA	x AA	33	3.7	1.2
10.	F	x Ayr	13	4.3	0.8
11.	Sim	x Ayr	11	4.8	0.6

Differences in flavour and juiciness were very
much smaller and rarely significant

Sex

     The male animal can produce less fat and more lean from a given
bulk of feed than either the female or the castrate and the elimination of castration of bull calves and boar pigs would be the
simplest, cheapest and most immediate method of increasing the lean
meat supply and reducing the demand for imported feedingstuffs in
the UK (see 32). In present circumstances it is really quite
indefensible that the producing and retailing industries virtually
ignore this immediately available saving in input costs and
reduction of loss from waste, and the official side remains modest
in its encouragement of the production of the sort of meat the
consumer prefers. The arguments based on quality may be summarised
as follows.

     Bull beef is alleged to be less tender than steer but, among
the many carefully controlled experiments comparing the two, as
many can be quoted where no difference was found as the reverse
(34). Two factors may account for this, firstly, the fat
cover is thin in a bull carcass and cooling is much more rapid
which may induce toughness due to cold shortening in the muscles;
secondly, the lack of intramuscular fat will influence tasters who
like fat to discriminate against a leaner sample. No more than
10% of male cattle are castrated in northern Europe, especially in
Germany, and Weniger et al (31) examined a large sample of German

slaughter animals, including old cows, and assigned 10-20% of variability in tenderness to sex, and 0-10% to juiciness.

Meat from boars of up to 100 kg carcass weight has been found to be indistinguishable in texture and overall eating quality to that of castrates in both laboratory and large consumer tests in this country and elsewhere, and there are no valid reasons for the continued castration of pigs except that of a mistaken apprehension about boar odour in the fat; a subject discussed adequately elsewhere (35) and not relevant to this meeting.

Feedstuffs and Flavour

Flavorous compounds in plants are generally fat soluble, oxygenated and unsaturated and such compounds are readily hydrogenated in the powerfully reducing conditions of the rumen; in all animals they are subjected to the normal digestive and metabolic processes which provide an efficient filtering mechanism preventing access to the carcass fats. Substances from the indigenous forages and the standard supplementary feeds which do appear in meat are presumably regarded by the consumer as contributing to the normal flavour and are, therefore, desirable. Major changes in raising practice need examination, however, since myths rapidly appear; for example, intensive rearing on barley or other grain was denigrated at one time on grounds of insipid flavour: we made two large trials using side-by-side taste panel comparisons of meat from Hereford x Friesian raised to 12 months on barley or 24 months on grass, and from Friesian x Ayrshire heifers raised intensively on grass to 18 or to 24 months. In no case was a significant difference in flavour detectable (36).

We have also looked critically at the meat from animals fed various exotic supplements; for example, field beans, brewers grains, lucerne pellets or dried poultry waste were all quite innocuous in beef. Similarly three of the recently developed single cell proteins gave pork meat as good as, if not better than, fish meal as a supplement for pigs (37). An acid-treatment silage made from fish waste did introduce a rancid greasy odour into pork fat when fatty fish was used. In Australian work (38,39) rape, oats, lucerne and white clover were found to impart unpleasant tones into the flavour of lamb compared to standard grasses; but we have found no influence of red clover or barley concentrates on lamb flavour. In a recent paper from Australia a rich source of energy for animal production was identified in Government waste paper; unfortunately no measurements of eating quality were made on the animals raised (40).

Tissue Composition

Eating quality of meat results from the chemical and physical

modifications of the components of tissues brought about by heating
during cooking. Texture and juiciness result mainly from
denaturation of the contractile and soluble sarcoplasmic proteins
to give a solid precipitated coagulum on the one hand, and the
heat induced collapse of the tertiary structure of the connective
tissue collagen, producing a weak gel or solution of gelatin on
the other. Reactions producing meat flavour are the extremely
complex pyrolytic degradations of all sorts of tissue components,
soluble and insoluble, which produce a vast range of compounds,
volatile and odoriferous as well as those soluble in the water
phase.

Tissue proteins. The biochemical mechanism of the contractile
process in all land animals is the same and, in healthy animals, the
molecular organisation and amino acid composition are identical.
At the macroscopic level there is variability in the amount of
connective tissue according to the function and size of the muscle
and this accounts for the gross differences in tenderness between
muscle groups and the necessity for using different degrees of time/
temperature treatments in cooking to render the grosser elements
soft enough for mastication. The amount of collagen does not
change greatly with age in individual muscles of the mature animals,
but intermolecular links form which reduce the proportion which
degrades to a soluble gelatin on heating. This factor is not,
however, a great determinant of tenderness, for example in a
comprehensive study using cattle up to 14 years old Cross et al
(16) found that collagen content, soluble collagen content, and
elastin content accounted for less than 5% of the variability in
taste panel tenderness or shear force value measured within each
of the major meat muscles. Severe undernutrition followed by
compensatory growth also had no effect on tenderness of beef (41).
In the normal carcass tenderness is dependent very much more upon
the state of contraction of the muscle elements and this is
determined by the interaction of the rate of fall of temperature
during cooling, the transformation of glycogen to lactic acid
inducing a progressive drop in pH, and the exhaustion of ATP as
the living processes decay. Too rapid cooling - well within the
possibilities of modern refrigerating plant - produces a 'cold-
shortening' which can turn a tender lamb carcass into quite
unacceptably tough meat (42), and severely affect a beef side (43);
cold-shortening is not, however, a problem in pork.

Tissue fats. From the point of view of eating quality the
intracellular lipids may be ignored; they are small in quantity
and are integral components of the biochemical structures; the
major concern is with the depots of triglyceride fat lying between
muscle bundles and between the muscles themselves. Fat has provided a major source of calories in the human diet and only recently
has central heating, heated transport and the substitution of fuel
for muscle-power rendered the consumption of fat calories

unnecessary or, as a growing body of opinion on the causation of circulatory diseases concludes, positively undesirable. Because of these factors and the influence of nutrition education, in general consumer preference is moving steadily toward leaner meat, a trend intensified by rising prices. Today more than half of our population do not wish to eat fat and actively reject it by leaving it on the plate (Table 4). Yet the industry as a whole is moving only slowly toward the leaner carcass and our fatstock marketing system and the regulations governing it still apply traditional yardsticks of 'finish' to beast and carcass which have little or no rational justification. Some of this resistance derives from the belief that tenderness, juiciness and flavour in beef depend primarily upon the level of fatness, a belief strongly supported by the USDA quality grade system which bestows the accolade of prime grade on the carcass containing gross quantities of visible fat. It is encouraging to note that the USDA standards are at present undergoing a reappraisal with the admission that such levels of fatness do not contribute significantly to eating quality and that the lavish waste of feedingstuffs required to produce them cannot be justified.

What evidence is there concerning the effect of fatness on eating quality? Firstly, the question becomes academic for that part of the population that likes to eat fat as a part of meat; they will wish to purchase fat and enjoy it. For those who do not, it is important to know whether the advice in the cookery books and magazines, from official sources and from the trade, that eating quality in the lean depends upon excess fat is or is not correct: the thesis is not supported by experiment or observation.

Backus (44) found a negative correlation between marbling and tenderness wihin 84 Hereford steers while Larmond el al (45) concluded that tenderness is not associated with intramuscular fat. Henricksen and Moore (46), in a comprehensive study of steer meat from 6 to 92 month old beasts, found fatness to be unimportant up to 18 months but in 42 and 92 month old meat it had a significant influence on tenderness, juiciness and flavour. Norris et al (10) examined meat at three maturity and two fatness levels and found no influence of either factor in shear value, tenderness or palatability. Similarly, in 60 heifer or cow carcases, marbling level was not significantly related to texture although it did affect juiciness and flavour (7) and, in Herefords, McKee (19) concluded marbling was of little importance in determining tenderness, juiciness, or flavour. Weniger and Steinhauf (31) in their examination of German beef production, reported low correlation coefficients (0.13 - 0.38) for tenderness and fat content in various muscles. Moody el al (47) also found that fine or coarse marbling has no discernible effect on tenderness. In a study of 24 pure Friesian steers raised to different levels of fatness (15 - 45% arbitrary units) the correlation coefficients between

Table 4

Personal preferences among 504 people for meat fat

| | % of sample who | |
	EAT fat	LEAVE on the plate
Men	55	45
Women	40	60
Boys	39	61
Girls	24	76
Adults	48	52
Children	31	69
Whole sample	46	54

fatness and texture of roasts or grilled steaks by taste panel or objective measurements of texture were below 0.24 (48). Much higher correlation coefficients have been obtained in similar panel studies when a wider range of fatness levels is included. McBee et al (8) examined samples from USDA marbling grade 1 to 9 inclusive but the ranges of sensory tenderness, flavour or juiciness covered less than one unit on a 10 point scale, though each was highly correlated with marbling level.

These and many other papers have been analysed exhaustively by Jeremiah et al (49) who combined the results into overall weighted correlation coefficients based on some 40 studies and two thousand carcasses. They concluded that only about 10% of the variability in tenderness (measured by machine or by taste panel) juiciness, flavour or overall eating quality was accounted for by the fat content of beef as assessed by chemical determination, or marbling score by USDA quality grade.

Dramatic reductions have been effected in the fat content of the pig carcass in the last 70 years by selection and the trend continues. Experimental comparisons of very lean pigs with the present average levels shows there is little danger of affecting consumer acceptability by continued selection for leanness (50) and in consumer studies using even leaner meat from boars, as pork or bacon, many favourable comments were received (51,52).

Producing fat in an animal carcass to provide calories in the human diet is an extremely inefficient way of using feedingstuffs; even if all the fat so produced were consumed, the system would not bear rational economic examination. In fact, most of the fat in a carcass is discarded in the abattoir, in the retail cutting, in the kitchen and on the plate and that sold at meat prices fulfils little useful purpose in meat quality. It is not wanted by a half of consumers and may also be harmful to health. Clearly we should be aiming to produce the minimum of body fat and exploring the possibilities of eliminating the huge deposits in the kidney knob, channel fat and the mesenterium of our meat species.

REFERENCES

1. Ann. Abs. Stats. (1972) H.M.S.O. (London).

2. McCance, R. A. and Widdowson, E. M. (1960) "The Composition of Foods", H.M.S.O. (London).

3. Wld. Health Org. (1973) Tech. Rpt. Series No. 522.

4. Bender, A. E. (1974) Rpt. Adv. Panel Med. Aspects Food Policy. H.M.S.O. (London).

5. Romans, J. R., Tuma, H. J. and Tucker, W. L. (1965) J. Anim. Sci. 24, 681.

6. Walter, M. J., Goll, D. E., Kline, E. A., Anderson, L. P. and Carling, A. F. (1965) Food Technol. 19, 841.

7. Breidenstein, B. B., Cooper, E. C., Cassens, R. G., Evans, G. and Bray, R. W. (1968) J. Anim. Sci. 27, 1532.

8. McBee, J. L. and Wiles, J. A. (1967) J. Anim. Sci. 26, 701.

9. Covington, R. C., Tuma, H. J., Grant, D. L. and Dayton, A. D. (1970) J. Anim. Sci. 30, 191.

10. Norris, H. L., Harrison, D. L., Anderson, L. L., Welk, B. von and Tuma, H. J. (1971) J. Food Sci. 36, 440.

11. Suess, G. G., Bray, R. W., Lewis, R. W. and Brungardt, V. H. (1966) J. Anim. Sci. 25, 1197.

12. Purchas, R. W. (1972) J. Food Sci. 37, 341.

13. Epley, R. J., Stringer, W. C., Hedrick, H. B., Schupp, A. R.,
     Kramer, C. L. and White, R. H. (1968) J. Anim. Sci. 27, 1277.

14. Murray, D. M. (1970) Proc. Aust. Soc. Anim. Prod. 8, 226.

15. Berry, B. W., Smith, G. C. and Carpenter, J. A. (1974).
     J. Anim. Sci. 38, 507.

16. Cross, H. R., Carpenter, Z. L. and Smith, G. C. (1973)
     J. Food Sci. 38, 998.

17. Thornton, J. W., Alsmeyer, R. H. and Davey, R. J. (1968)
     J. Anim. Sci. 27, 1229.

18. Hiner, R. L., Thornton, J. W. and Alsminger, R. H. (1965)
     J. Food Sci. 30, 550.

19. McKee, M. (1968) Dissert. Abs. B30, 1432.

20. Cartwright, T. C., Butler, O. D. and Cover, S. (1958)
     Proc. 10th Res. Conf. Am. Meat Inst. Found. 10, 75.

21. Bradley, N. W., Cundiff, L. V., Kemp, J. D. and Greathouse,
     T. R. (1966) J. Anim. Sci. 25, 783.

22. Cover, S., Cartwright, T. C. and Butler, O. D. (1957) J.
     Anim. Sci. 16, 946.

23. Kiefer, N. M., Henricksen, R. L., Chambers, D. and Stephens,
     D. F. (1958) J. Anim. Sci. 17, 1137.

24. Means, R. H. and King, G. T. (1959) J. Anim. Sci. 18, 1475.

25. Alsmeyer, R. H., Palmer, A. Z., Coger, M. and Kirk, W. G.
     (1959) Proc. 11th Res. Conf. Am. Meat Inst. Found. 11, 85.

26. Bryce-Jones, K., Houston, G. W. and Harries, J. M. (1963)
     J. Sci. Food Agric. 14, 637.

27. Palmer, A. Z., Carpenter, J. W., Reddish, R. L., Murphy, C. E.
     and Hallet, D. K. (1961) J. Anim. Sci. 21, 193.

28. Zinn, D. W. (1964) Proc. Recip. Meat Conf. 17, 43.

29. Gregory, K. E. (1961) Neb. Ag. Exp. Stat. Res. Bull. No. 196.

30. DuBose, L. E. and Cartwright, T. C. (1967). J. Anim. Sci.
     26, 203.

31. Weniger, J. H. and Steinhauf, D. (1969) In: "Meat Production from Entire Male Animals", p. 211 (Ed. D. N. Rhodes). (Churchill, London).

32. Bryce-Jones, K. (1965) Exp. Husb. No. 16, 25.

33. Rhodes, D. N. (Ed.) (1969) "Meat Production from Entire Male Animals". (Churchill, London).

34. Turton, J. D. (1969) ibid p. 1.

35. Rhodes, D. N. (1971) In: "Meat Res. Inst. Ann. Rept. 1970-71", p. 13 (Bristol).

36. Rhodes, D. N. (1973) Meat Res. Inst. Mem. No. 24, 10 (Bristol).

37. Rhodes, D. N. Unpublished observations.

38. Park, R. J., Corbett, J. L. and Furnivall, E. P. (1972) J. agric. Sci. Camb. 78, 47.

39. Park, R. J., Spurway, R. A. and Wheeler, J. L. (1972) J. agric. Sci. Camb. 78, 53.

40. Coombe, J. B. and Briggs, A. L. (1974) Aust. J. exp. Agric. Anim. Husb. 14, 68.

41. Morgan, J. H. L. (1972) J. agric. Sci. Camb. 78, 417.

42. Taylor, A. A., Chrystal, B. B. and Rhodes, D. N. (1972) J. Fd Technol. 7, 251.

43. Dransfield, E. and Rhodes, D. N. Unpublished observations.

44. Backus, W. R. (1969) Dissert. Abs B29, 2692.

45. Larmond, E., Petrasovits, A. and Hill, P. (1969) Canad. J. Anim. Sci. 49, 51.

46. Henricksen, R. L. and Moore, R. E. (1965) Oklahoma State Univ. Tech. Bull. T115.

47. Moody, N. G., Jacobs, J. A. and Kemp, J. D. (1970) J. Anim. Sci. 31, 1074.

48. Rhodes, D. N. (1973) Meat Res. Inst. Mem. No. 15 (Bristol).

49. Jeremiah, L. E., Carpenter, Z. L., Smith, G. C. and Butler, O. D. (1970) Texas Agric. Exp. Stat. Tech. Rept. No. 22.

50. Rhodes, D. N. (1970) J. Sci. Fd Agric. <u>21</u>, 572.

51. Rhodes, D. N. (1971) J. Sci. Fd Agric. <u>22</u>, 485.

52. Rhodes, D. N. (1972) J. Sci. Fd Agric. <u>23</u>, 1483.

DISCUSSION

Dr. Dickerson agreed with Dr. Rhodes that there had always been a lack of feed-back from the processor to the producer. He conjectured that there might be economic reasons why the processor should be unwilling to supply the necessary information to the producer. *Dr. Rhodes* replied that it was his belief that, since the retailer was provided with lean, fat and bone, it is to his advantage to sell fat at the price of lean in order to make a profit. He cannot, therefore, take the view that fat is undesirable, so perpetuates the myth that fat is a necessity for the eating qualities of meat. This is one of those economic situations which is impossible to resolve without re-education at all levels. *Prof. Lucas* thought that Dr. Rhodes had perhaps overstated the case against fat. In buying carcasses from the wholesaler, the retailer was free to choose those carcasses which he considered most desirable. He asked Dr. Rhodes if there was not some truth in the trade belief that a certain amount of fat cover was necessary if only to prevent moisture loss from carcasses. He recalled that some years ago this was the view of the New Zealand lamb processors, whose interest was primarily in frozen meat. *Dr. Rhodes* agreed that there was a measure of over-statement. He recognized that some subcutaneous fat was of importance to prevent moisture loss which was of considerable economic, though not of nutritional, importance, but this end could be achieved by as little as 3 mm of evenly distributed subcutaneous fat. *Prof. Ingram* commented that there were cheaper ways of preventing moisture loss from carcasses and that the requirements of the fresh and frozen meat trades should not be confused. Dr. Braude warned that, in reducing the subcutaneous fat depots of animals, there was a danger of simply moving lipid into other parts of the carcass. He also suggested that increasing the leanness of carcasses may simply result in an increase of water. *Dr. Dickerson* disagreed with Dr. Braude's suggestion that the water content of the lean tissue could be varied.

Dr. Moody suggested that the US system of yield grades had something to offer in combining quantity and quality of carcass. Mr. Mason noted that Italian housewives refused to buy beef with any trace of fat and saw no reason why the same consumer pressure should not be applied in this country. *Dr. Fuller* referred to the continuing resistance of the meat trade to the carcasses of intact males, and asked whether it would not be better to provide improved description of such commodities to allow consumers to exercise their own choice. *Dr. Fowler* commented that a component of the enjoyment of eating meat is psycho-social in origin. He asked if it was not likely that much of the effort currently expended on improving meat quality was dictated by rather frivolous and ephemeral fashions in eating. *Dr. Rhodes* agreed with this. He thought the only area for improving meat quality

was in the immediate pre- and post-slaughter treatment. He re-
emphasised that the most damaging myth was that fat was necessary
for the best eating characteristics of meat. The concepts of
finish, while imprecise and irrational, are perpetuated through
tradition in both education and commercial practice.
Dr. Duckworth suggested that collective agreement would be useful
on Dr. Rhodes' generalisation that the traditional concept of
finish was irrelevant. This would be useful even though past
experience suggested that the dissemination of such information,
within and between different sections of the livestock industry,
was extremely slow and could be expected to meet considerable
resistance.

The Efficiency of Meat-Producing Systems

THE RELEVANCE OF VARIOUS MEASURES OF EFFICIENCY

C. R. W. Spedding

Department of Agriculture and Horticulture

University of Reading

THE DEFINITION OF EFFICIENCY

Efficiency may be defined simply as Output per unit of Input, for specified outputs and inputs, over a specified period in some stated context (of environment, for example) (1). There is thus no "right" expression of efficiency (E), since any number of outputs (O) and inputs (I) may be selected. The sole criterion in this selection is whether the resulting ratio $(\frac{O}{I})$ accurately reflects the reasons for wishing to make the calculation.

The main difficulty is usually in having to select only one of the outputs and one of the possible inputs, since we are rarely interested in increasing the efficiency of one process without regard to the consequences to all the other processes involved. Any combination of groups of outputs or inputs must involve the use of weighting factors: the commonest device being expression in monetary terms. This is not really a solution to the problem, however. If a financial ratio is what is required, then all monetary outputs and inputs are involved from the start. If such a ratio is not of interest, then monetary expression of inputs and outputs has solved nothing. Exactly the same is true of 'energy'. If energetic efficiency is of interest, then all energy outputs and inputs are wanted: if it is not, then expressing outputs and inputs in energy terms will simply obscure the main interest.

So the choice of outputs and inputs and the choice of the units in which they are expressed must be based on the reasons for the calculation and the only legitimate test of an efficiency

expression is whether or to what extent it fulfils the purpose
underlying the calculation.

THE PURPOSES OF MEASURING EFFICIENCY

It may be worth recognising that efficiency is calculated
rather than measured, but this does not affect the argument at
this point.

The major purposes behind an efficiency calculation may be
grouped as follows:

(a)   in order to increase output per unit of input, or to decide
      whether efficiency needs to be improved;

(b)   in order to minimise the use of an input, without undesirable
      consequences;

(c)   in order to decrease unwanted outputs (e.g. pollution): the
      interest here is in negative efficiency;

(d)   in order to maximise the use of an input (as in the case of
      cheap or natural resources) in a production process.

It is clear that <u>maximising</u> the efficiency of one process
will rarely be the objective, but it is also clear that the aim
is frequently not even to <u>increase</u> efficiency. Of course,
constraints can be imposed, within which efficiency can be
increased, and this can be done systematically for a number of
different constraints.

In general, the objective is to improve our understanding of
a process, in order to improve our ability to manipulate it in a
desired direction. This practical purpose requires that the
ratio used allows us to define the factors that influence
efficiency. Thus, it must be possible to derive from any one
ratio, a list of these factors and a picture of the way in which
they operate.

For example, a ratio of meat output (M) to food consumed (F),
by a cow and its calf, could be expressed as:

$$\frac{M}{F}$$

but immediately expanded to:

$$\frac{\text{Calf Growth Rate x Time from birth to slaughter (t)}}{t_2(F_p) + t_3(F_L) + t_1(F_c)}$$

where F_p, F_L and F_c represent daily food intakes of the pregnant cow, the lactating cow and the calf, respectively, and t_1, t_2 and t_3 are the relevant periods of time. It is then a straightforward matter to continue the expansion process until all the important factors (such as cow size, food quality, disease incidence) have been included.

However, it is already obvious that the factors affecting Outputs and Inputs are not independent. It is not possible, for example, to guarantee improvement in efficiency by independently increasing O or decreasing I: indeed they cannot be varied independently at all in some cases. Quantities such as F_p, F_L and F_c cannot be varied without the possibility of consequent effects on calf growth rate (see Fig. 1 for an illustration relating to meat production in general). As the ratio is increased in complexity, so the interrelationships between input and output determinants become more complicated.

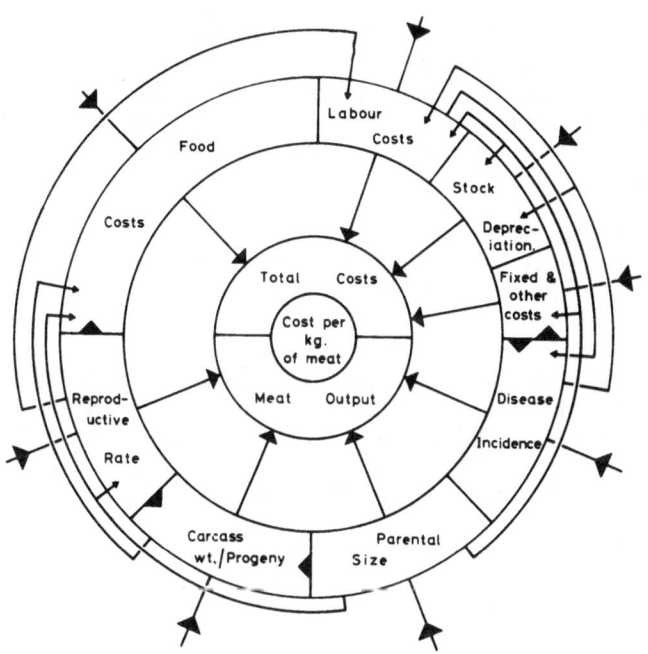

*1 ♀ + a proportion of the ♂ depending upon the ratio of ♀ : ♂ required

Fig. 1.   Meat production and costs (per parental unit*) - a diagram illustrating the relationships between factors influencing the inputs and outputs of an efficiency ratio.

The simple proposition, that measuring efficiency and describing the factors influencing it will make it possible to manipulate processes in a desired direction, is therefore to some extent misleading. It is valuable to know what the situation is, and a knowledge of relative efficiencies makes it possible to choose between processes or systems, but the understanding required for improvement is of a different order.

It is this kind of problem that leads to the use of mathematical models as tools in the understanding of the consequences of manipulation of agricultural systems (2,3). There has been considerable effort in this direction in recent times and a range of successful modelling has been undertaken (4,5,6,7,8) but there is scope for much further development (see 9), particularly in relation to bio-economic models.

Models of one kind or another are bound to be used at all levels of understanding, and they are no less useful in achieving insights into the biological processes underlying agriculture than they are in other disciplines. However, all the earlier arguments apply and models are simplifications, by definition, and must relate to some defined purpose. It is not possible to describe processes completely, at any level of detail, because any description is based on a point of view, a way of looking at things, and there are innumerable points of view that could usefully be adopted. This is true of quite ordinary objects. A moment's consideration will demonstrate that no one description of a cow, for example, can possibly encompass all external and internal features and all possible ways of looking at it, even for a stationary cow (or even for a dead cow).

It follows that total comprehension of systems is neither a useful nor a practicable goal, but we do not necessarily know enough to decide which particular views are most important. We may therefore need to explore many possible models in order to gain insights which will eventually help to decide on the models that are required for specified purposes.

Much of this is independent of any argument for assessing efficiencies but the latter do summarise the results of particular viewpoints and may serve to establish differences or similarities that may imply different or similar mechanisms (or similarities or differences in the efficiency of such mechanisms). In this way, the measurement of efficiency may draw attention to important factors influencing biological processes.

EFFICIENCY IN MEAT PRODUCTION SYSTEMS

The efficiency with which any of the numerous agricultural animals produces meat from given resources can obviously be expressed in many different ways. As already emphasised, these must relate to the purposes for which such calculations are made but there are some general points that are worth noting.

First, agricultural animals, including those that produce meat, may serve several purposes. Table 1 illustrates some of the most important of these, together with the associated roles and functions of such animals.

Table 1

The Functions of a Meat-producing Animal

Major purposes for which meat-producing animals are kept

1. The provision of a product (& by-products)
    (a) for direct consumption
    (b) for processing
    (c) for feeding to other animals (including pets)
    (d) for export

2. Monetary reward to the producer - this may be expressed in many ways, such as:
    (a) return on invested capital
    (b) profit
    (c) gross margins

3. Use of available resources for purposes other than production, such as:-
    (a) employment
    (b) a way of life
    (c) preservation of amenity
    (d) use of locally-produced inputs
    (e) use of resources imported for other reasons
    (f) import-saving and assistance to balance of payments

Major functions of meat-producing animals

1. Production.

2. Collection of plant feed - especially that which is otherwise unavailable, for physical or economic reasons.

3. Conversion of nutrients otherwise unsuitable for human consumption.

4. Concentration of nutrients present in the original food in low concentration.

5. Elimination of toxic materials present in the original diet.

6. Maintenance of continuity of human food supply during times when crop growth is negligible.

The efficiency of an animal can legitimately be assessed in relation to any of these purposes and functions and the relative efficiencies of different species will vary accordingly (10).

Secondly, the efficiency of animal species can be compared in one of two main ways: either in the same situation or in situations that are different and in some sense appropriate. Since feed accounts for a very high proportion of the total costs in animal production (See Table 2), the efficiency of feed conversion is an important ratio to assess. In some circumstances, the aim will be to assess the efficiency with which different species use the same kind of feed. This is true for comparisons of sheep, cattle, goats and horses, say, all eating grass. In other cases, however, the aim may be to compare efficiencies (often in terms of energy or protein) on those feeds that are preferred by the animals concerned (or on which they perform best). This might be so for comparisons of sheep and pigs. Theoretically, comparisons would be better based on a series of diets that

Table 2

The proportion of the total costs of meat production that is attributable to feed (these figures are quoted to indicate the order of magnitude only)

Enterprise	Feed costs as % of total costs	Sources of information Ref.No.
Beef Production		
18-month beef	48	20
Barley beef	60	21
Suckler beef	45	22
Sheep Production		
Suckled lamb	57	23
Artificially-reared lamb	50	24
Pig Production		
Bacon	47	25
Pork	38	25
Broilers	54 – 61	26

included those preferred by each species: however, there might be little point in feeding simple-stomached animals solely on very fibrous feeds.

The same argument applies to other resources and to aspects of the environment. It may be important to know that pigs use barley more efficiently than do sheep but also that this is not necessarily so in all climatic conditions.

This implies that, even for one expression of efficiency, there is no one value that states the efficiency of an animal species.

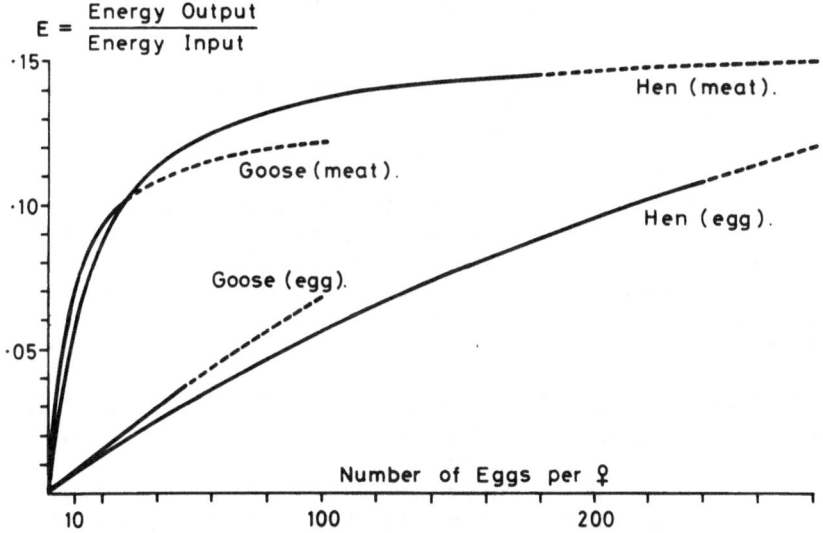

Fig. 2.   Efficiency of energy production as edible meat from progeny and as eggs, by the hen and the goose, related to the number of eggs produced per female per annum. Solid lines refer to levels of egg production per annum that are readily achieved on a flock basis. Efficiency is calculated as the energy produced per unit of feed energy consumed per annum by one female bird and a proportion of the feed required to support one male.
The female/male ratios used were:- 12:1 and 4:1 for egg production by hens and geese, respectively, and, for meat production, 10:1 (hens) and 3:1 (geese).

Thirdly, there are attributes of each animal species that influence efficiency and there are attributes of the particular population considered that may exert even larger effects. This is notably so for size of both male and female, especially in homeotherms, for reproductive rate and for longevity (1,11). Egg production in birds provides a simple example (see Fig. 2) and shows how the variation of efficiency with reproductive rate also depends on the form of product considered.

In the light of these relationships, it is clearly useful to state efficiencies, however calculated, not as single figures (even with a standard error or some other measure of variation) but as relationships with changes in some important factor. A series of such relationships might prove to be the most useful description, indicating ceiling values that cannot be exceeded. In this context, it is possible to say that one species is more efficient than another, using given genetic material.

One example of this is the contrast between efficiencies of individual animals and of the populations of which they form a part.

INDIVIDUAL AND POPULATION EFFICIENCIES

Every individual (at conception, birth, hatching or emergence) has already incurred a production cost, represented by the costs of maintaining its parents, spread over the number of individuals produced. This overhead cost, whether expressed as feed, energy or money, has to be balanced by the meat production of the progeny: that is why the reproductive rate (the number of progeny per female per year) is so important a determinant of efficiency of feed conversion.

The feed conversion efficiency of an individual by itself is therefore generally higher than that of the whole population of which it is a part. If the overhead cost can be reduced to negligible proportions (by a high reproductive rate, small female size and low ratio of males to females), then the efficiency of a population can approach that of the isolated individual. In these cases, therefore, the efficiency of the individual represents a ceiling value for the population and it is possible to judge the potential efficiency of the population on this basis. In this way, populations of two species can be compared, even though the current performance of each may represent a different proportion of its potential. However, it is necessary to consider not only whole populations but the production processes, enterprise and agricultural systems in which they operate.

One major problem here is the physical impossibility of experimentation on large-scale systems (12). It is extremely costly to

experiment with whole agricultural systems at all but to do so on
a large scale over substantial periods of time and to study the
results of change in all the important variables would be out of
the question.

It is therefore impossible to measure directly the change in
efficiency of such systems in response to experimental treatments,
quite apart from the inherent difficulties in terms of control.
All that can be done directly is to calculate the efficiency of
existing systems and try to deduce, from observation and the
monitoring of components, the reasons for the calculated result.
This makes mathematical modelling an indispensable tool in the
study of agricultural systems, since models can be constructed
from smaller parts of the whole and the predictions of the model
tested against reality. It is not a simple matter, however, to
decide which of the smaller parts of a large system are satisfactory units for separate experimentation, either to produce data
for model building or to use in experimental testing of the model.

It has been argued that each constituent process, surrounded
by its immediate ring of variables, should be the unit for
acquiring data for modelling (12), but the important decisions
have then already been taken in deciding how to construct the
model. Conceptually, it is clear how the efficiencies of constituent processes can be assessed and used but it is not so clear
how the importance of efficiency in one process could be judged
in relation to the operation of the whole system.

It has been argued (13) that "sub-systems" should be delin-
eated, as the minimal units for experimentation. The basis for
this argument is simply that, if the primary interest is in the
whole system, then the parts studied must be relevant to the
whole. The problem may be readily seen from a simple example. A
honeybee might be an important part of a crop production system
(e.g. for pollination) but it would be possible to take that constituent out of the system and study it separately without ever
adding to our knowledge of its role within the system. Before
that can be done it is necessary to define a relevant sub-system,
of which the bee is a part, but which also specifies the
particular bee/plant (and other) relationships that are important
in this particular context.

There are methods that can be used to help in defining such
sub-systems (13) but the important point here is that, if the
argument is correct, then an understanding of the efficiency of a
large system could be derived from the measurement of the
efficiency of its sub-systems.

Of course, it can be argued that there are always still
larger systems to be considered and boundaries are difficult to

draw. An illustration of this problem that also touches on an important aspect of efficiency is related to the use of "support" energy.

"SUPPORT" ENERGY USE

"Support" energy has been defined as energy other than that derived directly from solar radiation: it includes "fossil" fuels, the energy content of inputs and the energy required to produce them, and the energy needed to process and distribute outputs. There are many problems in deciding what to include and what to leave out but there is no question that support energy is an important agricultural input and that the efficiency with which it is used matters greatly (14,15,16,17,18,19).

In energetic terms, modern agriculture is extremely inefficient, using vastly more energy (in total) to produce a product than is contained in it. For most production systems, even the additional support energy employed greatly exceeds the energy content of the products (see Table 3).

Several points need to be emphasised. Firstly, the efficiency of support energy use may be important because of its decreasing availability or its increasing cost. Secondly, some systems are more efficient that others, both within and between products. So, until 'within-product' efficiency has been explored, it is difficult to make comparisons between production systems. Thirdly, the support energy used during the production process may be small compared to that used before and after it. It is therefore necessary to consider the wider system, including, for example, distribution costs , in order to assess the relative importance of an efficiency value relating to the production process itself.

As mentioned previously, the efficiency with which one resource is used can rarely be viewed in isolation, but support energy is probably a good example of a case where one particular resource can rapidly assume a high degree of importance and justify consider-able effort in assessing the efficiency with which it is used.

CONCLUSION

It has to be recognised that we are bound to be interested in how well meat-production systems achieve the purpose for which they are carried out, and one major way of judging this is to assess the efficiency with which these purposes are achieved in relation to the costs incurred (or resources used). Efficiencies are bound to be rather static summaries of system performance and their value should not be overestimated. Nevertheless, the main difficulties are not inherent in the concept of efficiency but reflect the problems of defining purpose or combining several purposes in one expression.

Table 3

Efficiency of use of support energy (MJ) in Meat
Production (as MJ or kg Protein)

| Meat Product | Efficiency of production | | | |
	MJ/MJ	Ref.	MJ to produce 1 kg protein	Ref.
Beef	0.33	27		
	0.4	29	326	29
Lamb	0.37*	28		
	1.0	29	117+	29
Broiler	0.11	17		
	0.3	29	167	29
Pig meat	0.3–0.5	29	234–268	29

The figures from references (17) and (29) relate only to <u>direct</u>
energy costs of food production: they exclude processing,
transport etc.

*   including wool

+   excluding wool

REFERENCES

1.   Spedding, C. R. W. (1973). In: "The Biological Efficiency
of Protein Production".(Ed.J.G.W.Jones)(University Press, Cambridge)

2.   Charlton, P. J. and Street, P. R. (1970). In: "The use of
models in agricultural and biological research".
(Ed. J. G. W. Jones) P.50 (G.R.I. Hurley).

3.   Dent, J. B. and Anderson, J. R. (1971). "Systems Analysis
in Agricultural Management." (John Wiley and Sons. Austr.).

4.   de Wit, C. T. and Brouwer, R. (1968). Angewandte Botanick
      42, 1.

5.   Holling, C. S. (1966). "The strategy of building models of
      complex ecological systems". Ch. 8 in 'Systems Analysis in
      Ecology". (Ed. K. E. F. Watt) (Academic Press, London and
      New York).

6.   Watt, K. E. F. (1968). "Ecology and Resource Management".
      (McGraw-Hill, New York).

7.   Van Dyne, G. M. (Ed.) (1969). "The Ecosystem Concept in
      Natural Resource Management". (Academic Press, London and
      New York).

8.   de Wit, C. T., Brouwer, R. and Penning de Vries, F. W. T.
      (1971). "A Dynamic Model of Plant and Crop Growth". Ch. 8
      in "Potential Crop Production". (Eds. P. F. Wareing and
      J. P. Cooper) (Heinemann Educational Books, London).

9.   Morley, F. H. W. (1972). Proc. Aust. Soc. Anim. Prod. 9,
      137.

10.   Spedding, C. R. W. and Hoxey, A. M. (1974). The potential
      for conventional meat animals. Proc. 21st Nottingham Easter
      Sch. Agric. Sci. (Butterworth, London).

11.   Large, R. V. (1973). In: "The biological Efficiency of
      Protein Production". (Ed. J. G. W. Jones). p. 183. (University
      Press, Cambridge).

12.   Spedding, C. R. W. and Brockington, N. R. (1975). Agric.
      Syst. 1, 1.

13.   Spedding, C. R. W. (1974). "The Biology of Agricultural
      Systems". (Academic Press, London and New York).

14.   Odum, H. T. (1971). "Environment, Power and Society".
      (Wiley, New York).

15.   Black, J. N. (1971). Ann. Appl. Biol. 67, 272.

16.   Lawton, J. H. (1973). In: "Resources and Population".
      (Eds. B. Benjamin, P. R. Cox and J. Peel). p. 59. (Academic
      Press, London and New York).

17.   Leach, G. A. (1973). In: "The Man-Food Equation". (Ed.
      A. Bourne) (Academic Press, London and New York).

18. Leach, G. A. (1974). In: "Human Food Chains and Nutrient Cycles". (Eds. A. A. Duckham and J. G. W. Jones) (Elsevier, Amsterdam). In Press.

19. Slesser, M. (1973). J. Sci. Fd Agric. <u>24</u>, 1193.

20. Joint Beef Production Committee (1967). Handbook No. 1. (Meat and Livestock Commission, Bletchley).

21. Joint Beef Production Committee (1968). Handbook No. 2. (Meat and Livestock Commission, Bletchley).

22. Joint Beef Production Committee (1972). Handbook No. 3. (Meat and Livestock Commission, Bletchley).

23. Young, N. E. (1971). Unpublished data.

24. Penning, P. D. (1970). Quoted in: "Sheep production and grazing management". 2nd Ed. (Ed. C. R. W. Spedding). (Baillière, Tindall and Cassell, London).

25. Norman, L. and Coote, R. B. (1971). "The Farm Business". (Longmans, London).

26. Walsingham, J. M. (1974a). Pers. comm.

27. Jones, J. G. W. (1974). Pers. comm.

28. Walsingham, J. M. (1974b). Pers. comm.

29. Holmes, W. (1974). Assessment of alternative nutrient sources. Proc. 21st Nottingham Easter Sch. Agric. Sci. (Butterworth, London).

THE INFLUENCE OF REPRODUCTIVE RATE ON THE EFFICIENCY OF MEAT PRODUCTION IN ANIMAL POPULATIONS

R. V. Large

The Grassland Research Institute

Hurley, Berks

INTRODUCTION

At least half the cost of producing meat is accounted for by the food consumed by the animals; it is important that this food should be used efficiently. The many ways of expressing efficiency and the reasons why the efficiency of animal production processes should be carefully considered have already been discussed (1).

Before delving into the detailed biochemistry of muscle and fat formation, later in the programme, it is worth looking at the relatively simple relationship of the amount of meat produced per unit of food consumed. It is certain that in some of our species of domestic animals it is possible to achieve increases in efficiency purely by manipulating the structure of the animal population. One way is by improving the reproductive rate.

There are many reports in the literature (2,3,4,5,6) of the relative efficiencies of different species for the production of human food. Reference will be made to several different species (domestic fowl, pigs, rabbits, sheep and cattle) but it is not the purpose of this paper to compare them. It is preferable to consider each species in turn and assess its performance in relation to the factors that may influence its efficiency for meat production. The reasons for this are that (a) it is not always easy to compare the value of the products from different species and (b) it is difficult to equate the inputs of different types of food required by species of animals having quite different digestive attributes. Financial comparisons are outside the scope of this paper.

If efficiency is defined as the amount of meat (carcass) produced per unit of food consumed then the simplest example to take
is that of a young weaned animal, where efficiency will be a
straightforward ratio of meat output to food input, for a given
period of time. This calculation takes no account of the cost of
producing the young animal and growing it to a point where it is
independent. A more meaningful assessment of efficiency can be
made by including the cost of maintaining the breeding female
required to produce the young animal. If the female normally only
reproduces once a year (e.g. a cow or a ewe) then the cost of
keeping the breeding female for the whole year must be included.
An expression for the annual efficiency of a breeding female may
be derived as follows:

$$(1) \qquad E = \frac{W \times N}{F_D + (F_p \times N)}$$

where   W = Wt of the progeny at slaughter

       N = No. of progeny per year

      F_D = Food consumed by the dam in a year

      F_p = Average food consumption of the progeny from birth
          to slaughter

This expression contains all the factors that will have a
marked effect on efficiency. The carcass weight (W) can be
manipulated but is governed, to a certain extent, by consumer preference. The number of young (N) produced per year represents the
number of young born per litter and the frequency with which litters
are produced. The total food consumed by the progeny (F_p) is
related to the number of progeny and their rate of growth. Food
consumption by the dam (F_D) will be influenced by the total foetal
burden, the amount of milk produced and her own maintenance
requirement which will be related to her size.

When considering a whole population of animals, allowance
must be made for the replacement of breeding stock. This will
reduce the level of output by the young animals but output of meat
from the older females that are culled out will restore the balance
to some extent. Additional food will be required by these replacement animals to grow on to breeding age and allowance must be made
for a proportion of male animals to be maintained. The numbers of
young required as replacements will be influenced by the longevity
and mortality rate of the breeding stock.

An expression for the biological efficiency of a population
of animals is:

$$(2) \qquad E = \frac{W_1 \times (N_1 - N_2) + (W_2 \times N_3)}{F_D + (F_p \times N_1) + (F_R \times N_2) + F_{S/N_4}}$$

where W_1 = Weight of progeny at slaughter

W_2 = Mature wt of cull dams

N_1 = Number of progeny reared to weight W_1

N_2 = Number of replacements reared to weight W_2

N_3 = Number of dams culled

N_4 = Number of dams per sire

F_D = Food consumed by dam during the year

F_P = Food consumed by progeny to weight W_1

F_R = Food consumed by replacements in growing from W_1 to W_2

F_S = Food consumed by sire during the year

Wassmuth and Beuing (7) have extended this expression for sheep, by including factors to allow for differences in the quality of food inputs and of the meat output, between animals of different ages. One way of taking into account the quality of food input and output is to use financial terms; it is then a relatively simple step to convert this to an economic model by inserting current prices (see 8).

As Spedding (1) has already pointed out, the highest level of efficiency attainable is given by the direct conversion of food to product by the individual which excludes the overheads of maintaining the breeding stock. Examples of these maximum values are shown in Table 1.

The general relationship between the efficiency of the individual and that for a female producing n progeny per annum is illustrated in Figure 1. The efficiency of the individual may be changed by increasing its rate of growth, but similar increases in growth rate will also change the relative efficiency of the dam and her progeny.

Table 1

Efficiency (E) of individual animals

$$E = \frac{\text{carcass (kg)}}{\text{G.E. (MJ)}}$$

Domestic fowl	18.9
Pig (pork)	18.6
Rabbit	13.3
Sheep	5.2
Cattle	6.6

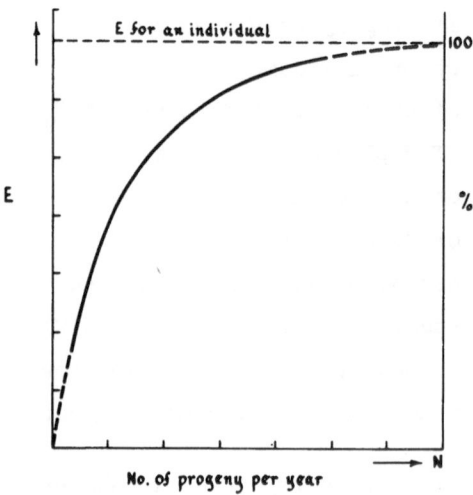

Fig. 1.   The relationship between the efficiency of meat production
(E) by an individual and that of a female producing N progeny per year

From the formula (1) it can be seen that when reproductive
rate is low most of the food input is accounted for by the dam
(i.e. her maintenance requirement); this proportion can be
influenced, to some extent, by the size of the product. The
product output relative to dam size is shown in Table 2 for the
five species mentioned. With increasing reproductive rate a
greater proportion of the total food intake is accounted for by
the progeny.

RESULTS OF CALCULATIONS

The efficiency of the dam and her progeny, of each species,
has been compared, over a range of reproductive rates, with the
values given for the individuals of each species (see Table 1).

The units used in these calculations are, for output, the
weight of dressed carcass produced and, for input, the gross
energy of the food consumed. The average reproductive rate and
the suggested potentials of each of the five species are shown in
Table 3.

The domestic fowl has the highest reproduction rate of our
meat producing animals. Figure 2 shows the maximum value for
efficiency as given by the value for an individual chick growing
to broiler weight. The effect of including the cost of maintaining the hen is shown to depress efficiency to 55% of the maximum
value where only twenty chicks are produced per year. If the

Table 3

Reproductive rate of domestic animals (Average values derived
from the literature. Figures in parenthesis
are suggested potentials).

Species	Gestation length	No. of parities per annum	Mean litter size	Total no. of progeny per annum
Domestic fowl	-	-	-	120 (240)
Pigs	112 days	2	8 (15)	16 (30)
Rabbits	32 days	6 (10)	8 (10)	48 (100)
Sheep	5 months	1 (2)	115 (3)	1.5 (6)
Cattle	9 months	1	1 (2)	1 (2)

current level of production is in the order of 120 chicks per
hen, this will result in a level of efficiency of about 95% of
the maximum value, and therefore increasing the reproductive rate
of the hen further will not improve efficiency to any great extent.

Figure 3 shows the results of a similar calculation for pigs
with the value of the individual pig, taken to porker weight,
compared with that of a sow and her progeny. If a current average
production of 16 pigs per year is taken then the efficiency level
is 83% of the maximum value and increasing the rate to 30 per
year would raise the efficiency to just over 90%.

Rabbits are considered in the same way in Figure 4 and the
maximum value is approached quite early on in the scale of reproductive rate. Although breeders claim that the new hybrid rabbits
can produce up to 80 young per year, 91% of the maximum value for
efficiency is reached with numbers of young less than 50: a
reproductive rate that may be achieved without much difficulty.
Thus, although higher rates of reproduction may be achieved in
rabbits, there would seem to be little advantage in doing so.

The curve for sheep, as shown in Figure 5, has been extended
well beyond the practical possibilities. The average level of
reproduction in sheep is about 1.5 lambs per ewe per year,
resulting in an efficiency of about 35% of the maximum value.
If the reproductive rate of sheep could be improved, to produce
6 lambs per year as suggested in Table 3, the efficiency would be

Table 2

Production attributes of domestic animals

Species	Wt of dam kg	Wt of product (carcass) kg	Product as % of dam's wt	No of progeny per annum	Total product as % of dam's wt	% of total food to dam	% of total food to progeny
Domestic fowl	3.0	1.3	45	120.0	5.200	10	90
Pigs (pork)	150.0	45.0	30	16.0	480	33	66
Rabbits	4.0	1.0	25	48.0	1,200	28	72
Sheep	75.0	17.0	23	1.5	34	72	28
Cattle	400.0	248.0	62	1.0	62	52	48

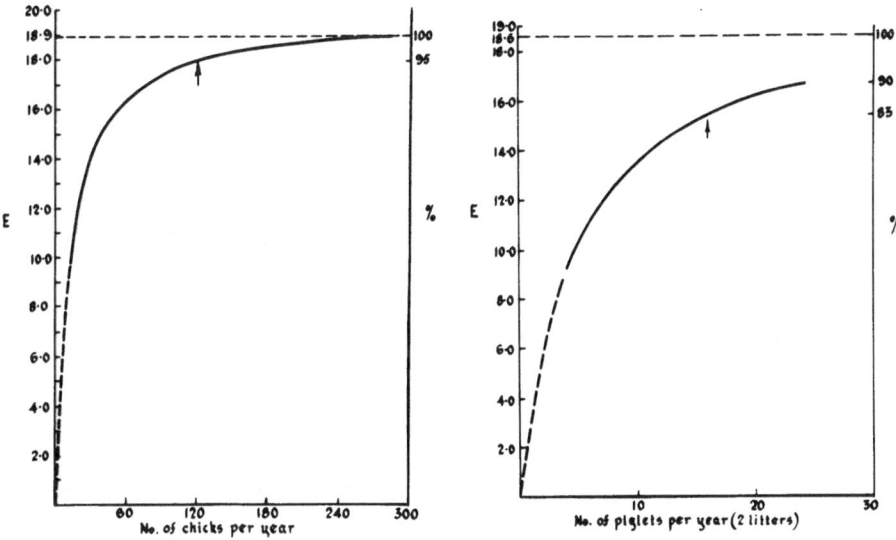

Figs. 2 and 3.   The effect of reproductive rate on the efficiency of meat production by domestic fowls (Fig. 2) and pigs (Fig. 3).

raised to 75% of the maximum value. However, this is not the only factor that can have a marked effect on efficiency (see later).

The potential for cattle, in terms of reproductive rate, is limited but Figure 6 indicates the efficiency that may be achieved by a cow producing two calves and, although very theoretical, a value for a cow with three calves has been included. The beef suckler cow with one calf gives an efficiency which is 35% of the maximum value. Using data for cows with two calves the efficiency is raised to 52% of the maximum value.

It has already been shown that the maximum value for efficiency of the individual was reduced by including the cost of maintaining the breeding female. In considering the influence of reproductive rate on the efficiency of a self-contained population the values will be further reduced by the necessity for maintaining the numbers of breeding animals and replacing, from the numbers of young produced, those breeding females that die or are culled out when their reproductive performance becomes unsatisfactory.

Obviously those animals that die are a complete loss to the system and will reduce the overall efficiency. Animals that are culled out have some value which can be added to the input side. The carcass weight of the culled animals may be used in the calculations but the meat may be of inferior quality commercially and

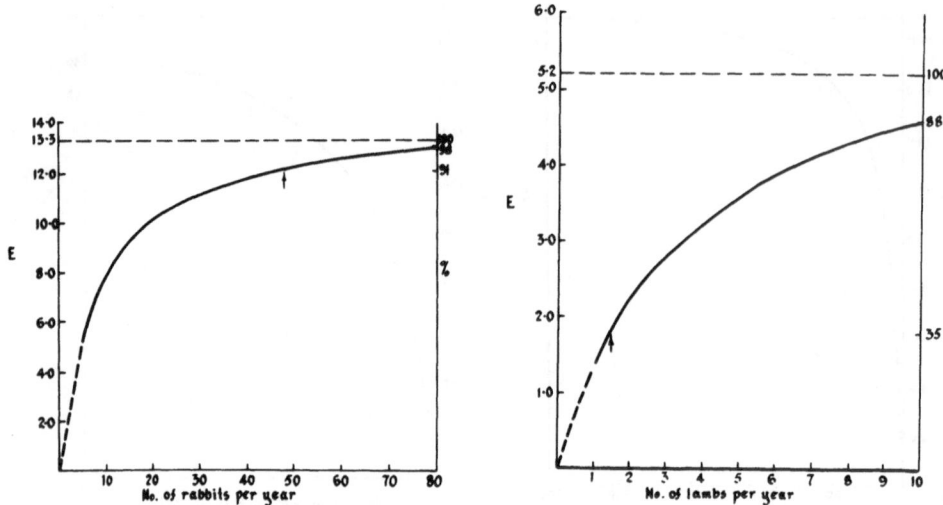

Figs. 4 and 5.   The effect of reproductive rate on the efficiency
of meat production by rabbits (Fig. 4) and sheep (Fig. 5).

Table 4

Factors affecting the population structure of domestic
animals (Average values from the literature).

Species	Ratio of males to females	Age to first parity	Average length of breeding life	Mortal- ity rate of breeding stock (%)	Replace- ment rate (%)
Domestic fowl	1 : 10–20	24 weeks	1 year	10	100
Pigs	1 : 20	1 year	2½ years	4	40
Rabbits	1 : 15–20	4–5 months	2 years	20	50
Sheep	1 : 30–40	2 years (1 year)	5 years	5	20
Cattle	AI	2 years (1½ years)	4 years	5	25

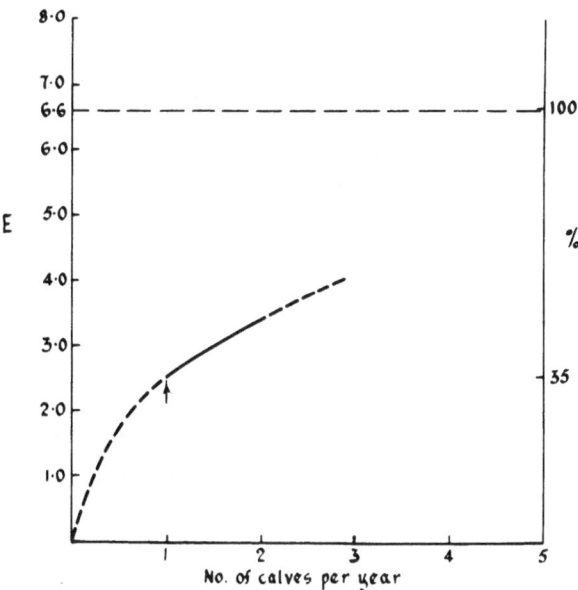

Fig. 6.   The effect of reproductive rate on the efficiency of
meat production in cattle.

Wassmuth and Beuing's approach (8) is probably more realistic.
The number of animals that are culled, over and above those that
die, will depend on a management decision on the number of replacements to be made each year.   Some idea of the factors that
can affect the population structure and its efficiency are shown
in Table 4.

     There are many calculations that could be made incorporating
some or all of these factors.   Two examples have been taken from
species having widely differing reproductive rates.   Certain
assumptions have been made about some of the factors, mainly
because of the lack of the appropriate data to build into the
calculation.

     The result of a calculation on the effect of reproductive
rate on the efficiency of a population of rabbits is shown in
Figure 7.   The assumptions made here are that the variation in the
number of young produced per year is related to the number of
litters born per year; an average litter size of eight was used.
A variable mortality rate has also been included on the basis that
the more pregnancies a doe has in a year the greater is the risk
of her dying.   However, Figure 7 shows that the efficiency of the
population differs very little from that calculated for a doe and

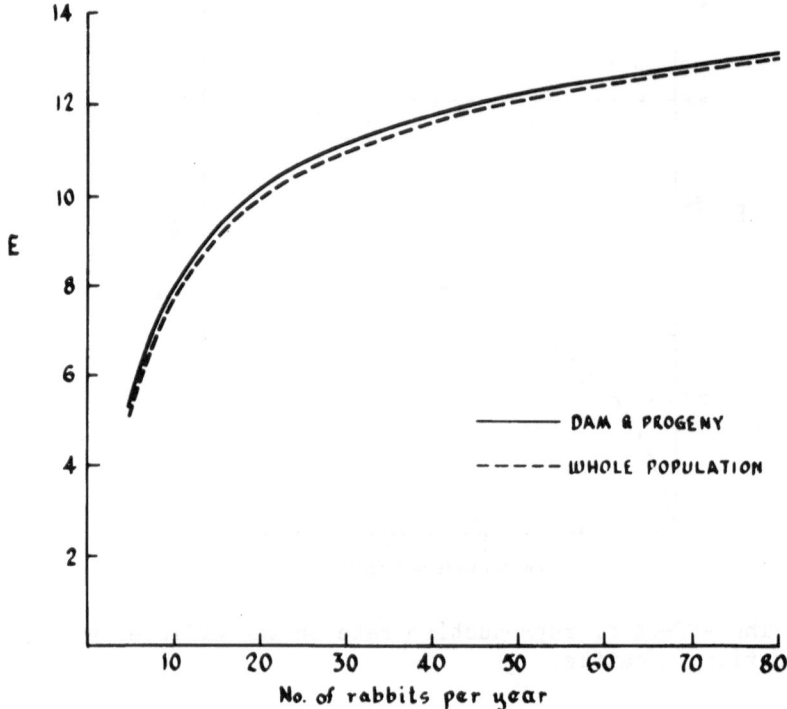

Fig. 7.   The efficiency of meat production by a whole population
of rabbits.

her progeny. This is because, at the lower end of the reproductive
scale, the mortality rate of the females is low and the culled
animals are balancing the numbers of young required for replacements and at high levels of reproduction the proportion of young
required as replacements is a very small percentage of the total
number produced.

For sheep, a calculation has been made to show the effect, on
efficiency, of the number of years that a ewe is kept for breeding.
Here it has been assumed that the longer the ewes are kept for
breeding (i.e. the lower the replacement rate) the higher will be
the mortality rate and hence the fewer will be the ewes culled for
sale. Figure 8 shows that efficiency decreases the longer ewes
are kept. Another factor that has been included is that ewes
bearing triplets and twins will have a higher mortality rate
than ewes bearing singles. However, from Figure 8 it is clear
that the effect on efficiency of varying a factor like the replacement rate in a flock, is small compared with that likely to be

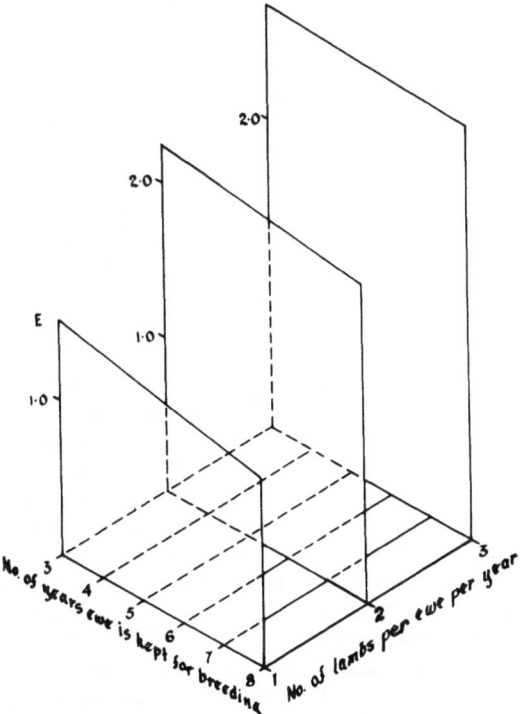

Fig. 8.   The effect of the replacement rate of breeding ewes on the efficiency of meat production by a flock.

to be achieved by increasing the reproductive rate of the ewes.

     In general, it would seem that the effects of the population structure are relatively small, compared with those that affect the individual animal or a dam and progeny, due largely to the balancing effect of production from culled animals (9); the main factor likely to upset this balance being the mortality rate of the breeding stock.

CONCLUSIONS

     The efficiency of meat production by five species of domestic animals has been compared, over a range of reproductive rates, with that of the individual animal for each species: the efficiency of the individual representing the maximum value attainable.

     For those species with an inherently high reproductive rate, such as domestic fowl, rabbits and to some extent pigs, current levels of production achieve an efficiency of around 90% of the maximum value and, biologically, there would seem to be little value in trying to improve on this, but if some other measure of efficiency was being considered the conclusion might well be different. Cattle offer limited scope for improvement in reproductive rate although on the basis of the calculations made here, the production of twins would increase the actual value for efficiency by some 30% over that for singles. Sheep seem to offer some scope for increasing the reproductive rate with a reasonable return in terms of increased efficiency. Although considerable progress has been made in this field (10), biological possibilities must be tempered with the recognition of the additional inputs required both in terms of management and economics.

Table 2 shows the proportions of food that are eaten by the dam and the progeny respectively. Where the reproductive rate is high, a high proportion of the total food is consumed directly by the young animals and the maximum value for efficiency is approached. Therefore the main factors affecting the overall efficiency, in these instances, are those that affect the efficiency of the individual, e.g. the growth rate of the progeny. Conversely, for sheep, where the proportion of food consumed by the lambs is much smaller than that for the ewe, the growth rate of the lamb is relatively unimportant. The more important factor here is the amount of food consumed by the ewe which may be reduced by using a smaller ewe, but then the relative level of product output must be maintained by crossing the small ewe with a large ram. Thus, for sheep, there is the possibility of improving efficiency by increasing the reproductive rate and by reducing the size of the breeding female (11). Cattle have some advantage over sheep (12) in that, despite their low reproductive capacity, their product size is large relative to the size of the dam; in fact it is the greatest, per individual, of all the species considered.

Consideration of some of the factors contributing to the structure of a population of animals suggests that they are less important than those factors that already have a marked effect on the efficiency of the individual and of the dam and her progeny.

REFERENCES

1. Spedding, C. R. W. (1974) In:"The Control of the Growth and Productive Efficiency of Meat-Producing Animals." p. 29 this volume.

2. Dickerson, G. (1970). J. Anim. Sci. 30, 849.

3. Holmes, W. (1971) In:"Potential Crop Production. A Case Study." (Ed. P. F. Wareing and J. P. Cooper) p. 213 (Heinemann, London).

4. Wilson, P. N. (1973) Phil. Trans. R. Soc. Lond. B. 267, 101.

5.   Blaxter, K. L. (1968) Proc. 2nd Wld. Conf. Anim. Prod.
     Maryland p.31 (University of Maryland).

6.   Reid, J. T. (1970) Proc. 1970 Cornell Nutr. Conf. p.50
     (Cornell University).

7.   Wassmuth, R. and Beuing, R. (1973) Wld. Rev. Anim. Prod.
     9, 48.

8.   Wassmuth, R. and Beuing, R. (1974) Livestock Prod. Sci. 1, 67.

9.   Large, R. V. (1973) In: "The Biological Efficiency of Protein
     Production", (Ed. J. G. W. Jones) p.183, (University Press,
     Cambridge).

10. Robinson, J. J. (1974) Proc. Brit. Soc. Anim. Prod. 3, 31.

11. Large, R. V. (1970) Anim. Prod. 12, 393.

12. Baker, R. D., Large, R. V. and Spedding, C. R. W. (1973)
     Proc. Brit. Soc. Anim. Prod. 2, 35.

THE OPTIMUM SIZE AND STRUCTURE OF ENTERPRISE

P. N. Wilson

BOCM SILCOCK

Basingstoke, Hampshire

INTRODUCTION

Five interacting variables jointly determine the optimum size of any meat-producing enterprise. The optimum structure will depend upon technological and managemental innovation, but the variables which affect size, also contribute, though at a different order of magnitude.

The five variables are:-

1. The profit required from the enterprise as a whole.

2. The number of animals.

3. The profit per head or the level of productivity per head.

4. The land area required per animal.

5. The chosen strain, breed, species or genus.

With the passage of time, there will be a desire to increase the profit from an enterprise partly because producers will expect a higher reward and partly because inflation lowers the value of 'fixed returns'. Social pressures restrict the availability of land for agricultural use, its value rises and in time the area available per animal tends to fall nearer to a minimum than to a theoretical optimum. As stocking rates or carrying capacities increase, the total number of livestock will also tend to increase, to a limiting point after which there will be a tendency for the number of livestock to stabilise. At this point the aim

Table 1

Size and numbers of Feedlots in United States (1)

	No. of Head ('000)	1962		1967	
		No. Lots	Cattle ('000)	No. Lots	Cattle ('000)
Major 'beef' States	1-2	801	900	960	1.200
	2-4	385	800	510	1.500
	4-8	195	1,100	313	2.000
	8-16	106	1,500	153	2.200
	16-32	26	900	59	2.100
	>32	5	300	13	1.100
All States	<1	234,646		209,505	
	>1	1,518		2,008	
	Total	236,164		211,513	

will be to increase production, and hence, profit, per animal
rather than per unit area of land.

TEMPORAL CHANGES IN THE SIZE OF LIVESTOCK HOLDINGS

Table 1 illustrates the dramatic changes which are currently
taking place in the sizes of feedlots in the major beef producing
areas of the USA. Thus the number of feedlots carrying more than
32,000 head increased $2\frac{1}{2}$ times during the period 1962-1967.
Similar changes are taking place within both the UK and the EEC,
although the initial size of the enterprises and the scale of the
increase are very much smaller. Table 2 presents data for the 9
member states of the EEC, and shows that the average size of dairy
and beef cattle herds in the UK at 57.8 head is roughly twice that
in the remaining 8 member states. Within the EEC, mean herd size
is rising at a figure varying between 1 and 4 cows/herd/annum, and
this trend is likely to continue for many years. Table 3 presents
somewhat dated information relating to beef cattle holdings in the
former 6 member states of the EEC. It will be noted that the
greatest percentage of cattle are to be found in herd sizes of

Table 2

Distribution (%) of herds by size relative to all herds in each of the countries of the EEC (2)

No. in herd	Year	1-5	5-10	10-20	20-50	50-100	>100	Mean size of herd
Belgium	67	4	7	19	48	22	–	22.6
Luxemburg	70	1	2	8	41	39	9	31.5
*France	69	11	28	43	17	1	–	18.6
*W Germany	69	22	34	31	12	1	–	14.1
Italy	61	28	22	22	16	6	6	6.2
**Holland	70	6	11	20	45	19	6	33.0
Denmark	71	1	3	12	47	28	9	28.5
*Eire	60	33	26	23	–	19	–	20.4
U.K.	71	–	1	21		29	49	57.8

* Cows only   ** Cows for breeding

Table 3

Number of Holdings with Fattening Cattle
by Herd size in the six: 1966-7 (2)

Number in Herd	% of all Holdings	% of all Cattle
1	27	5
2	22	9
3-5	26	19
6-9	12	17
10-14	6	14
15-19	3	9
20-29	2	10
30-49	1	8
50-99	0.4	6
>99	0.2	5

between 3 and 15 head, and that only 0.2% of the holdings, and only 5%
of the total cattle, are to be found in herd sizes greater than
99 head.

RELATIONSHIP BETWEEN SIZE OF HOLDING AND LIVESTOCK PERFORMANCE

Tables 4 and 5 summarise data on beef farms recorded by and
recently analysed by the Meat and Livestock Commission. The
correlation coefficients have been calculated between the size
of the livestock unit on the one hand, and various parameters of
livestock productivity on the other. Although information from
1900 farms has been analysed in this survey the correlation coefficients are in all cases extremely low and, in several cases,
negative. The uncorrected average correlation coefficients for
the 3 factors examined are positive, but on average they are not
significantly different from zero. There are therefore no grounds
for postulating that an increase in herd size will, per se,
increase production per head. Perhaps the most useful interpretation however, is the negative one, namely that increasing
herd size will not necessarily decrease production per beast.
This might be assumed by virtue of the fact that increasing herd
size of necessity must decrease the individual care and attention
given by stockmen to each individual breeding cow and finishing
steer.

It therefore follows that the well known 'economies of scale' can be expected to accrue as beef herd size increases. In other words, fixed costs per head, particularly of labour and management, can be effectively reduced by spreading them more efficiently over a larger number of productive animals.

APPLICATION OF THE LAW OF DIMINISHING RETURNS TO THE ECONOMICS OF ENTERPRISE SCALE.

Clearly any meat producing system, and any given level of management competence, has its own specific economic optimum size. Total enterprise profit, expressed by such parameters as return on capital investment, will be normally distributed around such optimum size, with the two tails of the curve indicating sub-optimal performance at herd sizes which are larger or smaller than optimal. In practice, however, the actual data on which such a distribution curve is based are likely to cause the curve to be skewed rather than normally distributed, for it is unlikely that data will be available from many farms which are markedly larger than the optimal size of enterprise. For this reason, and taking a small sample of farms ranging from very small to optimal, or marginally above optimal, size, it follows that one might expect a spurious correlation, of a low order of magnitude, between economic efficiency and enterprise size. Just such a situation is revealed in the data tabulated in Tables 4 and 5. Farm size, both in the USA and the EEC, is currently increasing closer to, and rarely past, optimal enterprise size and hence the data are necessarily skewed. Theoretically, if there were as many herds in excess of optimal size as there were below it, the regression of economic efficiency on size of enterprise would be curvilinear, and not linear.

SIZE OF ENTERPRISE AND THE SPREADING OF DIRECT COSTS

As already described, most of the so called 'economies of scale' are derived from the more equitable spreading of fixed costs over a larger number of productive outputs. This is illustrated in Table 6, which relates to beef farms in the UK in the early 1970's. Two levels of annual wage are included in the calculation, one relating to a high paid specialist stockman (£4,000 per annum inclusive of overtime) and one relating to a part time stockman (calculated at £1,000 per annum inclusive of overtime). The technological efficiency of the enterprise is accounted for in terms of calving interval (18 v 12 months) and calf mortality (20 v 0%) from which the number of calves produced per annum can be calculated and the labour costs apportioned to each calf sold. Four representative herd sizes have been chosen, spanning herds from 25 to 200 cows, so that the type of beef

Table 4

Correlation between size of Unit and Performance (3)

System	Age at slaughter (months)	Number of Cattle and daily live weight gain (r)	Number of cattle and mortality % (r)	Number and stocking rate (r)
Cereal Beef	<12	0.019	0.098	N/A
Grass/Cereal	15	0.069	-0.069	0.147
Grass/Cereal	18	-0.101	-0.045	0.223
Grass/Cereal	24	0.043	-0.133	0.214
Grass	20	0.121	-0.098	0.168
Grass	>24	0.097	-0.007	0.207
Suckled Lowland	>24	0.040	-0.060	0.092
Suckled Upland	>24	-0.108	-0.005	0.143
Suckled Hill	>34	0.146	0.182	0.193

Table 5

Correlations between size of Unit and Performance (3)

System	Number in herd and daily liveweight gain	Number in herd and mortality	Number in herd and stocking rate
Overwintering of suckled calves	-0.056	0.138	N/A
Overwintering of stores	0.184	0.102	N/A
Finishing suckled calves	0.207	-0.041	N/A
Finishing Lowland stores	0.076	0.137	N/A
Finishing Hill stores	0.107	-0.048	-
Grass Finishing Lowland stores	-0.205	-0.008	0.032
Grass Finishing Hill stores	0.272	-0.016	0.141
Grass Finishing Irish stores	0.008	-0.083	0.002
Overall Correlation	0.036	0.002	0.142

enterprise currently to be found in the UK falls within the middle
of this range. It will be noted that the 'direct labour cost per
calf sold' ranges from a preposterous figure of £360/steer to an
acceptable £5/steer. Although it may be considered to be highly
unlikely that a small beef breeding herd of 25 cows would ever
justify the employment of a full time stockman earning £4,000 per
annum, such herds do exist, but they rely on the expectation that
the income from fat cattle will be supplemented by the income
from the sales of pedigree breeding stock. If, in any year,
there are no sales of stock for breeding, then the labour costs
per finished steer would indeed be as high as the level shown in
Table 6.

ENTERPRISE PROFIT

As mentioned in the introduction, the size of a given enterprise is often fixed in such a manner that the total enterprise
profits produced therefrom is acceptable. Table 7, accordingly,
calculates the size of enterprise required to produce any given
level of total enterprise profit, at varying net margins per
animal sold. It will be noted, therefore, that the expected total
profit from a typical U.K. enterprise expecting a margin between
£8 and £16 on 50-100 head lies between the limit of £44 and £1600
per annum. Assuming that this range of total enterprise profit
is, or is likely, shortly to become totally unacceptable it follows
that either the size of the enterprise must increase in future or
the net margin per head must improve dramatically. Relevant
interpolation in Table 7 can determine the required number of
cattle, at a given net margin/head, assuming a known requirement
for total enterprise profit.

CONCLUSION

This paper has shown that there is no proven correlation
between size of enterprise and animal productivity within the
beef industry. However, current mean sizes of beef enterprises,
both within the UK and even more so within the EEC, are extremely
low with the consequence that the total enterprise profit from
such enterprises is unacceptably low. In the past, the profit
from the beef enterprise has been counterbalanced by profit from
other farm enterprises in a mixed farming situation, but it is
highly likely that agricultural systems will become more specialised, and hence 'mixed farming' will decline. Various calculations
are presented which show the inter-dependence of total enterprise
profit on factors both within the farmer's control (size of
enterprise) and only partially within his control (net margin per
head).

It is concluded that the trend towards larger units will

Table 6

Labour costs (£) per calf reared and sold on beef breeding farms
(For details see text)

Wages (£ p.a.)	High (4000)				Low (1000)			
Calving interval (mth)	18		12		18		12	
Calf Mortality (%)	20	0	20	0	20	0	20	0
Number of cows in herd								
25	360	240	200	160	75	60	50	40
50	160	120	100	80	38	30	25	20
100	80	60	50	40	19	15	13	10
200	40	30	10	20	10	8	7	5

Table 7

Calculation of Enterprise Profits

Net Margin per head (£)	Number in herd					
	10	25	50	100	500	1,000
0.5	5	13	25	50	250	500
1	10	25	50	100	500	1,000
2	20	50	100	200	1,000	2,000
4	40	100	200	400	2,000	4,000
8	80	200	400	800	4,000	8,000
16	160	400	800	1,600	8,000	16,000
32	320	800	1,600	3,200	16,000	32,000

continue to increase for some considerable time to come, although
it is highly improbable that Europe will emulate the very large
units typical of the feedlots in the Southern states of the USA.

REFERENCES

1.   Preston, T. R. and Willis, M. B. (1970). "Intensive Beef
     Production". (Pergamon Press, Oxford).

2.   Anon. (1973). "U.K. Farming and the Common Market: Beef".
     (N.E.D.O., London).

3.   Records of Meat and Livestock Commission, Bletchley.

DISCUSSION

Dr. van Es asked which of the various measures of efficiency
discussed by Prof. Spedding would be used by farmers and which by
advisory workers. *Prof. Spedding* replied that farmers would
always be concerned with profit and would, therefore, be concerned
with efficiency in the monetary sense. When they or advisers
were concerned with analysing the components of that efficiency,
then the problem must be examined at a different level. In
research one may take this analysis to an even more detailed level.
Dr. Fowler asked what allowance Mr. Large had made for the value
of the dam at the end of her breeding life; this was particularly
important in the case of once-bred animals when the offspring
could in some sense be regarded as a by-product. In reply
Mr. Large said that no allowance had been included in his presen-
tation, but approximate calculations suggested that in such
systems, efficiency would approach the maximum value. The
maximum value quoted might be increased if it is true that pregnancy increases the efficiency with which food is utilised for the
production of maternal body tissue.   *Prof. Spedding* noted that
the production of pigs from the once-bred gilt was no different
in principle from any other agricultural system in which there was
more than one product, for example the beef calf produced by the
dairy cow. *Dr. Braude* commented that financial return was
determined, not only by the efficiency of the individual animal,
but by the turnover per unit of production. Prof. Spedding had
implied that it was possible to look at only one dimension of a
problem at one time; *Dr. Fuller* wondered if it was not possible by
linear programming techniques to look at all relevant dimensions
simultaneously. *Prof. Spedding* replied that although simulation
techniques and the use of computers would extend one's capacity to
grapple with complexity, the more complicated the system one tries

to examine, the less efficiently one examines any particular
component of it. To construct the ultimate model for all purposes
would result in a replica of the original which, by definition,
is no longer a model.

Asked to what extent the optimum size of an enterprise is
dependent on the mixture of enterprises of which it is a part,
Dr. Wilson pointed out that there are conflicting interests
between the flexibility of a mixture of small enterprises and the
economies of large-scale operations. *Dr. Fowler* thought that
economic forecasting in the beef industry was hindered by the
lack of structure in the marketing organisation. *Dr. Wilson*,
while sympathising with Dr. Fowler's view, thought it significant
that, where the greatest technological progress had been
achieved, for example in broilers, there is no national marketing
agency. *Prof. Lucas* thought that the intensification of animal
production envisaged by Dr. Wilson would inevitably lead to
competition for the produce of arable land and under-use of the
productive potential of marginal land. *Dr. Wilson* foresaw the
increased use of cheap transportable arable by-products, such as
chemically-treated barley straw, to allow greater intensification
of upland areas. He regretted that the indiscriminate application
of Government grants had tended to provide for the survival of
uneconomic units rather than an incentive to greater efficiency.

The Efficiency of Food Conversion

COMPARISON OF BIOLOGICAL MECHANISMS FOR CONVERSION OF FEED TO MEAT

D. G. Armstrong

Department of Agricultural Biochemistry
University of Newcastle upon Tyne

INTRODUCTION

The carcases of domestic livestock are made up of muscular tissue, fatty tissue and bone. Muscular tissue, from which meat is essentially derived, comprises protein with its associated intramuscular fat. As the percentage of fatty tissue in a carcass increases, due for example to an increase in plane of nutrition, the percentage of intramuscular fat also tends to increase (1,2). Thus in cattle with 20% fatty tissue in the carcass the L. **dorsi** muscle contained 4.5% intramuscular fat; with 39% fatty tissue present this value had risen to 11.1%.

Meat is primarily regarded as a protein food and clearly protein biosynthesis is of over-riding importance. It is the intention in this paper to refer to the processes of digestion and subsequent metabolism within the body whereby the constituents of food are converted to the proteins of muscle tissue. Particular emphasis will be laid on comparative aspects of these processes as they relate to the pig and ruminant animal. No reference will be made to aspects of fat digestion and metabolism since this will be dealt with in a later contribution.

It should be noted that there are a number of biophysical and biochemical changes commencing in muscle tissue at the time of death which result in differences between meat and the muscle from which it is derived (3); these changes will not be discussed in this paper.

The processes and factors which govern the ability of a food to supply the body with the essential (and non-essential) amino

acids and additional non-amino, energy-yielding substrates necessary for protein biosynthesis will first be considered.

SUPPLY OF AMINO ACIDS TO THE BODY

The Simple-Stomached Animal

Digestion of proteins in the monogastric animal has recently
been reviewed (4). The stomach plays an important part in the
process not only by acting as a temporary storage organ and initiating proteolysis with gastric proteases following HCl denaturation but also by virtue of its involvement in regulatory control
of protein release into the small intestine (5) and hence regulation of the overall rate of protein digestion. The rate of
stomach emptying is inversely related to the dietary concentration
of protein (6) and is dependent upon the kind of protein (7). It
has also been shown that for a given protein, rate of gastric
emptying is affected by the nature of the dietary carbohydrate (7).
Heat-damaged proteins of lowered digestibility leave the stomach
more speedily than if given in the unheated or mildly heated form
(8). Porter and Rolls (5) conclude that, in general, as the
quantity of food given at a single meal increases, so does the
absolute rate of stomach emptying; the rate of emptying expressed
as the proportion of the food fed that leaves the stomach per unit
of time however declines. As Erbersdobler (4) points out, the
mechanisms controlling stomach emptying require further study but
some feedback mechanism depending upon the concentrations of free
amino acids in portal plasma or in the duodenum seems likely.

Within the small intestine the change in pH associated with
the influx of bile and the resulting action of the pancreatic
proteases and of the peptidases associated with the intestinal
secretions rapidly complete the digestion process. There is increasing evidence that uptake of peptides by the epithelial cells
of the small intestine is of major importance in the process of
protein digestion (9). These cells possess the peptidases to
complete the hydrolysis of such peptides to their constituent amino
acids and thus only free amino acids enter the portal blood (10).

Studies with animals fed protein supplements such as sunflower
or fish meal together with additions of lysine or methionine
reveal no differences in time of uptake of the amino acids as
between those ingested free or as protein (see 11). Porter and
Rolls (5) consider, however, that with poorly digested proteins
such as zein or severely heat-damaged fish meals, the rate of
intestinal proteolysis may be an important factor in limiting
their effective use for subsequent protein biosynthesis within
the body, and especially where supplementation of the limiting
amino acid is practised.

Most of the observations referred to above have been obtained
using the rat as experimental animal. In so far as conclusions so
drawn apply equally to the pig, it is apparent that the quantities
and proportions of amino acids presented to the tissues of the pig
are dependent upon the amount and amino acid make-up of the dietary
protein and the availability (i.e. digestibility within the stomach
and small intestine) of its constituent amino acids. Concerning
this last point, Eggum (12) has determined values for true digestibility (TD) of the individual amino acids in a considerable range
of feedstuffs when fed to pigs and some of the data are shown in
Table 1. Inspection of the data in Table 1 suggests that the
values for individual amino acids in barley are lower than in
wheat and that in the fish meal used the amino acids were almost
completely available. It should be noted that TD is not synonymous
with the term availability as defined above since it does not
exclude events occurring within the caecum and colon; in that TD
is, however, calculated from faecal loss of an amino acid after
correction for metabolic faecal loss of the acid, differences
between TD and availability, if any, are likely to be small. It
is interesting to note that the TD values determined by Eggum (12)
for pigs were not significantly different from those found for the
rat fed the same diet. Microbiological techniques for determining
availability of amino acids have recently been reviewed (13,14).

Table 1

True digestibility values[*] determined with young
pigs for four amino acids present in a
selection of feedstuffs (data of Eggum; 12)

	Lysine	Methionine	Valine	Leucine
Barley	72	78	83	84
Oats	73	77	79	82
Wheat	84	89	91	93
Maize	89	94	90	93
Fish meal	96	94	94	95
Soya bean meal	92	87	90	91
Groundnut meal	88	89	90	91

[*] In calculating these values the total faecal loss of an individual amino acid was corrected for the metabolic faecal loss of
that amino acid.

The Ruminant Animal

In marked contrast to the situation pertaining in the simple-stomached animal, the early intervention of microbial fermentation in the digestive system of the ruminant markedly affects the relationship between dietary supply of amino acids and the amounts subsequently available for absorption into the body.

N transformations in the rumen.   N transformations occurring within the reticulo-rumen have been reviewed (15,16). Essentially the amount of protein entering the small intestine is dependent upon the extent to which dietary protein escapes microbial fermentation within the reticulo-rumen and the amount of microbial protein synthesised therein. Protein breakdown in the rumen, at least as reflected by ammonia release in vitro (17), is a function of the solubility of the protein in rumen liquor; the higher the solubility the greater the degradation. Some recently determined (18) values for the percentage of protein escaping digestion within the rumen of sheep fed a partially purified diet are lupin meal 36, ground nut meal 37, soya bean meal 61 and fish meal 71. The extent of microbial cell synthesis is largely governed by the levels of energy and of N available to the micro-organisms, the last-mentioned being associated with the magnitude of protein breakdown in the rumen and the extent of urea recycling. N-energy relationships occurring within the rumen with particular reference to factors affecting microbial cell yield have been reviewed (19). Hobson (20) has stressed the importance of dilution rate (D) in the rumen as a factor in governing the yield of microbial cells arising from carbohydrate fermentation in the rumen. It is noteworthy that Harrison, Beever, Thomson and Osbourn (21) have shown that the intra-ruminal infusion of 4 1/d of artificial saliva containing 4% w/v polyethylene glycol into sheep fed a diet of flaked maize and dried grass induced a significant increase in D (from $0.039h^{-1}$ to $0.098h^{-1}$) in two out of three animals and that this increase in dilution rate was associated with a 23% increase in flow of microbial amino acids into the small intestine (an increase of 20% in total amino acid flow). In the third sheep, in which the D value was already high without the infusion (0.087^{-1}), the infusion of artificial saliva did not increase the D value further and nor was there any increase in total or microbial amino acid flow into the duodenum. Reference will be made later to aspects of volatile fatty acid (VFA) metabolism and in this connection it is noteworthy that in the two animals in which D was increased by the rumen infusion, the molar proportion of pro-pionate in the rumen liquor was reduced (mean values: 29.5% control; 20.2% infused); butyrate and, to a lesser extent, acetate were increased. From data obtained on a wide range of all-forage and high concentrate (cereal) diets fed to sheep it has been calculated (19) that mean values for microbial N entering the small intestine/100g organic matter fermented in the rumen were

respectively 3.3 ± 0.1 and 2.2 ± 0.2 g. In that the consumption
of forage diets tends to be associated with high D values and low
molar proportions of propionate in rumen VFA while high cereal
diets are associated with lower D values and higher proportions of
propionate, these findings are in agreement with those of Harrison
et al (21).

The extent to which effects such as the above are associated
with changes in the degree of N recycling within the rumen requires
elucidation. It is known that the protozoa obtain a major part of
their requirements for growth from engulfed bacteria. Nolan and
Leng (22) have shown that, in sheep fed chopped lucerne and in
which some 12.0g/d microbial N left the reticulo-rumen, 3.1g/d
microbial N were recycled within the rumen. Weller and Pilgrim
(23), in studies with sheep equipped with fistulas in both rumen
and omasum and given a variety of roughage diets, have observed
that the passage of protozoa from the reticulo-rumen to the omasum
in the fluid effluent ranges from only 6 to 29% of the amount
expected based upon concentrations of protozoa in rumen fluid.

The tendency towards uniformity in amino acid composition of
duodenal digesta noted by a number of workers (24,25,26) is not
surprising in view of the relative constancy of amino acid
composition of microbial protein (27) and the contribution that
such protein and endogenous protein secretions make to the total
protein entering the small intestine. However, when sheep were
fed a diet containing fish meal protein, which was only partially
degraded within the rumen, the amino acid composition of the
digesta entering the small intestine reflected in small part the
differences in composition between fish meal protein and microbial
protein (28).

Determinations of the biological value of rumen microbial
proteins (29,30) - albeit with the rat - give values which lie
in the range 78-82. Protein quality is clearly quite high and the
lack of variation in values is in keeping with the constancy of
amino acid composition referred to above. The upgrading of low
quality dietary proteins by conversion to microbial protein is of
benefit to the host animal; equally it is undesirable to allow
high-quality proteins such as fish meal to be used as N sources for
the rumen microflora.

Protein digestion within the small intestine. The
ruminant possesses the same pancreatic proteases (31) and, most
probably, intestinal peptidases as those present in the small
intestine of the simple-stomached animal. Data relating to both
sheep (31,32) and cows (33) suggest that protein digestion is
reasonably efficient (see Table 2). Appreciable amounts of nucleic
acids also enter the ruminant's small intestine and are efficiently
digested therein (32).

Table 2

Mean values for net disappearance of amino acid N
in the small intestine of sheep (31) and cows (33)

	Sheep			Cows	
	Grass (fresh, frozen)	Grass (dried, chopped)	Barley/ dried grass (7:1)	Hay/ Conc (4:3)	Grass (fresh)
Total amino acid N	65	79	71	70	75
Essential amino acid N	68	82	68	72	75
Non-essential amino acid N	62	76	74	68	74
Lysine	70	77	74	77	78
Methionine	64	88	81	75	73
Valine	60	82	71	68	77
Leucine	71	82	77	75	77

N intake and amino acid uptake from the small intestine.
      From data obtained with sheep fed dried and fresh forages the
following linear equation relating N intake to amino acid N uptake
from the small intestine has been derived (26):-

$$Y = 0.888 - 0.116x \quad \dots\dots\dots\dots\dots\dots\dots\dots \text{equation 1}$$

      where Y = g total amino acid N disappearing in small
      intestine/g N intake and x = % N in organic matter of
      forage (n = 16; r = 0.912).

Table 3 shows some values for amino acid uptake at different
dietary N concentrations, calculated using the equation. It is
emphasised that this equation relates to forage diets fed to sheep
and is based upon a limited amount of data; such relationships
are required for a wide range of feeds and for both cattle and
sheep.

Table 3

Amino acid N uptake (g/d) from the small intestine in
sheep fed a daily ration of 1kg organic matter containing varying levels of dietary N calculated from equation 1

% N in Organic matter fed	N intake g/24h	Amino acid N uptake g/24h
1	10	7.72
2	20	13.12
3	30	16.20
4	40	16.96

Comparative Aspects

From the foregoing it can be seen that in the pig the amount
of protein fed and its amino acid composition are the major
determinants of the amounts of essential (and non-essential) amino
acids available to the tissues for protein biosynthesis and their
ratios one to the other. The amount of dietary carbohydrate
present has little effect apart from contributing to energy
requirements for absorption of amino acids by the gut wall to which
further reference will be made later. In the ruminant, on the
other hand, the major determinants are the magnitude of microbial
protein biosynthesis - which is dependent upon the availability of
the dietary carbohydrate as well as the N (but not necessarily
protein) supply - and the extent to which dietary protein escapes
fermentation in the rumen. Amino acid composition of digesta
entering the small intestine of ruminants fed conventional diets
varies but little. Furthermore in the ruminant, in which flow of
digesta from the abomasum is virtually a continuous process, the
supply of amino acids being absorbed is more or less continuous.
In the pig the absorption of amino acids occurs in peaks associated
with interval of feeding. In the young pig the maximal concentrations of amino acids is reached between 1 and 1½h after the end of
a meal (34).

DIGESTION AND METABOLISM OF CARBOHYDRATES

In contrast to carbohydrate digestion in the pig, which
primarily takes place in the small intestine and results in the
uptake of monosaccharides, primarily glucose, the major end products
of carbohydrate digestion in the ruminant are the VFA absorbed prior
to the small intestine. On certain high-cereal diets appreciable

amounts of α-linked glucose can enter the small intestine and be
digested therein (35,36) and there is indirect evidence that such
digestion is associated with the uptake of glucose or glucose
precursors (see 36). The importance of these differences in
carbohydrate digestion between ruminant and non-ruminant on overall
protein metabolism has yet to be fully evaluated. Three areas of
possible significance are 1, the uptake of amino acids and their
immediate fate within the mucosal lining of the intestine, 2, the
reliance of the ruminant upon gluconeogenesis and 3, the specific
effect of glucose metabolism on protein biosynthesis.

Amino acid uptake and immediate fate within the intestinal
wall. Quantitative aspects of amino acid metabolism within the
bodies of sheep fed lucerne pellets have been studied by Wolff and
co-workers (37,38,39). Most of the amino acids, including all the
essential ones, were added to the portal blood in the highly
significant amounts to be expected of their disappearance from
digesta in its passage through the small intestine. Net uptakes
of glutamic and aspartic acids were, however, almost zero, implying
extensive metabolism of these two amino acids in the gut wall.
Glutamine was also extensively utilized in the portal-drained
visceral tissues. It has been postulated (31) that the extensive
metabolism of these amino acids and glutamine in the gut wall may
provide the substrates for oxidative metabolism needed to meet the
energy demands of this metabolically very active tissue. Amino
acid absorption is known to be an active, energy consuming process
and reference has already been made to the fact that with most
dietary regimes little carbohydrate capable of being digested
within the small intestine enters therein. Essential amino acid
levels in the portal blood of the pig appear to parallel those in
the dietary protein although there does appear to be some
metabolism of glycine, alanine and ornithine (see 11).

Gluconeogenesis. Understandably gluconeogenesis is a very
important and continuing process in ruminants, unlike the
situation in non-ruminants where it assumes significance only
under the stimulus of energy deprivation. The amino acids (with
the exception of lysine and leucine which are ketogenic, and
isoleucine, phenylalanine and tyrosine which are partially
ketogenic) in addition to propionate are the major glucose
precursors and the liver and kidney cortex the major sites of
gluconeogenic activity (40). From their studies with sheep fed a
lucerne diet, Wolff and Bergman (39) concluded that amino acids
may contribute between 11 and 30 per cent of the total glucose and
that the most important substrates were alanine, glutamate and
aspartate; the essential amino acids made little contribution.

The protein-sparing effect of carbohydrate. Since protein,
carbohydrate and fat can all serve as dietary energy sources,
the last two mentioned can spare protein from being used to
meet energy requirements. However it is known (41) that in the

monogastric, dietary carbohydrate has an additional protein-sparing
effect not shared by fat. The effect is mediated primarily through
stimulation of insulin secretion by the absorbed glucose resulting
in the increased uptake of plasma amino acids by skeletal muscle
(42); thus the process may have special significance in the meat-
producing animal. An additional effect of glucose, again mediated
through insulin, may lie in repression of a key gluconeogenic
enzyme (phosphoenolpyruvate carboxykinase) in the liver and
repression of a number of others, also in liver tissue associated
with catabolism of gluconeogenic amino acids, i.e. serine dehy-
dratase which catalyses the non-oxidative deamination of serine to
pyruvic acid, a key gluconeogenic metabolite, and ammonia (43).

    With conventionally-fed ruminants in which little glucose is
absorbed from the digestive tract the question as to whether
propionate, a glucose precursor, exerts a comparable carbohydrate-
sparing effect is not easy to answer. Certainly the experiments of
Potter and colleagues (44), in which changes in amino acids after
short-term, low-level arterial energy infusions into sheep that had
been fasted for 24h were followed, suggest that propionate, unlike
acetate or butyrate, stimulates amino acid uptake from plasma; it
was almost as effective as glucose. This finding is supported by
the experiments of Eskeland and colleagues (45) who evaluated the
three volatile fatty acids and glucose, each separately, as energy
sources by adminstering them intravenously to growing lambs and
observing changes in N balance. All energy sources increased N
balance over control as would be expected since dietary N was not
limiting. However, glucose proved more efficient than any of the
VFAs; propionate had a significantly greater effect than acetate
and, in some instances, than butyrate. According to Bassett (46),
while the C_3 and C_4 acids are potent stimulators of insulin release
(presumably when administered within the body), intraruminal
administration of these acids does not stimulate release of the
hormone. To this extent the observations of Potter and Eskeland
and their colleagues must be interpreted with caution in relation
to their significance for the fed animal.

Protein Biosynthesis

    The mechanisms of mammalian protein biosynthesis at the
cellular level have been reviewed (46,47,48) and no attempt will
be made to consider them here. From current knowledge it is clear
that the quantities of amino acids reaching the tissues and their
proportions, particularly with regard to the essential amino acids,
together with availability of energy-yielding substrates at the
tissue level are important nutritional factors governing protein
biosynthesis. The complex interactions leading to protein biosynthesis or degradation are largely under the control of endocrine
function. Thus insulin stimulates protein synthesis within the
muscle, growth hormone plays an important part in the continuous

regulation of protein and of energy metabolism by stimulating
protein synthesis, lipolysis and glucose oxidation. Androgens have
a marked anabolic effect on N metabolism while the corticosteroids
induce loss of protein from skeletal muscle and gain in that of
the liver; not surprisingly the skeletal muscle is the major con-
tributor to loss of carcass protein resulting from corticosteroid
administration. These aspects will be dealt with fully in later
contributions.

The liver, at least in the monogastric animal, plays an
important role in regulating overall amino acid metabolism in
extra-hepatic tissues by its ability to regulate the flow of amino
acids to the peripheral tissues. In response to the post-absorptive
rise in amino acids flowing to the liver via the portal blood
following protein digestion and absorption of the amino acids, there
is a marked uptake of amino acids by the liver associated with
increased synthesis of liver (42) and possibly plasma proteins (50).
In addition there is increased liver catabolism of amino acids (51).
The subsequent capacity of the liver to release the proteins
synthesised and thus augment amino acid supplies to the extra-hepatic
tissues under the influence of growth hormone has been outlined (52).

The studies of Wolff and his colleagues (37) indicate that
even in the sheep, in which the supply of amino acids via the
portal blood is more continuous, the liver is very effective in
taking up amino acids. As these authors point out, their findings
support the view of Elwyn (53), based upon observations in the dog,
that peripheral metabolism of plasma proteins synthesised within
the liver warrants consideration as a mechanism for supplying significant quantities of amino acids to extra-hepatic tissues.

Protein:Energy Relationships

In the pig, provided that neither quantity nor quality of the
amino acid supply nor the energy supply is limiting, the amount of
protein deposited daily throughout the growing phase is fairly
constant (see 11); it reaches a plateau of approximately 100g/24h
near to 30kg liveweight, rising by around 40kg liveweight to
110g/day and remains at this level until after about 130kg liveweight when it begins to decline. Unlike the relative constancy
of daily protein retention, daily lipid deposition increases rapidly
as growth proceeds, the quantity stored per 24h being considerably
dependent upon the amount of energy surplus to needs for maintenance
and protein deposition. In ruminants also similar relative constancy of N deposition is associated with increasing fat deposition
depending upon plane of nutrition.

There are therefore optimal daily supplies of amino acids both
in quantity and quality and an optimal ratio of amino acids to non-
anino acid energy sources that will ensure maximum protein depo-

sition with minimal deposition of energy as lipid. With the monogastric animal such ratios are capable of being translated into
dietary formulations and much of the success of present day feeding
practice in the pig industry can be attributed to the effectiveness
with which this has been done.

In the ruminant animal, on the other hand, the situation is
very different and some of the reasons for this have already been
discussed. The quantity and quality of the protein that eventually
supplies the host animal with its amino acids is as dependent on
events occurring within the reticulo-rumen as on the nature of the
dietary constituents. The ruminant makes continual demands upon
absorbed amino acids to meet the glucose requirements of its tissues.
Furthermore, the magnitude of this demand is likely to be related
to the amount of propionate absorbed from the rumen.

A further consequence of events occurring within the reticulo
rumen is a narrowing of the ratio of amino acid N to total energy
absorbed by the host animal. This is illustrated in Table 4. For
the pig a twofold difference in dietary N concentration is reflected
in a similar difference in amounts of metabolisable energy (ME)
absorbed/g amino acid N uptake from the small intestine. In the
sheep a similar difference in dietary N concentration has resulted
in little difference in the ME/g amino acid N uptake between the
two diets. It will be appreciated that part of the narrowing of
the ratio in this instance is due to the decline in ME content of
the dried grass with the lower N content; this grass was in fact
less digestible. Nevertheless, it can be seen from Table 4 that
even if the ME contents of the two grasses had remained the same,
the two-fold difference in dietary N concentration would still be
reduced by half because of the changes in N metabolism occurring
in the rumen. In a comprehensive study of protein-energy relationships in sheep fed forage diets, Egan (55) observed that while
dietary N intake extended over a thirty-fold range, N yields at the
duodenum showed only a twelve-fold difference.

FUTURE STUDIES

It cannot be emphasised too strongly that ruminant species,
by virtue of the early intervention of microbial fermentation in
their digestive systems, are able to utilise effectively the very
considerable amounts of solar energy stored annually within the
structural carbohydrates of plant tissues, and thus render available
to man a proportion of this energy. Nevertheless, there is little
doubt that the microbial process does impose limitations upon
the efficiency with which such species can convert feed energy into
animal products. This is well illustrated by the data of Black and
Tribe (56) who fed weaned lambs on lucerne pellets in amounts to
supply 38% of their daily energy intake and 43% of their daily N

Table 4

Comparison of ratios of metabolisable energy intake
to amino acid N uptake in diets fed to pigs
and ruminants

	Intake/kg feed organic matter		Amino acid N uptake in small intestine	Ratio
	N (g)	Metabolisable energy (MJ) (1)	g/kg OM fed (2)	$\frac{(1)}{(2)}$
Ruminant feeds[*]				
Dried Grass A	32.2	13.94	15.21	0.917
Dried Grass B	16.3	10.53	10.76	0.978
Pig feeds[†]				
Diet A	32	16.61	25.6	0.649
Diet B	16	16.61	12.8	1.297

[*] The N and metabolisable energy contents of the two dried
    grasses are those pertaining to ryegrass cuts 1 and 4 (54);
    amino acid N uptake was calculated using the equation
    derived for forages (29) (see text).

[†] For the two pig diets it was assumed that gross energy was
    20.50 MJ/kg organic matter and metabolisable energy was 81%
    of gross energy; amino acid N uptake from such small
    intestine was calculated as 80% of N intake.

intake. The remainder of the diet was infused in liquid form (as a
mixture of spray-dried cow's milk and whey, and butter oil) either
into the rumen or into the abomasum. Some of the data are shown
in Table 5 from which it can be seen that the lambs receiving the
abomasal infusions made daily gains in liveweight and in empty body
weight respectively that were 75% and 191% greater than those of
the rumen-fed lambs; daily N and energy retentions were superior
by 108% and 131%.

It is the author's view that future programmes of work should
be aimed at developing feed practices for the ruminant animal which,
while allowing the microbial fermentation to operate efficiently,
ensures that production in the host animal is not limited by
shortage of specific nutrients.

An effective method of achieving this is to utilize the
reticular groove (57) and thus by-pass the reticulo-rumen. Protection of nutrients from fermentation is another method by which
this can be achieved. Protection can occur naturally through the
presence in the feed source of tannins, result from heat treatment
as with fish meal or be induced by chemical means involving the
use of formaldehyde. Increases in the rate of wool growth in adult
sheep through the feeding of protected proteins is well documented
(58,59) but evidence for increased gain in bodyweight resulting
from their use is much more limited. Increased gains in weight of
lambs fed protected as compared with unprotected casein supplements
have been reported (60,61). However, studies with weaned calves
fed protected or unprotected peanut meal showed no significant
differences in rate of gain (62). Peanut meal protein has a low
biological value; thus gain in amino acid uptake in the small
intestine as a result of protection must have been offset by
lowered "quality" of the absorbed amino acids. The protection of
specific amino acids is an alternative approach and one which may
well receive increasing attention in the light of future knowledge
concerning amino acid requirements of the ruminant animal.

Table 5

Effect of administering part of the daily feed via
the rumen or abomasum in growing lambs (56)

		Site of infusion		Significance of Difference
		Rumen	Abomasum	
Liveweight gain	(g/24h)	72	126	P < 0.001
Empty body wt. gain	(g/24h)	45	131	"
Wool growth	(g/24h)	3.68	5.34	"
N. retention	(g/24h)	1.68	3.50	"
Energy retention	(MJ/24h)	1.39	3.21	"

REFERENCES

1. Callow, E. H. (1958) J. agric. Sci. Camb. 51, 361.

2. Callow, E. H. and Searle, R. L. (1956) J. agric. Sci. Camb. 48, 61

3. Lawrie, R. A. (1966)"Meat Science". (Pergamon Press, Oxford)

4. Erbersdobler, H. (1973) In:"Proteins in Human Nutrition"
    (Eds. J. W. G. Porter and B. A. Rolls) p.453 (Academic Press,
    London).

5. Porter, J. W. G. and Rolls, B. A. (1971) Proc. Nutr. Soc.,
    30, 17.

6. Gitler, C. (1964) In:"Mammalian Protein Metabolism" Vol.1
    (Eds. H. N. Munro and J. B. Allison) p.35 (Academic Press,
    London).

7. Buraczewski, S., Porter, J. W. G., Rolls, B. A. and Zebrowska,
    T. (1970) Br. J. Nutr., 25, 299.

8. Erbersdobler, H., Weber, G. and Gunsser, S. (1972) Z.
    Tierphysiol. Tiernahr. Futtermittelk., 29, 325.

9. Mathews, D. M. (1972) In:"Peptide Transport in Bacteria and
    Mammalian Gut" (Eds. K. Elliot and M. O'Connor) p.71
    (Elsevier, Amsterdam).

10. Mathews, D. M. (1972) Proc. Nutr. Soc., 31, 171.

11. Rerat, A. (1972) Nutr. Abst. and Rev., 42, 13.

12. Eggum, B. O. (1973) "A Study of Certain Factors Influencing
    Protein Utilization in Rats and Pigs" 406,beretning fra
    forsøgslaboratoriet, København, p.173.

13. Haenel, H. and Kharatyan, S. G. (1973) In:"Proteins in Human
    Nutrition" (Eds. J. W. G. Porter and B. A. Rolls) p.195
    (Academic Press, London).

14. Shorrock, C. and Ford, J. E. (1973) In "Proteins in Human
    Nutrition" (Eds. J. W. G. Porter and B. A. Rolls) P.207
    (Academic Press, London).

15. Armstrong, D. G. and Hutton, K. (1972) Proc. 2nd World Congr.
    of Animal Feeding, Madrid, Vol.4, 219.

16. Smith, R. H. (1975) Proc. 4th Internat. Symp. on Ruminant
    Physiology, Sydney, Australia (in press).

17. Hendrickx, H. and Martin, J. (1967) Comp. Rend. Rech. Inst. Rech. Sci. Ind. Agr., Bruxelles, 31, 9.

18. Hume, I. D. (1974) Aust. J. agric. Res., 25, 155.

19. McMeniman, N. P., Ben-Ghedalia, D. and Armstrong, D. G. (1975) Proc. 1st Symp. Protein Metabolism, E.A.A.P. (Nottingham Univ. July, 1974) (in press).

20. Hobson, P. N. (1972) Proc. Nutr. Soc., 31, 135.

21. Harrison, D. G., Beever, D. E., Thomson, D. J. and Osbourn, D. F. (1975) J. agric. Sci. Camb., (in press).

22. Nolan, J. V. and Leng, R. A. (1972) Br. J. Nutr., 27, 177.

23. Weller, R. A. and Pilgrim, A. F. (1974) Br. J. Nutr. 32, 341.

24. Hogan, J. P., Weston, R. H. and Lindsay, J. R. (1969) Aust. J. agric. Res., 20, 925.

25. MacRae, J. C. and Ulyatt, M. J. (1973) J. agric. Sci., Camb., 82, 309.

26. Armstrong, D. G. (1974) 5th General Meeting of the European Grassland Federation, Uppsala, 21-15 June, 1973.

27. Purser, D. B. and Buechler, S. M. (1966) J. Dairy Sci., 49, 81.

28. Ørskov, E. R., Fraser, C. and McDonald, I. (1971) Br. J. Nutr., 25, 243.

29. Reed, R. M., Moir, R. J. and Underwood, E. J. (1949) Aust. J. Sci. Res., Series B, 2, 304.

30. McNaught, M. L., Owen, E.C., Henry, H. M. and Kon, S. K. (1954) Biochem. J., 56, 151.

31. Armstrong, D. G. and Hutton, K. (1975) Proc. 4th Intern. Symp. Ruminant Physiol., Sydney, Australia (in press).

32. Armstrong, D. G. (1973) In: "Production Disease in Farm Animals" (Eds. J. M. Payne, K. G. Hibbitt and B. F. Sansom) p.43 (Baillière Tindall, London).

33. van't Klooster, A.Th. and Boekholt, H. A. (1972) Neth. J. agric. Sci., 20, 272.

34. Fauconneau, G. and Michel, M. C. (1970) In:"Mammalian Protein Metabolism" Vol. 4 (Ed. H. N. Munro) p.481 (Academic Press, London).

35. Armstrong, D. G. (1972) In:"Cereal Processing and Digestion",
     Technical Publ. U.S. Feed Grains Council, London Office, p.9
     (U.S. Feed Grains Council, London).

36. Armstrong, D. G. (1974) In:"Future Trends in the Supply and
     Utilization of Feed Grains", Technical Publ. U.S. Feed Grains
     Council, London Office, (U.S. Feed Grains Council, London)
     In Press.

37. Wolff, J. E., Bergman, E. N. and Williams, H. H. (1972) Amer.
     J. Physiol., 223, 438.

38. Wolff, J. E. and Bergman, E. N. (1972) Amer. J. Physiol., 223,
     447.

39. Wolff, J. E. and Bergman, E. N. (1972) Amer. J. Physiol., 223,
     455.

40. Leng, R. A. (1970) Adv. in Vet. Sci., 14, 209.

41. Munro, H. N. (1970) In:"Mammalian Protein Metabolism" Vol.
     Eds. H. N. Munro and J. B. Allison) p.381 (Academic Press,
     London).

42. Munro, H. N. (1970) In:"Mammalian Protein Metabolism" Vol. 4
     (Ed. H. N. Munro) p.299 (Academic Press, London).

43. Kaplan, J. H. and Pitot, H. C. (1970) In:"Mammalian Protein
     Metabolism" Vol. 4 (Ed. H. N. Munro) p.388 (Academic Press,
     London).

44. Potter, E. L., Purser, D. B. and Cline, J. H. (1958) J. Nut.,
     95, 655.

45. Eskeland, B., Pfander, W. H. and Preson, R. L. (1973) Br. J.
     Nutr., 29, 347.

46. Bassett, J. M. (1975) Proc. 4th Intern. Symp. Ruminant Physiol.,
     Sydney, Australia (in press).

47. Korner, A. (1964) In:"Mammalian Protein Metabolism" Vol. 1
     (Eds. H. N. Munro and J. B. Allison) P.178 (Academic Press,
     London).

48. Neuberger, A. and Richards, F. F. (1964) In:"Mammalian Protein
     Metabolism" Vol. 1 (Eds. H. N. Munro and J. B. Allison) p.243
     (Academic Press, London).

49. Heywood, S. M. (1970) In:"The Physiology and Biochemistry of
     Muscle as a Food" Vol. 2 (Eds. E. J. Briskey, R. G. Cassens
     and B. B. Marsh) p.13 (Univ. Wisconsin Press, Madison).

50. Hoffenberg, R. (1972) Proc. Nutr. Soc., 31, 265.

51. Munro, H. N. (1969) Proc. Nutr. Soc., 28, 214.

52. Turner, M. R. (1972) Proc. Nutr. Soc., 31, 205.

53. Elwyn, D. H. (1970) In:"Mammalian Protein Metabolism" Vol. 4
     (Ed. H. N. Munro) p.523 (Academic Press, London).

54. Armstrong, D. G. (1964) J. agric. Sci. Camb., 62, 399.

55. Egan, A. R. (1974) Aust. J. agric. Res., 25, 613.

56. Black, J. L. and Tribe, D. E. (1973) Aust. J. agric. Res.,
     24, 763.

57. Ørskov, E. F. (1972) Proc. 2nd World Congress of Animal
     Feeding, Madrid, Vol. 4, p.627.

58. Ferguson, K. A., Hemsley, J. A. and Reis, P. J. (1967) Aust. J.
     Sci., 30, 215.

59. Reis, J. A. and Tunks, D. A. (1969) Aust. J. agric. Res., 20,
     775.

60. Faichney, G. J. (1971) Aust. J. agric. Res., 22, 453.

61. Wright, P. L. (1971) J. Anim. Sci., 33, 137.

62. Faichney, G. J. and Davies, H. L. L. (1972) Aust. J. agric.
     Res., 24, 613.

EFFICIENCIES OF ENERGY UTILIZATION DURING GROWTH

A. J. F. Webster

Rowett Research Institute
Bucksburn
Aberdeen. AB2 9SB

INTRODUCTION

The efficiency with which the energy contained in animal feedstuffs can be converted into saleable animal produce depends on many factors which are usually considered, somewhat arbitrarily, under the following headings:

1.   The metabolizable energy (ME) of the food, which is that portion of the gross energy that can serve as a fuel for body functions.

2.   The ME requirement for maintenance of vital body functions, conventionally taken as the intake of ME (I_{ME}) which exactly equals metabolic heat production (H).

3.   The net efficiencies with which increments of ME are used below and above maintenance.

4.   The partition of retained energy, principally between protein and fat, which determines not only the chemical composition of the carcass but the amount of energy stored per kg of carcass gain.

5.   The physiological limit to the capacity of an animal to consume food energy and to store it as protein and as fat.

METABOLIZABILITY OF FOOD

The apparent metabolizable energy content of a food depends on

the proportion of the gross energy that is lost in the faeces and
urine or as combustible gases from the digestive tract. The processes of digestion and metabolism that determine the ME content
of any particular food will not be discussed. Hereafter food will
only be considered in terms of its ME content, i.e. that portion
which is oxidised with the liberation of heat, or stored in the
body principally as protein or as fat. Thus the gross efficiency
of retention of ME in the body of a growing animal is given by
$(I_{ME} - H)/I_{ME}$.

COMPOSITION OF CARCASS GAINS

The gross efficiency of conversion of ME into body mass (and
retail carcass yield) is influenced to a major extent by the composition and thus the energy content of the carcass gains. The
energy content of the dry matter of muscle protein is 23.5 kJ/g.
However each gram of muscle protein is associated with about five
times its weight of water. Thus the energy content of "wet" muscle
protein is about 4.7 kJ/g. Lipid has an energy content of about
39.2 kJ/g so that the energy retained in a gram of fat is theoretically eight times greater than that retained in a gram of wet
protein. Fortunately, lean meat is not just protein and water;
these merely act as a sponge for the really tasty constituents.
In practice the energy content of carcass gains ranges from about
8 kJ/g in very young animals growing slowly to about 32 kJ/g in
animals rapidly approaching slaughter weight(1) – still a four
fold range. Thus the relationship between energy retention and
the deposition of body tissue can only be established if the composition of body gains is known very precisely. It follows then
that any attempt to predict the efficiency of ME utilization from
body weight gains (or vice versa) without reliable measurements
of body composition is quite meaningless.

PREDICTION OF ENERGY RETENTION IN GROWING CATTLE

The main part of the paper is concerned with the gross
efficiency of retention of ME in the tissues of growing animals.
Control of this process has two distinct meanings. In one sense
it implies the prediction of output in relation to input and thus
the regulation of input to achieve the rate of output required.
In the other sense it implies a manipulation of the conversion
process so as to improve the ratio of output to input.

In any animal energy retention (R) = I_{ME} - H. The measurement of I_{ME} in growing cattle is laborious but presents no real
problems. The measurement of R in cattle by whole carcass analysis
on serially slaughtered animals is extremely expensive and time
consuming. The measurement of H using modern automated calori-

meters with computer links is relatively painless (2,3). Blaxter
and his colleagues(4) developed the following synthetic approach
to the prediction of H, and thus R, in a growing animal at a stated
intake of ME, and this approach has been adopted by the Agricultural
Research Council(1) as the basis of their recommendations concerning the energy requirements of ruminants.

Heat production is the sum of:

1.   Fasting metabolism (F) – heat production measured when
      food intake is zero and conventionally expressed as a
      function of "metabolic body size" ($kJ/kg^{0.75}$ 24h).

2.   Energy cost of activity – the increase in heat production
      arising from any activities additional to those occurring
      during the measurement of F. These include standing,
      walking and any metabolic response to environmental
      stress.

3.   The heat increment of feeding – the increase in heat
      production with increasing food intake is exponential
      in ruminants but usually expressed as two straight lines
      below and above the point of maintenance.

The heat increment of feeding varies a great deal in ruminants,
particularly above maintenance. Wainman and Blaxter (see refs. 4,
5,6) have measured the efficiency with which the ME of a variety
of foods can be used above and below maintenance by measuring the
heat production of mature sheep when fasted, or given amounts of
ME at about maintenance and about twice maintenance. The slopes
of the lines relating R to I_{ME} below and above maintenance are k_m
and k_f which are the <u>net</u> efficiencies of utilization of ME for
maintenance and for fattening respectively. For most practical
ruminant diets the range of k_m is 0.65 to 0.75 and k_f 0.30 to 0.60
(1,4).

The Metabolizable Energy system is a marked improvement on
the traditional Starch Equivalent system both conceptually and
practically; however it is still not so old and so respectable as
to be incapable of improvement. There are three strong conceptual
criticisms that may be made of the system.

1.   The physiological status of the growing animal is predicted
      from measurements made during a four-day period of starvation

This paradox is imposed by the system which arbitrarily
partitions heat production into that due to size (F), activity,
and food intake. In order to assess the effect of size the effect

of food intake must theoretically be removed. In practice this is not possible (4). In most conventions F in the growing animal is expressed per kg body weight $(W)^{0.75}$. This exponent of body weight was adopted by Kleiber(8) as being that which would approximately confer proportionality on measurements of F when comparing mature animals of different species. The change in F in a young animal as it grows in size is not however proportional to $W^{0.75}$ but to an exponent usually somewhere between $W^{0.5}$ and $W^{0.6}$ (9,10). Despite this the California Net energy system (11) and the Net energy, fattening system (12,13), predicting energy require- ments of growing cattle, both assume F is proportional to $W^{0.75}$. The Agricultural Research Council (1) recognises that this exponent does not confer proportionality but uses it anyway.

2.   The efficiency of utilization of ME for growth in cattle is assessed from experiments made with mature sheep which must not change in size

In this case sheep are used because they are more convenient and there is good evidence to suggest that adult sheep and cattle utilize most commercial diets with similar efficiency (14). Mature animals must be used because any changes in H due to changes in size during the experiment would distort the calculations of k_m and k_f (5).

3.   The term k_f describes the efficiency of utilization of ME for fattening rather than for growth

The proportion of energy retained as protein is greater in young growing animals than in adults depositing energy almost entirely as fat. This may affect k_f, although the variation over a large part of growth in the partition of retained energy between protein and fat is quite small. A calf weighing less than 100 kg and gaining only about 0.5 kg/24h still retains about 50% of its energy as fat. A 300 kg steer gaining 1 kg/24h retains about 85% of its energy as fat even though the contribution of fat mass to weight gain is only about 40% at this time.

To these three conceptual criticisms must be added the practical criticism that the ME system tends progressively to over- estimate weight gains in steers as they approach slaughter weight (15,16).

Recent experiments at the Rowett Institute have been designed to examine, in the light of these criticisms, the validity of the ME system as a predictor of the energy requirements for growth in cattle. The first and most direct approach was to measure, in a succession of calorimetric experiments, R in growing cattle given

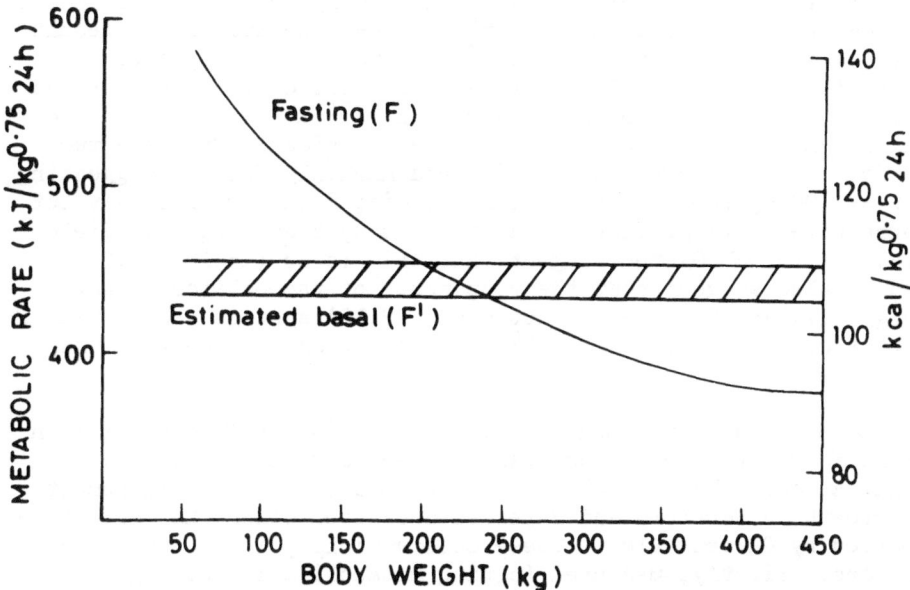

Fig. 1   Fasting metabolism (F) and predicted basal metabolism
(F') in growing steers - from Webster, Brockway and Smith (7).

diets for which the ME content, k_m and k_f had previously been
determined. (7).

According to the ME system

$$R = k_f(I_{ME} - F_{/k_m}) \text{\hspace{2cm}} (1)$$

In these experiments R was measured directly and F', the
predicted basal metabolism of the growing animal could be calculated from

$$F' = k_m(I_{ME} - R_{/k_f}) \text{\hspace{2cm}} (2)$$

No particular biological significance need be attributed to F'.
It is simply the empirical expression which when combined with
the best available description of a diet according to the conventions of the ME system most precisely describes energy
retention in the growing animal.

The first set of 61 trials (7) involving British Friesian and Aberdeen Angus castrate male cattle yielded the surprising result that log F' was proportional to $W^{0.73}$, Brody's interspecies exponent, precisely (17). Expressed in terms of the now accepted exponent, F' during growth was 440 kJ/kg$^{0.75}$ 24h, there being no suggestion of a difference between the individuals from the two breeds. Fig. 1 shows that F' differed markedly from F, suggesting that at 200 kg the ME System would predict R rather precisely but progressively overestimates R at increasing body weights. Field trials of the ME System suggest that this is so (15,16). The results also suggest that the proponents of the California Net Energy (11) and Net Energy Fattening (12,13) systems who assumed F to be a constant function of $W^{0.75}$ were rather more correct than they knew.

More recent experiments indicate that in Hereford x Friesian steers F' was reasonably constant between 200 and 600 kg body weight at about 370 kJ/kg$^{0.75}$ 24h. A reduced energy requirement for growth in Hereford and Hereford crossbred steers has also been reported by Garrett (18). These observations point to two con-clusions. Firstly, measured fasting metabolism is not a good base from which to predict the energy requirements for growth in cattle. Secondly, there may be physiological differences between individuals, attributable, for example, to sex or to genotype, which alter energy requirement in a way that is independent of the energy content of the food and therefore best considered within the predicted basal component of heat production. We shall return to this point later.

ENERGETIC EFFICIENCY OF PROTEIN AND FAT DEPOSITION

Protein and fat deposition in a growing animal are only a measure of the amount by which anabolism exceeds catabolism. The energy cost of the total anabolism of complex molecules like proteins can be considered in terms of the ATP equivalents required for their synthesis and the ME required to generate them (4,19).

Synthesis of a molar peptide bond requires 4 moles of ATP. To this one must add a further energy cost for the maintenance of the synthetic machinery (mainly tRNA and mRNA) which has not been determined precisely but which probably increases requirement to about 5 moles ATP per molar peptide bond. The energy required to synthesise one mole of ATP from ADP is about 79 kJ in monogastric species and 96 kJ in ruminants (20). From this it may be calcu-lated that the energy cost of protein synthesis in monogastric species is about 390 kJ/molar peptide bond or 4.46 kJ/g protein. The energetic efficiency of protein synthesis is then

$$\frac{\text{energy content of 1 g protein (23.8 kJ)}}{\text{energy content of precursors (23.8 kJ +}}_{\text{energy cost of synthesis (4.46 kJ)}} = 85\%$$

The energetic efficiency of fat synthesis, calculated using the same approach, is rather lower, about 70%. Values for k_f in monogastrics are close to 0.7 which suggests that the energetic efficiency of fat deposition measured by calorimetry is not very different from the theoretical efficiency of total fat anabolism.

There is however an impressive body of evidence to suggest that the energetic efficiency of protein deposition in a growing animal is lower than the energetic efficiency of fat deposition even though the efficiency of synthesis of protein is higher than that of fat. This conclusion was derived from the pioneering work of Kielanowski and Kotarbinska (21,22). In brief the approach is to relate ME intake to maintenance requirement (a $W(kg)^n$) and to the amount of energy deposited as protein (E_p) and as fat (E_f) in the growing animal.

$$I_{ME} = a\ W^n + b\ E_p + c\ E_f + d \ \text{————————}\ (3)$$

Kielanowski has always made clear the statistical limitations of this approach. To be effective the variables in a multiple regression analysis must be independent. During normal growth, however, the increase in body weight, and thus in maintenance requirement (a W^n) is inevitably linked to a decrease in the proportion of energy retained as protein. Moreover the energy stored as protein and fat is small compared with that required for maintenance, and small variations in b and c are likely to be overwhelmed by uncertainties attached to the calculation of the maintenance component. However Kielanowski (20) recently critically summarised the more reliable of the papers using this approach and concluded that the ME required for the deposition of 1 g protein in monogastric species was between 45 and 65 kJ, or that the energetic efficiency of protein deposition was 37 to 53%.

Recently Pullar and I (23) approached the same question from a different angle. We conducted energy and nitrogen balance trials using the Zucker rat (24) in which obesity of juvenile origin appears as a homozygous recessive trait. When lean Zucker rats and their fat siblings were pair-fed the ratio of energy retention in protein and in fat in the two groups was as follows.

	Body weight	Energy in protein (%)	Energy in fat (%)
Lean rats	330	75	25
Fat rats	320	14	86

This approach enabled us to vary the partition of retained energy between protein and fat in a way that was much more extreme than that observed during normal growth and also was largely independent of the effect of body size. Our estimates of the energetic efficiency of protein and fat deposition were 43 and 65% respectively, values which agree closely with those of Kielanowski (20).

All that this really means is that the heat production of a growing animal is directly related to the rate at which it is depositing protein. Whether one can justify using the rate of protein deposition, principally in muscle, as a predictor of heat production and ME requirements is quite another matter. Table 1 is an attempt to estimate the total rate of protein synthesis in a growing pig at 80 kg live weight. The protein contents of different tissues were provided by Dr. V. R. Fowler. The estimates of fractional protein synthesis rates are derived from several sources and several species (25,26,27,28) and can only be approximate. Total protein synthesis rate (g/24h) is the product of g protein x fractional synthesis rate/24h.

Table 1

An estimate of protein synthesis rates in a
growing pig weighing 80 kg

	Lean tissue weight (kg)	Tissue protein (kg)	Fractional synthesis rate ($24h^{-1}$)	Protein synthesis (g/24h)
Skeletal muscle	37.5	5.9	0.02	118
Liver	1.5	0.3	0.8	240
Gut Wall	4.5	0.7	0.6	420
Kidneys	1.5	0.3	0.5	150
Heart	3.0	0.5	0.1	50
Other viscera	3.0	0.5	0.3	150
Connective tissue	3.0	0.6	0.01	6
Total	54.0	8.8	-	1134

Ratio of protein synthesis in muscle to total
protein synthesis = 10%
Ratio of total protein deposition to total
protein synthesis = 7%

Energy cost of protein synthesis at 4.5 kJ/g protein = 5.1 MJ
Total heat production with ad libitum feeding   = 16.8 MJ

Ratio of energy cost of protein synthesis/total heat
production = 30%

The energy required for protein synthesis is here about 30% of total metabolic heat production. However the ratio of protein deposition to total protein synthesis is only 7% which illustrates the danger of predicting heat production from protein deposition since in this example the attempted correlation is between 7% of protein synthesis and 30% of heat production.

By far the greater component of the energy cost of protein synthesis relates to factors only indirectly related to rapid growth, such as rapid turnover of proteins in the liver and gut wall, and may be considered to be part of the energy requirement for so called maintenance. It is not therefore surprising that apparent maintenance requirements in the lactating cow (29,30) and in the rapidly growing lean steer (7) should be higher than those measured in animals neither growing nor lactating. Kielanowski (20) recently reached a similar conclusion on statistical grounds and suggested that ME requirement should be adjusted according to the rate of nitrogen retention according to the following equation

$$I_{ME} = E_{M.fp} + b\ E_p + c\ E_f \quad\text{————————}\quad (4)$$

Here $b\ E_p$ and $c\ E_f$ are the same as in equation 3 but the energy requirement for maintenance (E_M) varies as a function of protein retention (f_p). However since protein deposition is small in relation to total protein synthesis the value $b\ E_p$ may perhaps be discounted.

At a strictly practical level equation 4 may be re-expressed and simplified to

$$R = k_f(I_{ME} - A) \quad\text{————————————————}\quad (5)$$

which is a simpler form of the basic equation of the ME system (equation 1). Quite apart from its simplicity this equation has two biological points in its favour. The conversion of animal food to animal product depends on the interaction between the energy content of the food, and the size and innate capacity for growth of the animal. In this equation both the size and capacity for growth of an animal are incorporated in the intercept term A which can be determined empirically for a variety of strains and sexes. Much of the information needed to describe A to a degree of precision sufficient for practical purposes is available. It is, for example, F'/k_m, a term which probably predicts energy requirement more precisely than does F. Moreover, since the energy cost of protein synthesis is included within A, values for k_f determined in the conventional way with mature animals can be applied with confidence to the growing animal. Equation 5 could therefore form the basis of a modification to the ME system which

would amend the present description of the interaction between
food and animal to place more emphasis on the animal.

THE LIMIT OF APPETITE

The gross efficiency of retention of ME ($^R/I_{ME}$) is principally
a function of the amount by which I_{ME} exceeds maintenance, or in a
word, appetite. Theories of appetite control abound but are for
the most part not relevant to the present discussion since they
pose the question why - why does an animal start or stop eating?
The only point that relates to the present quantitative argument
is when does an animal decide it has had enough.

Pullar and I (23) observed that when fat and lean Zucker rats
were offered food to appetite, the fat rats consumed about 40%
more ME and energy retention was more than 100% greater. Nitrogen
retention, measured by balance, and heat production were however
the same in the two groups. When fat and lean rats were pair-fed,
nitrogen retention and heat production were less in the obese
individuals. This observation suggested that both groups regulated
their intake during growth to sustain the same rate of protein
deposition, or the same rate of heat production, since in this
experiment the two were closely correlated. Radcliffe (31), using
the comparative slaughter technique, recently confirmed the
remarkable constancy of protein deposition in fat and lean Zucker
rats allowed free access to diets varying in protein content
(Table 2). If this speculation is extended to domestic animals it
would imply that, given free access to a food that is not lacking
in any essential nutrient, an animal will eat to sustain an optimal
rate of growth of lean body mass. However, since deposition of
protein in lean body mass is such a small part of total protein
synthesis and the energy cost thereof (Table 1) it is perhaps more
realistic to consider that intake in a growing (or lactating) animal
is limited by the rate at which the animal can carry out the work
involved in synthesis of body constituents (perhaps 30% of which
is protein synthesis)

Table 2

Intake of ME, nitrogen and energy retention in lean and fat
Zucker rats offered free access for 64 days to semi-synthetic,
isoenergetic, diets containing 15 or 30% casein

	Casein content of diet (%)	ME intake (MJ)	N retention (g)	Energy retention (MJ)
Lean	15	15.0	4.73(±0.26)	2.65(±0.23)
	30	16.4	5.41(±0.21)	3.27(±0.33)
Fat	15	25.0	5.11(±0.45)	10.53(±0.58)
	30	26.0	5.21(±0.25)	11.38(±0.40)

Table 3 summarises a selection of energy balance trials con-
ducted with animals receiving food to appetite, or very nearly.
The results for veal calves and growing cattle are from the Rowett
Institute. Other sources are acknowledged in the Table.

Heat production at ad libitum intake is remarkably similar
(780-807 kJ/kg$^{0.75}$.24h) in rapidly growing monogastric animals
(pigs and veal calves), barley fed steers or even store calves
growing slowly on a diet having a metabolizability of only about
40%, although energy retention varies, according to the quality
of the diet from 104 to 673 kJ/kg$^{0.75}$.24h. This offers strong
support to the suggestion that there is a rather rigid upper limit
to the rate at which a growing animal can do work to support
tissue synthesis. In the very high-yielding dairy cow, this limit
appears to be reset at a higher level, but in the mature (2-4 year
old) sheep which has effectively finished growth the limit to
appetite and metabolic heat production is reduced. In all cases
however the limit appears to be remarkably insensitive to changes
in diet.

To summarise these observations: In animals given free access
to a balanced diet, protein deposition and heat production appear
to have rigidly defined upper limits which are set by the intrinsic
capacity of the individual for synthesis of milk or lean body mass.
These limits can only be adjusted by genetic selection or more

Table 3

Energy balance in different species of domestic
animals given food and libitum

	Body weight (kg, approx.)		Energy balance (kJ/kg$^{0.75}$.24h)		
			ME intake	Heat	Retention
Cattle					
Veal calf	100	Milk sub.	1265	800	465
Steer	150	Barley conc.	1060	780	280
	450	"	1172	807	365
Store calf	250	Roughage	897	793	104
Dairy cow(12)	550	Alfalfa	1550	905	645
(32)	550	Mixed	1470	882	588
Pig(33)	35	Barley conc.	1464	792	673
Sheep, Mature(6)		Chopped			
	60	dried grass	733	620	113
	60	Grass pellets	832	607	225

direct physiological interference. ME intake and energy retention
as fat are less well regulated by the animal and thus more amenable
to nutritional control. It follows therefore that by appropriate
nutrition one should be able to produce an optimal carcass from any
genotype. It also follows that attempts to improve appetite directly
by selection or pharmacology will probably only tend to increase the
conversion of ME to fat, which is an inefficient way to produce an
undesirable product. On the other hand, if the absolute capacity
for protein deposition can be improved by genetic or other means
then the appropriate adjustments in appetite should ensue.

REFERENCES

1. Agricultural Research Council (1965). "The Nutrient
    Requirements of Farm Livestock." No. 2 Ruminants.
    (Agricultural Research Council, London.)

2. Blaxter, K. L., Brockway, J. M. and Boyne, A. W. (1971).
    Q. Jl. exp. Physiol.<u>57</u>, 60.

3. Moe, P. W. and Flatt, W. P. (1969). In: 4th Symposium,
    "The Energy Metabolism of Farm Animals." (Eds. K. L. Blaxter,
    J. Kielanowski and G. Thorbek). Publ. Eur. Ass. Anim. Prod.
    No. 12, p.463.

4. Blaxter, K. L. (1967). "The Energy Metabolism of Ruminants",
    2nd impression. (Hutchinson, London).

5. Blaxter, K. L., Clapperton, J. L. and Wainman, F. W. (1966).
    J. agric. Sci., Camb. <u>67</u>, 67.

6. Wainman, F. W., Blaxter, K. L., Smith, J. S. and Dewey, P.
    J. S. (1970). In: 5th Symposium. "The energy metabolism
    of farm animals." (Eds. A. Schürch and C. Wenk). Publ. Eur. Ass.
    Anim. Prod. No. 13, p.17.

7. Webster, A. J. F., Brockway, J. M. and Smith, J. S. (1974).
    Anim. Prod. <u>19</u>, 127.

8. Kleiber, M. (1961). "The Fire of Life." (Wiley, New York.)

9. Blaxter, K. L. (1972). In: "Festskrift til Pr. K. Brierem".
    p.19. (Mariendals Boktrykkeri., Gjøvik).

10. Mount, L. E. (1968). "The Climatic Physiology of the Pig."
    (Arnold Ltd., London).

11. Lofgreen, G. P. and Garrett, W. N. (1968). J. Anim. Sci.
    <u>27</u>, 793.

12.   Nehring, K. (1969). In: 4th Symposium, "Energy Metabolism of
       Farm Animals." (Eds. K. L. Blaxter, J. Kielanowski and
       G. Thorbek). Publ. Eur. Ass. Anim. Prod. No. 12, p.5.

13.   Nehring, K. and Haelein, G. F. W. (1973). J. Anim. Sci.
       36, 949.

14.   Blaxter, K. L. and Wainman, F. W. (1961). J. agric. Sci.,
       Camb. 57, 419.

15.   Alderman, G., Morgan, D. E. and Lessells, W. J. (1970).
       In: 'Energy Metabolism of Farm Animals. (Eds. A. Schürch and
       C. Wenk). Publ. Eur. Ass. Anim. Prod. No. 13, p.81.

16.   Kay, M., Massie, R. and MacDiarmid, A. (1971).
       Anim. Prod. 13, 101.

17.   Brody, S. (1945). "Bioenergetics and Growth." (Reinhold
       Pubs., New York).

18.   Garrett, W. N. (1971). J. Anim. Sci. 32, 451.

19.   Buttery, P. J. and Annison, E. F. (1973). In: "The Biological
       Efficiency of Protein Production" (Ed. J. G. W. Jones)
       p.141. (University Press, Cambridge).

20.   Kielanowski, J. (1975). Proceedings 1st Symposium Protein
       Metabolism. (in press).

21.   Kielanowski, J. (1965). In: 3rd Symposium, "Energy Metabolism
       of Farm Animals." (Ed. K. L. Blaxter). Publ. Eur. Ass. Anim.
       Prod. No. 11, p.13.

22.   Kielanowski, J. and Kotarbinska, M. (1970). In: 4th
       Symposium, "Energy Metabolism of Farm Animals."
       (Eds. A. Schürch and C. Wenk). Publ. Eur. Ass. Anim. Prod.
       No. 12, p.145.

23.   Pullar, J. D. and Webster, A. J. F. (1974).
       Br. J. Nutr. 31, 377.

24.   Zucker, L. M. and Zucker, T. F. (1961). J. Hered. 52, 275.

25.   Arnal, M., Fauconneau, G. and Pech, Renee (1972).
       Ann. de Biologie Animale. 12, 91.

26.   Lobley, G. E. and Nicholas, G. A. - personal communication.

27.   Millward, D. J. and Garlick, P. J. (1972). Proc. Nutr.
       Soc. 31, 257.

28.   Perry, B. N. (1974).   Br. J. Nutr. 31, 35.

29.   Flatt, W. P., Moe, P. W., Munson, A. W. and Cooper, T. (1969).
      In: 4th Symposium, "The Energy Metabolism of Farm Animals".
      (Eds. K. L. Blaxter, J. Kielanowski and G. Thorbek).   Publ.
      Eur. Ass. Anim. Prod. No. 12, p.235.

30.   Moe, P. W., Tyrrell, H. F. and Flatt, W. P. (1970).   In: 5th
      Symposium, "Energy Metabolism of Farm Animals".   (Eds.
      A. Schürch and C. Wenk).   Publ. Eur. Ass. Anim. Prod. No. 13,
      p.65.

31.   Radcliffe, J. D., Webster, A. J. F., Dewey, P. J. S. and
      Atkinson, T. E. (1975).   Proc. Nutr. Soc. 34, 54A.

32.   Van Es, A. J. H. and Nijkamp, H. J. (1969).   In: 4th Symposium,
      "The Energy Metabolism of Farm Animals".   (Eds. K. L. Blaxter,
      J. Kielanowski and G. Thorbek).   Publ. Eur. Ass. Anim. Prod.
      No. 12, p.209.

33.   Verstegen, M. W. A., Close, W. H., Start, I. B. and Mount, L. E.
      (1973).   Br. J. Nutr. 30, 21.

EFFICIENCY OF PROTEIN UTILIZATION

D. Lewis, K. N. Boorman and P. J. Buttery

University of Nottingham
School of Agriculture
Sutton Bonington, Loughborough LE12 5RD

It is hardly possible to consider all aspects of the
efficiency of protein utilization in all species of meat-producing
animals under all individual physiological circumstances: instead
a few specific concepts will be discussed. A simple concept of
the efficiency of protein utilization is that it is determined by
how closely the dietary supply of amino acids corresponds to the
need of the animal. Surpluses above a perfect balance can be
regarded as contributing to catabolic processes and loss of
nitrogen. It is also necessary to assess the supply of amino
acids in relation to other nutrients: an inadequacy of energy-
yielding constituents could for example contribute to the
catabolism of gluconeogenic amino acids with consequent inefficiency in terms of nitrogen utilization.

This simple concept of protein utilization suggests that the
value of the ingested protein is determined by the level of the
limiting amino acid. Any surpluses of other amino acids contribute
to catabolism and inefficient utilization. There are however
other factors contributing to inefficiency, and their effects
represented in schematic form in Fig. 1. The supply of each amino
acid is represented as a percentage of the individual requirement:
level A represents the average percentage, and is thus a measure
of the overall protein supply; were all the amino acids in perfect
balance they would all be at level A. The simple concept of the
limiting amino acid implies efficient utilization only up to level
B, all quantities of amino acid above this being subjected to
catabolism and nitrogen wastage.

It must however be recognised that there are other losses of
nitrogen contributing to inefficiency. There is an inevitable

103

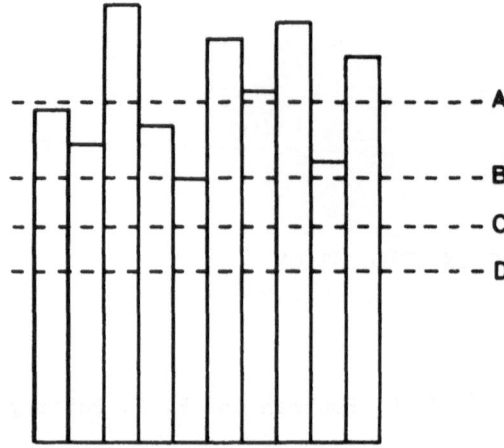

Fig. 1.   Schematic representation of the efficiency of
protein utilization.   For details see text.

component which can be regarded as contributing to maintenance or
turnover. After taking this into account the element remaining
for efficient utilization can be regarded as being at level C.
Inevitable losses include those of protein in hoof, hair and mucus.
There are also losses of protein from the gut wall by sloughing or
desquamation and by secretion. Some of this is reutilized lower
in the alimentary tract; losses from the gut wall are therefore
more of a penalty the lower in the tract they occur. There are
also inevitable metabolic losses of amino acids other than in the
form of protein, of histidine as histamine or of tyrosine as
thyroxine, melanin or adrenaline, for example, These are
physiologically useful processes but do not contribute directly to
incorporation of amino acid nitrogen in synthesised protein, the
usual measure of efficient protein utilization.

The process of protein turnover constitutes an inevitable
loss to the animal though it is more obviously a loss of energy-
yielding nutrients than of nitrogen (1,2). Protein turnover in
muscle is a more rapid process than protein deposition. In a 100
g rat 713 mg of muscle protein is synthesised per day while 564 mg
is degraded (3): in the rapidly growing broiler chick (900 g),
9.1 g of muscle protein is synthesised per day with a net protein
in crease in the muscle tissue of 3.9 g (1). Protein turnover
continues even in the mature animal: in a 50 kg wether, 85 g of
protein is synthesised in skeletal muscle per day while protein
deposition would be expected to be only a few grams (4). Hydroxy-

proline and the methylated amino acids, methyl histidine and methyl
lysine, released during the process of protein degradation cannot
be reincorporated into protein and are excreted (5,6). These
losses of histidine, lysine and proline might be expected to
unbalance the supply of amino acids for the re-synthesis of protein
and this might be expected to result in some catabolism. In
addition it would be surprising if there were no loss of an amino
acid in its passage through an intracellular amino acid pool
during the process of catabolism and resynthesis of protein.

Beyond the limit to utilization imposed by these inevitable
losses, represented by level C, there is a contribution of amino
acids to gluconeogenesis and lipogenesis, forming another limit
at level D. A surplus of dietary protein relative to other
nutrients could lead to this diversion of amino acid carbon
involving a loss of nitrogen and reduced efficiency. In the case
of the ruminant, which is particularly vulnerable to fluctuations
in glucose supply, there can be additional demands for gluconeo
genesis leading to reduced efficiency of nitrogen utilization. If
little hexose escapes breakdown in the rumen and the proportion of
acetate is high in rumen volatile fatty acids a breakdown of
gluconeogenic amino acids would tend to be encouraged.

Within this schematic representation of the efficiency of
protein utilization (Fig. 1) it is possible to visualise an
approach to achieve improvement and to minimise losses. This
essentially involves supplying amino acids in a pattern which is
matched as closely as possible to the animal's needs and which
simultaneously ensures balance between nitrogen-supplying and
energy-yielding nutrients.

To consider protein utilization in more realistic quantitative terms a system of evaluation is needed. Standard concepts of
the efficiency of protein utilization are usually based upon
measures of protein quality involving retention of nitrogen in
relation to total intake or absorbed nitrogen. This does not,
however, relate to useful tissue formation nor to individual
amino acids. Such measures are of little use in predictably pre-
paring or improving a diet. Functional definitions can be
considered which are able to take account of a specified objective
whether it be protein synthesis, lean meat production, liveweight
gain or even profit expressed in economic terms. It is also
possible to consider several alternative terms, for example the
efficiency of incorporation of a particular dietary amino acid
into useful meat amino acid or efficiency of conversion of food
nitrogen into meat nitrogen, lean meat or growth.

The efficiency of protein utilization can be considered in
terms of the incorporation of the limiting amino acid into meat
as influenced by protein level or the supply of energy-yielding

nutrients for example. Some values are given in Table 1 from an
experimental programme carried out by Hardy (7) to examine the
effects of energy-protein balance on lean meat production in the
growing pig. The factorial design included diets with four
concentrations of protein and four concentrations of digestible
energy. It was intended that the amino acids in the protein
should be ideally balanced but from improved knowledge of the
threonine requirement of pigs obtained by Taylor (8) it is clear
that threonine was the limiting amino acid in all diets.

The values presented in Table 1 show that approximately 25%
of dietary threonine, the limiting amino acid, is incorporated into
the lean of the sides (approximately 50% of the total protein of
the animal), the efficiency tending to be maximal at higher protein
concentration, the higher the energy concentration. Since the
amino acid composition of the dietary protein was kept constant
the percentage incorporation of dietary protein would be expected to
follow the same pattern.

Experiments which allow calculation of the efficiency of
incorporation of an individual amino acid, given at different
concentrations in the diet, have yielded some interesting results.
Taylor (8) fed pigs on a 12.5% protein basal diet to which graded
amounts of threonine were added and the threonine requirement for
growth was found to be met when the average daily threonine intake
was 9 g (Fig. 2). The efficiency of incorporation of threonine
into the lean of the sides also tended to be maximal when the
daily threonine requirement was met (Fig. 3).

In a similar experiment to estimate the lysine requirement of
the growing pig, Cooke (10) fed diets of differing lysine content.
The responses of growth, food utilisation and carcass lean content
to lysine intake were variable but calculation of the gain in lean
at each lysine content produced a relationship which showed a
clear asymptote and suggested a lysine requirement of about 18 g
per day (Fig. 4). The change in the efficiency of utilization of
dietary lysine with increasing lysine intake is shown in Fig. 5.
It appears that the efficiency of lysine utilization decreases
continuously with lysine intake from about 19%, at the lowest
intake, to 12% with the highest intake. The relationship does
not, therefore, show a distinct inflection in the region of the
requirement, which is unexpected and in contrast to the findings
for threonine. Although a plausible explanation for each type of
response, and for the difference between the two, can be offered
more examples are needed before such conjecture is justified.

It is surprising that the percentage incorporation into body
protein of the limiting amino acid in the growing pig does not
even approach 100% but is nearer 40% (assuming that lean protein
in the sides accounts for approximately 50% of total body protein).

Table 1

Percentage of dietary threonine incorporated
into the usable carcass lean of pigs

Crude Protein (%)	Digestible energy (MJ/kg)			
	13.6	14.2	14.5	15.3
14.6	22.2	26.0	25.3	24.2
16.6	22.3	25.6	25.4	24.1
19.4	25.0	26.5	26.4	25.9
21.3	22.4	22.4	24.4	23.4

Each value represents the mean of four pigs growing from
25 to 55 kg. It was assumed that lean protein contains 5.1%
threonine and lean meat contains 18% protein (9).

This inefficiency is depicted by the gap between levels D and B in
Fig. 1. The magnitude of this inefficiency is surprising especially in the case of lysine which is only involved in protein
synthesis. Threonine is a gluconeogenic amino acid: however, the
extent of gluconeogenesis from amino acids in the diets fed by
Hardy (7), Taylor (9) and Cooke (10) would be expected to be quite
small.

Within the concept of improving efficiency of protein utilization by bringing the pattern of ingested amino acids closer to
the needs of the animal it is necessary to continue the search for
more precise information on amino acid requirements. This is
usually possible by recording a production pattern in relation to
a predicted change in amino acid intake. There are some circumstances, however, under which this type of approach is not
possible. In the case of the human subject, for example, or the
pregnant sow, it is hardly possible to select a production parameter that can be related to graded intakes. In the case of the
adult ruminant, although it is possible to identify a suitable
production parameter, it is not feasible to achieve predictable
graded dietary intakes of particular amino acids. An alternative
approach has therefore been considered in the case of the ruminant
(11,12). This basically involves establishing a normal dietary

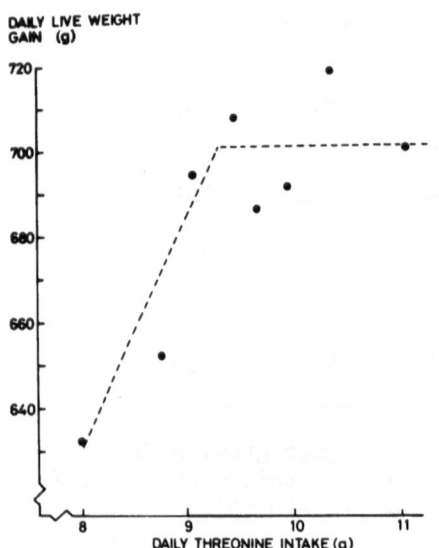

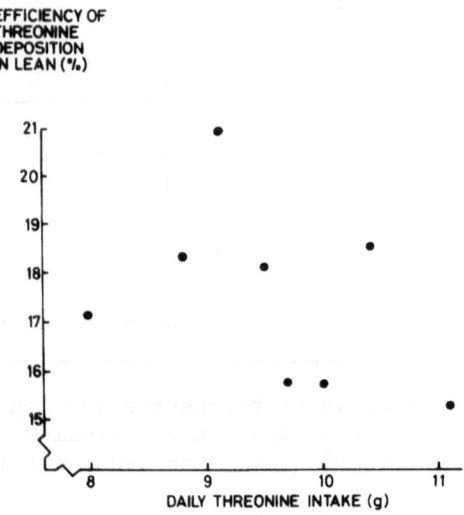

Fig. 2.   Daily live weight
gain of pigs given diets of
varying threonine content.
Each point represents the mean of
8 animals growing from 25 to
55 kg. Data from Taylor (8).

Fig. 3.   The change in the
efficiency of threonine deposition in the usable lean of
pigs given diets of increasing
threonine content.
Pigs grew 25 to 55 kg bodyweight. The middle joint was
dissected and the total lean in
the carcass calculated from the
whole body composition data
obtained by Hardy (7).   The
threonine content of lean
protein was assumed to be 5.1%
and the protein content of lean
was taken as 18% (9).   Data
from Taylor (8).

pattern by avoiding certain of the physiological consequences that
are the result of an unsuitable dietary balance of nutrients.

It can be considered that supplementation of a diet with an
amino acid in short supply, without changing the intakes of others,
would encourage protein synthesis. The plasma level of that amino
acid might thus not be expected to increase whereas the level of
other essential amino acids might be expected to fall slightly.
Beyond the point at which the dietary supply of the limiting amino
acid becomes adequate the plasma level would probably increase to
an extent determined by the effectiveness of pathways to deal with
surpluses, for example catabolic or excretory routes. According
to this concept, an inflexion point in the curve relating plasma
level to dietary supply can be taken to identify the point of most
efficient utilization or balanced input. Beyond this point of
dietary balance one would expect increased catabolism or excretion
as a contribution to inefficient nitrogen utilization. The

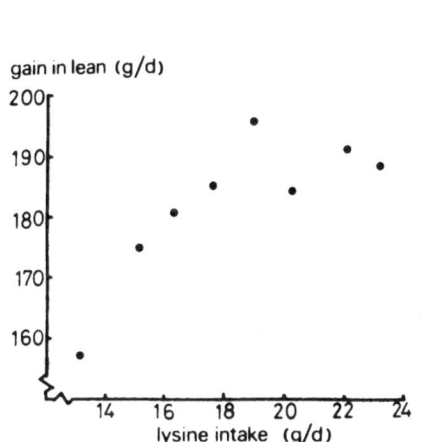

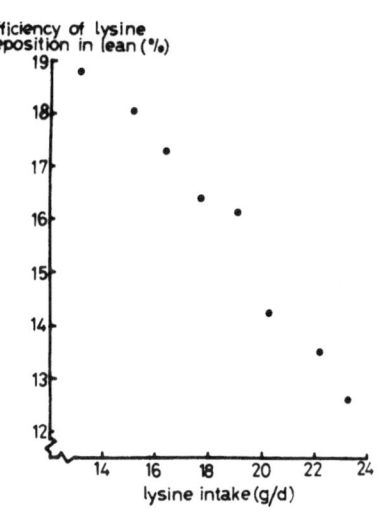

Fig. 4.   Gain in lean of
usable carcass of pigs fed on
diets of increasing lysine
content.
Each point represents 8 pigs
growing from 26 to 91 kg. Gain
in lean was calculated from data
of Cooke (10), assuming that a
26 kg pig contains 8.64 kg
usable lean (7).

Fig. 5.   The change in the
efficiency of lysine deposition
in the usable lean of pigs fed
on diets of increasing lysine
content.
Each point represents the mean
of 8 pigs growing from 26 to
91 kg. Values were derived
from Fig. 4, assuming that the
protein content of lean is 18%
and the lysine content of lean
protein is 7.8% (9).

increased catabolism can be recognized by a greater respiratory
loss of labelled carbon dioxide when a labelled dietary supplement
was offered; at the same time it might be possible to observe an
increase in plasma circulating urea levels. Should the excretory
route of disposal of surplus be important, for example in the case
of an amino acid with few active catabolic pathways available, an
examination of the urinary loss of amino acid could be fruitful.

A particularly interesting approach was proposed by Mercer
and Miller (13) for establishing the methionine requirement of
lambs. They used an abomasal injection of ^{35}S —methionine to
introduce a graded supply to the animal and recognised an
inflexion point in catabolic rate by monitoring the urinary
excretion of ^{35}S . Unfortunately such a procedure does not
appear possible with other amino acids since no such simple
labelling procedure seems to exist.

An approach was derived by Wakeling, Lewis and Annison (11)

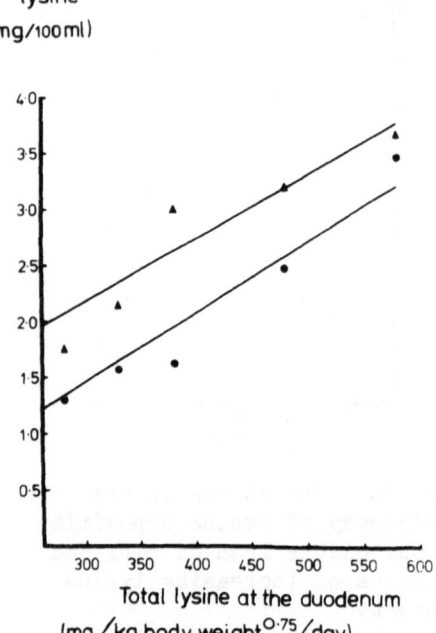

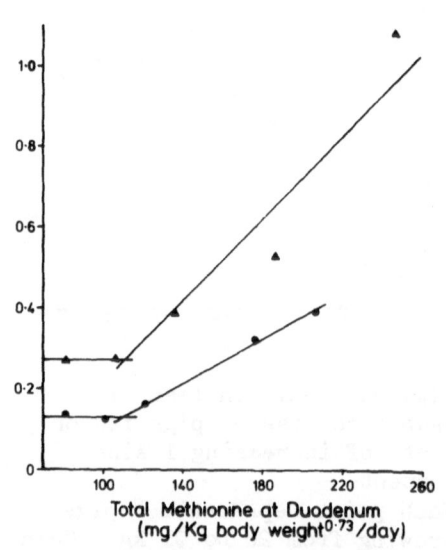

Fig. 6. Plasma lysine concentration in two sheep with
increasing passage of lysine
at the duodenum. Data from
Mitchell (15).

Fig. 7. Plasma methionine
concentration in two sheep with
increasing passage of methionine
at the duodenum. Data from
Wakeling, Lewis and Annison
(11).

to establish the identity of the limiting amino acid in a
particular diet and then to determine the need for that amino acid
under a particular set of physiological circumstances. Sheep
fitted with duodenal re-entrant cannulae were employed. In the
first instance the duodenal material was supplemented with graded
quantities of lysine. When plasma samples taken under standard
conditions were analysed for lysine it was found that the concentration increased steadily (Fig. 6). This was interpreted as
suggesting that lysine was not the limiting amino acid at the
level of the duodenum and that the unsupplemented quantity was
sufficient: the added amounts led to elevated plasma levels and
therefore probably to lower efficiency.

When comparable supplements of methionine to duodenal
material was made a different pattern of change was seen in plasma
methionine, (Fig. 7) with an inflexion point that was considered
to identify dietary balance or requirement. A similar pattern was
shown by Brookes, Owens, Brown and Garrigus (14) who monitored
expired $^{14}CO_2$ when graded levels of labelled lysine were administered. The studies of Wakeling et al (11), were extended by

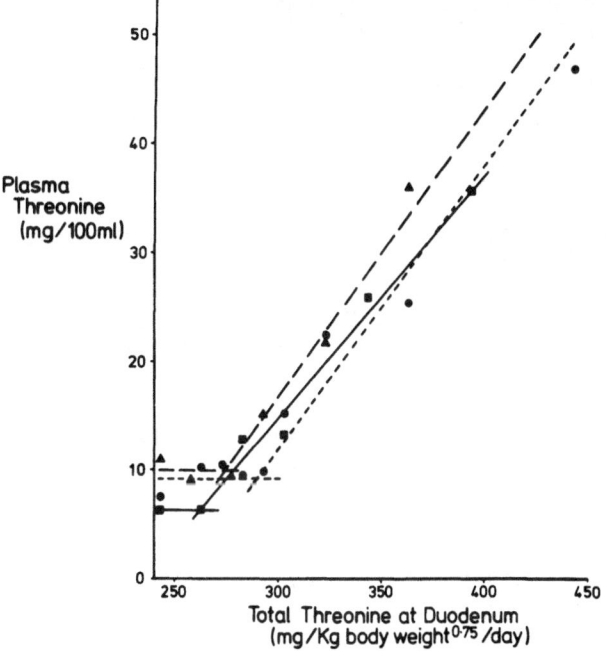

Fig. 8.   Plasma threonine concentration in two sheep with
increasing passage of threonine at the duodenum.
Data from Mitchell (15).

Mitchell (15) to identify subsequent limiting amino acids and the level at which these are in balance with the others. In one experiment a duodenal supplement of methionine was given which was somewhat in excess of that shown to be required. Graded supplements of lysine were added to the duodenum. A plasma lysine curve was obtained without a point of inflexion, suggesting that lysine was not the limiting amino acid. When this was repeated with threonine, the results suggested that this was the second limiting amino acid and that a point of balance or requirement could be recognised (Fig. 8). Several other investigations have recently been developed to establish a suitable pattern of amino acid supply for ruminant animals (16,17). It is clear that in the case of the ruminant as with the non-ruminant, as the effective supply of amino acids comes closer to a physiological balance, the major area in which improvements in protein utilization are possible is between levels B and C in Fig. 1.

With a balanced supply of amino acids the best efficiency of protein utilization can be expected. However, absolute requirements for particular amino acids are clearly determined by other factors, especially the energy-yielding nutrients, age and body size. The first need is therefore to identify for a particular set of physiological circumstances an ideal balance of amino acids, both essential and non essential. Beyond this 'physiological' balance, improvements in efficiency can be expected by adapting this pattern to changing animal circumstances (age or potential production), overall nutrient supply (energy-protein balance, nutrient 'density') or even economic issues.

REFERENCES

1.   Buttery, P. J., Boorman, K. N. and Barratt, E. (1973).
      Proc. Nutr. Soc. 32, 80A.

2.   Buttery, P. J. and Boorman, K. N. (1975). In: "Proceedings
      1st International Symposium on Nitrogen Metabolism in Farm
      Animals," E.A.A.P., in Press.

3.   Millward, D. J. (1970). Unpublished results quoted by
      V. R. Young (1970). In: "Mammalian Protein Metabolism"
      Vol. 4. (Ed. H.N. Munro). p. 623 (Academic Press, London and
      New York).

4.   Beckerton, A., Buttery, P. J. and Annison, E. F.
      Unpublished observation.

5.   Munro, H. N. (1970). In: "Mammalian Protein Metabolism"
      Vol. 4. (Ed. H. N. Munro). p.90 (Academic Press, London and
      New York).

6.   Young, V. R., Alexis, D. D., Baliga, B. S. and Munro, H. N.
     (1972). J. Biol. Chem. <u>247</u>, 3592-3600.

7.   Hardy, B. (1974). Ph.D. Thesis, University of Nottingham.

8.   Taylor, A. J. (1975). Ph.D. Thesis, University of Nottingham.

9.   Schweigert, B. S. and Payne, B. J. (1956). Amer. Meat Inst.
     Bull. No. 30, quoted by Lawrie, R. A. (1974) in "Meat Science,"
     2nd edition, p.347 (Pergamon Press, Oxford).

10. Cooke, R. J. (1969). Ph.D. Thesis, University of Nottingham.

11. Wakeling, A. E., Lewis, D. and Annison, E. F. (1970).
     Proc. Nutr. Soc. <u>29</u>, 60A.

12. Lewis, D. and Mitchell, R. M. (1975). In: "Proceedings 1st
     International Symposium on Nitrogen Metabolism in Farm
     Animals", E.A.A.P.   in Press.

13. Mercer, J. R. and Miller, E. L. (1973). Proc. Nutr. Soc.
     <u>32</u>, 86A.

14. Brooks, I. M., Owens, F. N., Brown, R. E. and Garrigus, U.S.
     (1973). J. Anim. Sci. <u>36</u>, 965.

15. Mitchell, R. M. (1974). Ph.D. Thesis, University of
     Nottingham.

16. Williams, A. P. and Smith, R. H. (1974). Proc. Nutr. Soc.
     in Press.

17. Armstrong, D. G. and Annison, E. F. (1973). Proc. Nutr.
     Soc. <u>32</u>, 107.

DISCUSSION

Dr. van Es asked if there were any differences between sheep
and cattle in the microbial activity of the rumen and substrate
transformations. *Prof. Armstrong* replied that most of the
available information had been obtained with sheep, although some
comparative information was now becoming available. These
results suggest that, with long forage diets, digestion is similar
in sheep and cattle, but with high concentrate diets, more starch
escapes fermentation in the steer.

Dr. Rérat reported that transamination in the gut wall
occurred in pigs as well as ruminants; recent work with pigs in
his own laboratory indicated that alanine and glycine in the
portal blood were increased at the expense of glutamic and
aspartic acids. *Dr. Rérat* made the further comment that gastric
emptying was controlled not only by the concentration of metabolites in the portal blood, but by the volume of the contents of
the small intestine. *Prof. Armstrong* thanked Dr. Rérat for the
information on the deamination of glutamic acid, which suggested
that the older concept of the specific gluconeogenic significance
of glutamic acid in the ruminant may not be valid.
Prof. Kielanowski pointed out that, because of the degradation to
ammonia of amino acids in the large intestine, the determination
of amino acid digestibility should be made at the end of the
small intestine.

Dr. van Es remarked that there were several reasons why the
metabolism of an animal changes with increasing age; amongst these
he noted decreasing activity and an increasing ratio of fat to
lean tissue. He enquired whether protein turnover was faster in
younger than in older animals, and wondered if this represented
the elimination of incorrectly synthesised proteins. He cited the
case of the dairy cow where the energy cost of protein synthesis
was less than in the growing animal. *Dr. Buttery* replied that
protein turnover was faster in the younger animal and also in the
smaller as opposed to the bigger species. *Dr. Webster* said that
the Zucker rat was a model for the extremes of some of these
interactions. Direct observation comparing lean and obese individuals during growth suggested no difference in activity, the
sluggishness of the fat adults was considered a consequence rather
than a cause of their obesity. Clearly, as the animal approaches
maturity, net protein deposition becomes a decreasing fraction of
total protein turnover and the efficiency of protein deposition
falls to zero. *Dr. Dickerson* asked to what extent in the Zucker
rat it was possible to maintain maximum lean-tissue growth while
modifying fat deposition by dietary manipulation. *Dr. Webster*
quoted his collaborative work with Mr. Radcliffe at the Rowett
Institute showing that, with diets varying widely in protein
concentration, fat and lean rats apparently varied their food

intakes so as to achieve a similar rate of nitrogen retention
though with vastly different rates of fat deposition.
Dr. Dickerson continued by asking whether it was possible to
prevent the fat genotype from becoming obese while still maintaining its maximum rate of lean-tissue deposition. Dr. Webster
replied that this was, to some extent, possible by increasing the
protein content of the diet, leading to a reduction in food intake
and in fat deposition, while maintaining the lean-tissue growth
rate.

Prof. Armstrong asked what correction Professor Lewis had
made in his calculations for the difference in the efficiency of
uptake between amino acids flowing from the rumen and those
infused into the duodenum, which he assumed would be completely
absorbed. Prof. Lewis replied that no correction had been made:
in his view not only was the absorption of amino acids in protein
flowing from the rumen less than 100%; so also was the absorption
of the infused amino acids which were available only in spurts.
He felt that existing information did not permit meaningful
corrections.

Dr. Buttery reported from his own results that, with sheep,
muscle accounted for over 50% of total protein synthesis, the
liver 15-23%; these results fitted with the compartmental model
of Nolan & Leng; however, in most sutdies it had been difficult
to estimate the contribution of protein turnover in the gut.
Dr. Webster referred to his own finding that the gut and liver
contributed 30-35% of total metabolic rate of animals fed almost
ad libitum, and he supposed that this reflected the high contribution which visceral protein turnover makes to total turnover.
Dr. Fowler commented that the high protein turnover of the gut
wall and liver was explicable in terms of the flux of nutrients
across these organs. He enquired whether high turnover in muscle
might be explained by the rejection of incomplete proteins by an
immunological screen. Dr. Buttery thought that, although protein
catabolism was in general a random process, there were elements
which were not random. There was evidence, for example, that
some of the subunits of structural proteins were broken down after
synthesis.

The Development of Muscle

The Development of Muscle

TOWARDS MORE EFFICIENT MEAT ANIMALS: A THEORETICAL CONSIDERATION

OF CONSTRAINTS AT THE LEVEL OF THE MUSCLE CELL

I. G. Burleigh

Agricultural Research Council, Meat Research Institute

Langford, Bristol, BS18 7DY, U.K.

I have chosen to consider the overall problem in the
following terms:-

1.   Should meat animals grow bigger or grow faster?

2.   Leanness and its side-effects.

3.   Muscle growth at the cellular level: introductory comments.

4.   The problem of what causes differences in the size of mammals
     and their muscles.

5.   A comment concerning the efficiency of muscle growth.

6.   A design for a hypothetically more efficient animal.

7.   Some diagnostic properties of the proposed animal's muscula-
     ture.

8.   Unresolved fundamental questions.

9.   Harnessing the answers to practical ends.

     1.   Should Meat Animals Grow Bigger or Grow Faster?

     I am concerned with processes that might operate within
muscle cells to regulate the rate and extent of accumulating protein.   This is research at a very fundamental level.   However, a

practical aim is the eventual development of techniques which will
further improve the growth efficiency of meat animals after present procedures have begun to yield diminishing returns.

For example, in the United Kingdom there is currently some
interest in producing meat from larger breeds of cattle.   At
present, there may be some advantages of growth efficiency to be
gained but presumably limits will be reached since cattle have not
evolved to the size of elephants.   One can think of several
reasons: (a) At the cellular level, the larger terrestrial animals
tend to grow relatively slowly (see below).   (b) In the absence
of compensatory adaptations, such animals will tend to expend more
of their metabolic energy in combating gravity, since body mass
increases with the cube of linear dimensions while the strength of
the structures which support that mass should increase only with
the square (1).   In fact, on an evolutionary scale, the bones of
large terrestrial animals are disproportionately thick (2) but
that will hardly commend itself to meat producers.   (c) Genetic
variability is said to decrease with increasing size of animal.
Thus, A. V. Hill (1) has pointed out that there are about 6,000
species of rodent but only three species of elephant (Walker (3)
cites only two species of elephant).   This is partly because
large terrestrial animals replicate themselves rather slowly and
partly because very few biological load-bearing structures can be
evolved to carry particularly heavy weights on land.

In passing, the restrictions of gravity on growth are less
important in marine animals which are buoyed up by the water they
displace.   Whales reach weights of over 100 metric tons and there
are at least nine species of whale whose weight (3) is more than
twice that of the two known species of elephant.   Genetic diversity
thus seems greater in the sea.

As regards my immediate brief, therefore, there must be some
limit beyond which making terrestrial animals bigger by present
processes of selection becomes uneconomic.   If British cattle are
considered in isolation, the scope for increasing the size of
several breeds, the Hereford for example, is obviously significant
since the South Devon, as well as several breeds of cattle on the
European mainland, are a good deal heavier.   More problematical
is whether the size of pigs can be increased since the musculature
of pigs is physiologically less developed at birth and may be less
able to support increased body weights in the neonatal period.

Other problems may be encountered with particularly large or
rapidly growing animals.   Thus, Miss Christine Gibbs of the
University of Bristol's Veterinary School has encountered Charolais
and Simmental cattle with malformations of the hips and of the
fetlocks (personal communication).   Both these conditions appear
to be associated with rapid weight gain in immature cattle (4).

The birth problems of large cattle are well-known (5,18,19), though there is some debate as to whether this is because of the animals' shape or their size.   One breeder of Herefords has recently commented that larger members of the breed are the least adaptable and the most unpredictable breeders (6).

To begin seeking means of obviating these problems, or potential problems, one might think of accelerating muscle growth, in the sense of taking less time to produce the same amount, or even more muscle, from animals whose birth weight does not alter. If one could shorten the time taken to produce a meat animal, prospective savings emerge on the amounts of food and physical energy that are consumed.

This is essentially to invoke Blaxter's suggestion (7) that the faster an animal is made to grow, the less energy is used up simply in maintaining the animal.   However, instead of using nutrition to manipulate rates of growth, I am here concerned with doing so genetically.   The problem, as will be mentioned again shortly, is that a major evolutionary mechanism for achieving faster rates of growth on land seems to involve more inefficient growth per unit time, indeed to the point that prospective gains in efficiency are cancelled out by the losses.

2.   Leanness and its Side-Effects

There is the additional aspect of making animals leaner i.e. increasing their complement of muscle relative to fat and bone. Lean breeds of animal do exist and a well-known example is the Pietrain pig.   However, pigs of this sort tend to yield pale and watery meat and they tend to die prematurely.   Lister and his co-workers consider that these disadvantages are a side effect of hormonal changes which diminish rates of accumulating fat, rather than accelerating muscle growth (8,9).

Holmes and Ashmore have advanced a different explanation of stress-susceptibility in lean, double-muscled cattle (10).   They have found that the muscular hypertrophy of such animals partly involves the accumulation of a large type of fibre which is rich in enzymes catalysing the catabolism of glycogen to lactic acid. The authors consider that this is one of several factors combining to induce stress-susceptibility by elevating the concentration of lactic acid in the blood.

I shall not attempt to decide which view of stress-susceptibility is correct, and they are not necessarily incompatible. Together, however, they advance two further specifications for inclusion in our ideal animal for current market requirements. (a) The animal should be lean because of genetic alterations to a

Table 1

Amounts of muscle and fat in pigs genetically
selected for fatness and leanness

| | Backfat thickness | | |
	High	Medium	Low
Age of pig at slaughter (days)	185	173	169
Muscle in half carcass (kg)	14.63	15.12	16.21
Fat in half carcass (kg)	9.54	7.95	6.70
Weight of half carcass (kg)	31.81	31.43	31.17

Calculated from Standal et al. (11)

mechanism which accelerates the growth of muscle, rather than
diminishing the accumulation of fat and (b) that mechanism should
not of itself produce stress-susceptibility.

Such a mechanism will be difficult to detect in operation.
This is because fat and muscle protein both require energy for
their synthesis and because some amino acids can potentially be
employed in the synthesis of both fat and protein.   Therefore, it
is theoretically possible to accelerate muscle growth by suppressing the accumulation of fat and diverting energy sources and certain amino acids into the protein-synthesizing pathway.   Consider
some data on the pigs which Standal et al. (11) bred to high and
low thicknesses of backfat (Table 1).   If the rate of accumulating
fat is not affecting muscle growth, then lean and fat animals of
the same age should have the same amount of muscle, expressed in
absolute terms.   However, the lean pigs produced about 10% more
muscle in about 91% of the time that fat pigs did (Table 1).
Adipose tissue isolated from the lean pigs also mobilized fat
relatively actively when adrenaline was added to it.   Like the
Pietrain, such pigs may have been stress-susceptible, though this
was not established.

One's chances of unambiguously detecting a growth-regulatory mechanism operating within muscle are probably best in very young animals, or even foetuses, since it is then that the ratio of muscle to fat is greatest.

3.   Muscle Growth at the Cellular Level:
Introductory Comments

In placental mammals, the major events in determining muscle size occur before birth.   Most of the cells in the musculature have formed by then and, depending on whether one is speaking of a cow or a rat, 30-40 generations of cells are produced before birth and the equivalent of only 2-4 generations accumulate thereafter.   Broadly speaking therefore, muscle growth falls into two phases;   one in which cells destined to form muscle are actively replicating for a significant portion of the animal's gestation period, and a succeeding phase in which amounts of muscle protein per cell increase in circumstances where cells replicate progressively more slowly.

Skeletal muscles are composed of fibres in which reside the specialized proteins concerned with contraction.   These fibres are essentially long multinucleate cells formed, as is now generally believed, by the fusion of elongated precursor cells called myoblasts.   Myoblasts have single nuclei and they originate from actively dividing precursor cells.   In vertebrates, at least, cells that are becoming myoblasts cease to divide by mitosis and, to my knowledge, nobody has provided evidence that nuclei divide within muscle fibres.   However, some cells with a capacity to replicate themselves remain associated with muscle fibres after they have formed.   These cells are called satellite cells and several authors consider them to be a reserve of myoblasts that can fuse with the fibres.

Once cells begin to accumulate the characteristic proteins of skeletal muscle they do not appear to divide again and, in addition, cells that have not become specialized within a developing muscle divide more slowly with age (12).   Two mechanisms combine, therefore, to produce the overall decline in the rate at which cells multiply in a muscle as it grows.

4.   The Problem of What Causes Differences in the Size
of Animals and their Muscles

The grosser variations in muscle size i.e. those between mammalian species, obviously correlate with the number of cells and nuclei which each muscle possesses, as well as with the number of fibres.   Also, in placental mammals, most or all the muscle fibres have formed by birth (12-14,37).   Thus, at least in regard

to differences between species, cell and fibre number represent
parameters which begin to meet the requirement for an index of
muscle growth potential that is established early in life.

Widdowson (15) has proposed that humans are larger than rats
because human cells continue to divide for a longer period before
birth, although they divide more slowly.   She may well be
describing the major mechanism for determining the size of animals
on land although other mechanisms probably exist.   Thus,
Kihlström (16) found that gestation periods become more protracted
with increasing body weight among species of terrestrial mammal
with placentas of similar morphology.   This supports Widdowson's
idea since the length of gestation is presumably some measure of
the overall time for which cells continue to divide actively.
However, Kihlström also observed that quite significant variation
in the length of gestation occurs between animal species of similar
size and that the differences reflect the type of placenta.   So
there is presumably some scope for the size of animals to correlate
directly with rates of cell replication, rather than inversely as
proposed by Widdowson.

However, large animals on land do seem to be so at least
partly because their cells have divided longer, albeit more
slowly, before birth.   As metabolic processes occur more slowly
in large terrestrial animals than in small ones (17), I recently
suggested (12) that the events which cause cells to stop dividing
proceed more slowly during the pre-natal growth of animals
destined to be large.

As regards post natal growth, this is a period in which muscle
fibres are principally growing in size rather than in number.
Relative to smaller mammals on land, the larger species perhaps
possess somewhat broader muscle fibres, but such variation is minor
compared with differences of several orders of magnitude in the
number of fibres (see above).   Indeed, fibre breadths that have
been quoted for several muscles in mice and cattle fall in the
range 40-50 μm.   What matters here is not so much differences in
the width of fibre as in the time required for the fibres to grow
to roughly similar widths.   Thus, mice and rats take about 15-18
weeks to achieve fibre breadths in the above range, cattle take
over 2 years and humans require at least 12 years (12).

There is probably a trend for muscles to grow more slowly in
animal species of increasing size.   This can be inferred from
post natal growth in body weights, in the absence of suitably
comprehensive data on muscle itself.   I have calculated the time
taken for various species of mammal to increase their body weight
by a constant factor, arbitrarily set at 5-fold.   To encompass the
very wide variation in these times within a single graph, the
logarithms have been taken and in Fig. 1 they are plotted against

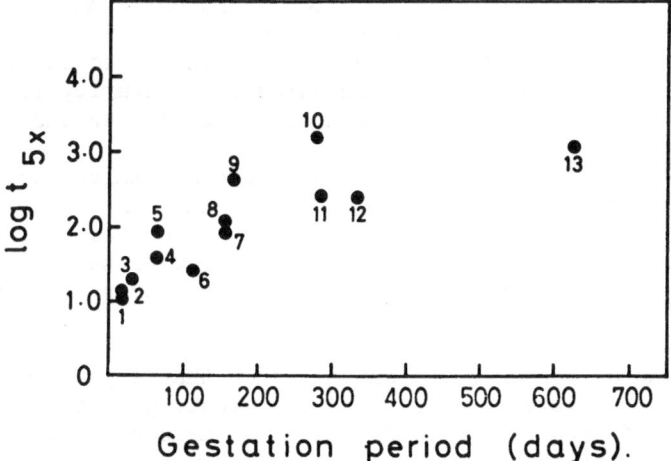

Fig. 1.   The rate at which terrestrial mammals grow postnatally
(log t_{5x}) plotted against their gestation period (for explanation,
see text). t_{5x} is the time in days needed to multiply an
animal's birth weight by 5-fold.

1 = mouse,   2 = rat,   3 = rabbit,   4 = cat,   5 = guinea pig,
6 = pig,   7 = sheep,   8 = goat,   9 = rhesus monkey,   10 = man,
11 = ox,   12 = horse,   13 = African elephant.

Calculated from data to be found in references 15,45,49.

the gestation periods for the same species.   There is obviously
a direct relationship, indicating a trend for the rate of prenatal growth to be reflected in the rate of postnatal growth.
It can be seen that gestation tends to be longer in the larger
species also.

    Some common process would therefore seem to regulate rates
of growth before and after birth.   Thus, if animal breeders
speed up this mechanism in postnatal life, they may accelerate
the processes by which cells stop dividing before birth.   The
result could be a smaller animal.   It is interesting to speculate
that manipulations of this sort have caused the trend for certain

breeds of British cattle, Herefords for example, to mature
earlier in recent times and to have become smaller in the process.
The opposite trend appears to have occurred in the so-called
exotic breeds of European cattle in which late maturity is
accompanied by large size and by longer gestation periods (18).

Consequently, breeds like the Charolais and Maine Anjou,
which have rather large birth weights (18), may simply have
acquired more cells by protracted and somewhat slower cell division
before birth.   This will make the animals grow faster overall
after birth though individual muscle cells or fibres are not
necessarily growing faster.

For example, consider two animals such that a given muscle in
animal A has 10^7 fibres at birth and its counterpart in animal B
has 1.1×10^7 fibres.   This difference has arisen because precursor cells have divided somewhat more slowly but for a longer
period overall in animal B.   Assume that at birth the fibres are
of the same average breadth, 10 μm say, in both animals and that
they subsequently grow at the same rate to breadths of 30 μm in
the next 300 days.

The cross-sectional area of each fibre, here taken as proportional to its mass, will increase from 78.5 μm^2 to 705 μm^2.
The total cross-sectional area of the two muscles will increase
by 10^7 (705-78.5) μm^2 in animal A and by 1.1×10^7 (705-78.5) μm^2
in animal B, i.e. by 62.7 cm^2 and 68.9 cm^2 respectively.   In
other words, the muscle in animal B appears to be growing faster
but at the cellular level it is not.   In fact this muscle would
still appear to grow faster overall were its fibres to grow up
to 10% more slowly than those in animal A.

There is, in fact, some evidence that muscle fibres grow
comparatively slowly in double muscled cattle (19,20), although
there are some observations to the contrary (10).   Lean Pietrain
pigs are said to have somewhat enlarged fibres (21-23) but the
last findings are possibly an artefactual effect of a phenomenon
described by Lister (8,9) whereby lean Pietrain and fat Large
White pigs accumulate muscle at the same rate.   The Pietrain
accumulates fat more slowly, however.   Thus, if it is allowed to
achieve the same body weight as the Large White, it will possess
more muscle and, presumably, broader muscle fibres but only because
it has grown for longer.   In terms of total body weight the
Pietrain, of course, grows comparatively slowly.

As regards large or prospectively leaner animals that grow
rapidly, the above discussion suggests that they could simply
possess more muscle fibres of unaltered growth potential.   In
turn, the latter could arise because the replication of precursor
cells has been prolonged and is perhaps even somewhat slower in

Table 2

Average birth weights for calves
of different breeds (52)

Breed	Sex	No.	Mean (lb)
Aberdeen Angus	Bulls	350	63
	Heifers	310	59
Beef Shorthorn	Bulls	108	72
	Heifers	72	63
Belted Galloway	Bulls	64	68
	Heifers	51	61
Charolais	Bulls	370	101
	Heifers	310	94
Devon	Bulls	280	82
	Heifers	212	77
Galloway	Bull	38	71
	Heifers	54	67
Hereford	Bulls	1,856	75
	Heifers	1,513	71
Lincoln Red	Bulls	47	81
	Heifers	38	78
South Devon	Bulls	79	102
	Heifers	68	98
Sussex	Bulls	210	83
	Heifers	187	78
Welsh Black	Bulls	164	79
	Heifers	178	77

pre-natal life.   Diagnostic of such a mechanism, if only as a
first approximation, should be an increase in the gestation period
given that, on an evolutionary scale, this is more protracted in
the larger terrestrial animals.   As muscle fibres contain significant quantities of contractile protein, even at birth, the animals'
weight at birth should probably increase somewhat.   Gestation
periods may not be prolonged by very much in the larger or leaner
types of meat animal since small differences in the average rate
and duration of cell division can have disproportionate effects on
the final number of cells (Section 6).

    Bulls are a case in point.   They are particularly lean but
they weigh more at birth than heifers (Table 2).   MacKellar (19)
studied several British breeds and established that bull calves
were born after a slightly, but statistically significantly,
longer gestation period than heifers.   Gestation periods are also

said to be extended and birth weights increased in lean cattle
which have inherited the condition of double muscling (5,19), as
well as in the large 'exotic' breeds of cattle (18,19,24).   For
example, a recent estimate placed the gestation periods of
Simmental, Charolais and Limousin cattle at between 285 and 287
days, relative to 282 days for the British Hereford (24).   The
average length of gestation for the largest British breed, the
South Devon, has been found to be 290 days in a survey by MacKellar
(19).   This was seven days more than the average for several
British breeds.

For examples of animals in which a rapid rate of postnatal
muscle development reflects an accelerated rate of cell replication in prenatal life, one may have to turn to the sea.   As
pointed out by Kihlström (16), gestation periods among cetaceans
(porpoises, whales, etc.) increase little with body weight, in
contrast to those of terrestrial mammals whose placental morphology
is of the type found in cetaceans.   It is tempting to suppose that
a diminished need to combat gravity allows aquatic animals to
divert energy into making cells faster during gestation.

5.   A Comment Concerning the Efficiency of Muscle Growth

Proteins are continually being broken down and replaced in
animal cells and it is commonly believed that the rate of their
degradation is proportional to the amount or concentration of each
protein that is present at any given time.   The rate of synthesizing protein is considered not to vary in this way (25,26) and for
theoretical purposes it has been treated as constant in the absence
of factors which perturb growth.   Therefore as the amount of a
given protein increases within the animal cell, the rate of
degrading that protein should increase also.   Eventually the
rates of synthesizing and catabolizing protein should become
equal and growth should cease.

Calculations based on the above concepts have successfully
explained the different rates at which liver enzymes accumulate in
response to hormonal and nutritional manipulation (25,26), and
the rate and extent to which multiple forms of the enzyme, lactic
dehydrogenase, accumulate in cardiac muscle (27).

Similar principles have been applied to the growth of protein
in skeletal muscle, though on incomplete evidence whose status
has been reviewed previously (12).   It is reasonable to accept
that protein is continually being degraded and resynthesized in
skeletal muscle.   However, although many authors express the rate
of its catabolism in terms of half-lives, indicative of the
assumption that the rate of catabolism varies with the amount or
concentration of protein, firm evidence for the assumption is

lacking.   If there is a relationship of the above sort, then the
rate of catabolizing a muscle fibre's contractile elements, the
myofibrils, probably varies with their surface area rather than
with their mass (12).   As myofibrils are increasingly segmented
or partially segmented during growth, one cannot presently say
how their surface area is varying.

As a working hypothesis, therefore, and to generate growth
curves which at least approximate those of muscle growth with age,
I proposed a simple model in which the growth of a muscle fibre
represents a balance between a constant rate of protein synthesis
and a rate of protein degradation that increases linearly with
time (12).   I then compared a situation in which individual muscle
fibres in mice and cattle grow to the same final breadth but at
rates which differ markedly between the two species (see Section 4).
As a result, not only did the rate of synthesizing appear to be
less in fibres of cattle but the rate of degrading unit quantity
of protein was inferred to be slower also.   The implication was
that the total amount of protein that is degraded and replaced
during growth of the two types of fibre tends to be the same.
The overall efficiency of growth should also tend to be similar
given that the processes of protein degradation and synthesis,
including replacement synthesis, should liberate heat and consume
energy.

Fig. 2A-D takes the argument further.   In each case, the rate
of protein synthesis is initially identical and the initial rate
of protein degradation is zero.   Rates of synthesis and degradation vary linearly and always meet at the same time T.   The total
amount of protein formed (or the fibre's final cross-sectional
area) is constant and is proportional to the area between the
ordinate and the lines representing synthesis and degradation (S
and N).

In Fig. 2A, protein synthesis increases with time and in
Fig. 2B it is constant as in (12).   Synthesis slows with age in
Fig. 2C and D, in the last case to zero when growth stops at
time T.   It can be seen that identical growth curves are obtained
in each case, a prediction which I have verified graphically and
mathematically.

However, the total amount of protein that is catabolized
($\Delta OXT = n$) obviously varies between situations A-D.   When T is
constant, n varies in proportion to β, the ratio of the final rate
of protein synthesis to its starting value (s_0), since it can be
shown that $n = \frac{1}{2}\beta s_0 T$.   The final quantity of protein, P_{max} can
also be shown to equal $\frac{1}{2}s_0 T$, so that $n/P_{max} = \beta$.   Consequently,
when P_{max} is constant, the total quantity of protein that is catabolized and replaced during the production of P_{max} is constant for
any constant value of β.   It is therefore independent of whether

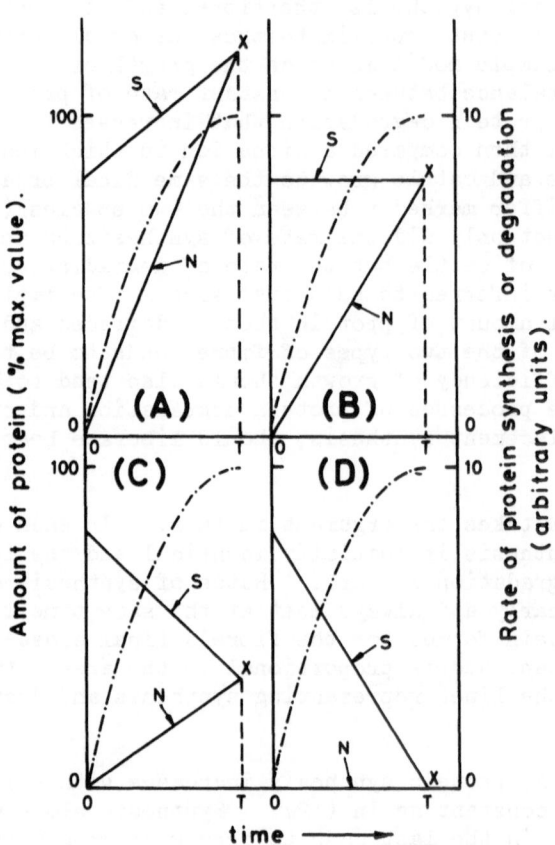

Fig. 2.   Constancy of the growth curve when rates of protein
synthesis and degradation vary linearly with time to different
extents.   Growth curve: S = rate of protein synthesis;
N = rate of protein catabolism.

protein synthesis increases or decreases with age and of the time taken to produce P_{max}, i.e. the total amount of protein degraded should be the same whether P_{max} accumulates in around 15 weeks as in mice or over a much longer period as in cattle. This, of course, would also be true if muscle proteins were completely stable during growth. Thus, all forms of the model in Fig. 2 tend to support my earlier and more tentative conclusions (12) which can also be adduced from a model of muscle growth in which the rate of protein synthesis is constant and the rate of degradation is proportional to the amount of protein that has accumulated at any given time (see Appendix). Faster growth thus seems to mean more inefficient growth, a prediction which accords with the fact the basal metabolic rate of rapidly-growing mice is several times greater than that of slow-growing animals the size of cattle (2). In fact, one has arrived at Kleiber's (28) conclusion that the overall efficiency with which animals convert food to body mass is independent of body size, the faster growth of smaller animals tending to be counterbalanced by greater inefficiency and heat output per unit time.

6.   A Design for a Hypothetically More Efficient Animal

In what follows I have assumed a model of muscle growth of the form shown in Fig. 2B. Whether Fig. 2A, B or C is chosen does not materially affect the argument that it may be possible to reduce the amount of protein that is degraded during the production of a given quantity of protein. Rather the ambiguity lies in the efficiency with which protein is accumulated by unimproved animals. If this is high as in Fig. 2C, scope for further improvement will be less and it will be zero if muscle growth occurs according to Fig. 2D. On biochemical grounds the last mechanism currently seems unlikely and whatever the mechanism of growth proves to be, shortening its duration should still diminish the amount of energy wasted on contractile activity by the muscle, and perhaps by other physiological processes.

Consider again what will be termed a basal amount of muscle protein, the aim being to accelerate the growth of this protein without making the animal unduly big, unduly small and without decreasing the efficiency of growth per unit time. In Fig. 3 growth of this protein is represented by triangle CDE which is here taken in a wider sense to describe the growth of a single muscle fibre of standard final breadth, or growth of a standard quantity of protein representing multiple fibres. Triangle ABC represents a situation in which the same amount of growth has been achieved in less time by the undesired mechanism which involves an increased rate of protein turnover. In triangle ACF we have a potentially better situation where protein synthesis has doubled without alteration to the rate at which a unit amount of protein is

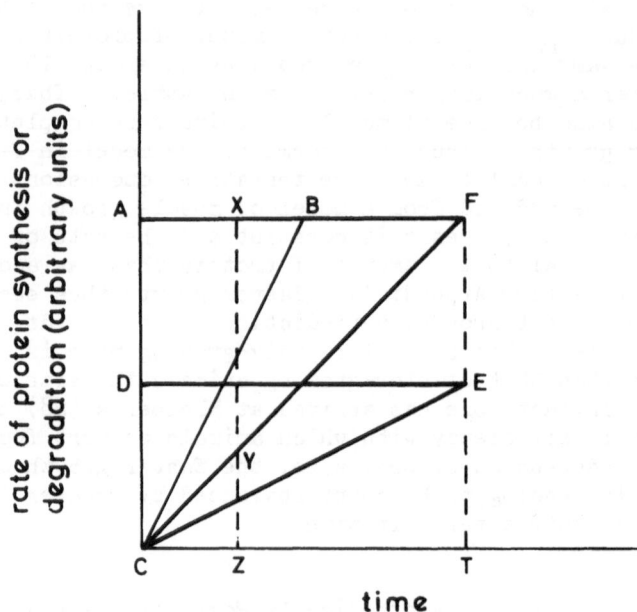

Fig. 3.   Hypothetical scheme to illustrate the amount of protein
that is degraded during the growth of muscle at different rates
and by different mechanisms.

degraded.   Twice as much protein (Δ AFC) is obtained in the same
time (CT) as the basal quantity (Δ CDE) took to grow and an
amount of protein (area AXYC) equal to the basal quantity accumulates in a shorter time (CZ).   At this stage the total amount of
protein that has been broken down (Δ CYZ) is nearly six times less
than that which would normally be catabolized (Δ CET) in forming
this basal quantity of protein.

In Fig. 3, the rate of protein synthesis could be increased
by doubling the activity of individual ribosomes or by producing
twice the number of ribosomes.   The second mechanism could simply
involve a doubling in the number of muscle-forming cells.   If the
number of these cells increases selectively, energy and branched
chain amino acids should be diverted from the production of fat
and the animal should also be leaner.

However, when the animal is killed, several physiological
properties of the muscle may well be underdeveloped.   The muscle
may be paler since myoglobin, the red pigment of muscle, accumulates

progressively after birth (29).   So also do the enzymes concerned
with degrading glycogen and generating energy for contraction by
the anaerobic route.   These should be present in comparatively
low concentration also.   Conversely, the concentration of mitochondria will probably be higher since it tends to decline as
muscles grow with age (12).   The last two factors should thus
combine to make the metabolism of the muscle more aerobic. Stress-
susceptibility will therefore be reduced if it is related to the
capacity of muscle fibres for anaerobic metabolism (Section 2).

To some extent, the slower and more protracted replication of
precursor cells might be harnessed to producing the extra cells
required in Fig. 3.   Given that a quite significant increase in
the final number of cells can theoretically result from a small
increase in the overall duration of replication, then on the
argument of Fig. 3 greater overall efficiency will result if the
extra nutrient consumed by the embryo, foetus and mother is less
than that saved by slaughtering the animal, when its fibres are
less developed.   Reduced food conversion ratios attributable to
double-muscled cattle, exotic cattle and bulls may have such an
explanation, at least in part.

There will almost certainly be limits to the increased
efficiency that can be achieved in the above manner.   Assuming
that gestation periods are a rough measure of the duration of
active cell replication, gestation has to be increased to 625
days in order to produce a new-born elephant weighing 120 kg.
Cattle weigh twice as much, or even more, at the same age from
conception.   Thus, at some stage in an attempt to produce more
cells in cattle by the above mechanism, any advantages of growth
efficiency will be offset by the economic disadvantages of slower
muscle production.

However, the immediate barrier to further manipulation of
present mechanisms for making cattle larger or leaner genetically,
probably lies in the fact that birth problems will increase more
than cattle were designed for, with attendant calving difficulties.
Circumventing this problem will not just be a matter of altering
rates of cell replication.   Even if we were to increase these in
embryonic life and if the resulting extra cells begin to form
muscle protein at the same stage in prenatal life and at the same
rate as in normal animals, the amount of muscle will still be
increased at birth, with possible problems of dystocia.   If we
continue active cell division and delay the process of cell
specialization till nearer birth, less muscle protein should have
formed by then and the birth weight should increase by rather less.
However, in so doing, we may have made the muscle fibres physiologically underdeveloped at birth and weaker.   Thus, there is evidence that the speed and total force with which a muscle contracts
increases during postnatal development.   Whether the immature

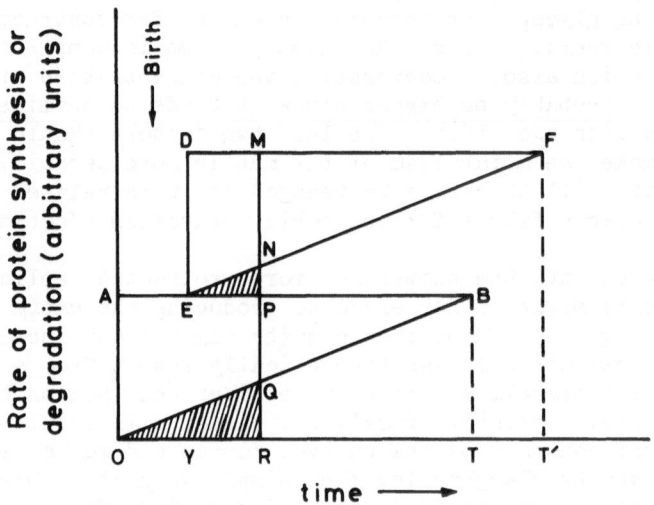

Fig. 4.   Extension of the scheme in Fig. 3 to encompass
asynchronous differentiation of muscle cells.

muscles exert less force per unit of muscle weight is unclear (30)
and that seems to be the more meaningful question in the present
context.   There is indirect evidence to the effect that neonatal
muscles are intrinsically weaker.   This takes the form of morphological observations that the myofibrils of immature muscle are
surrounded by comparatively few and relatively disorganized
elements of the sarcotubular system which are deficient in the
ability to take up calcium (31-33).   Whether the ATPase activity
of the myofibrillar proteins increases during postnatal development
is presently open to debate (12).

     Regarding the clinical effects of underdeveloped fibres in
neonatal animals, very young pigs are subject to a condition known
as splay leg in which the muscle fibres are poorly developed (34).
Such pigs are liable to be crushed by the sow.   Cattle might
possibly tolerate more immature fibres at birth since they are not
subject to crowding in large litters.

     Fig. 4 illustrates a hypothetical design for an animal which
might obviate these potential problems.   In this animal, a normal

basal amount of muscle protein (Δ ABO) starts to accumulate at
the same stage of prenatal life as in normal animals. The muscle
develops twice as much growth potential but expression of the
extra potential is delayed until nearer to, or after, birth
(Δ DEF).   Clearly, there are still prospective benefits of growth
efficiency to be gained if the rate at which individual fibres
degrade protein increases with age.   If the animal is slaughtered
at time OR to yield a basal amount of protein (the sum of areas
DMEN and APQO), five times less protein will have been catabolized
than in the normal animal i.e. the sume of the areas of Δs ENP and
OQR is five times less than that of Δ OBT.

     A more usual assumption is that the rate of protein degrada-
tion in muscle increases with the mass of protein rather than
linearly.   Fig. 5 illustrates that savings on protein turnover
are also expected on this basis in an animal whose doubled growth
potential has been delayed as in Fig. 4.   When expressed in
suitable units, the curves representing increments in total protein

Amount of protein
(arbitrary weight units)

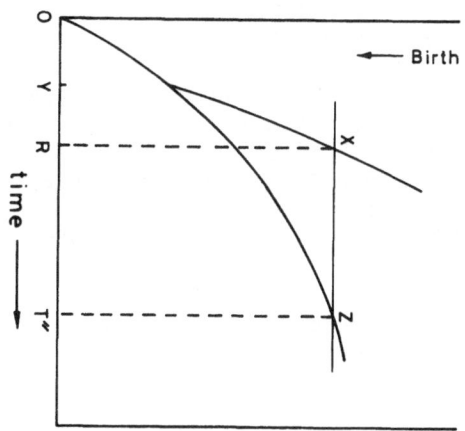

Rate of degrading protein
k.(arbitrary weight units)

Fig. 5.   Rates of accumulating and degrading protein in the
animal of Fig. 4 on the assumption that protein catabolism follows
first order kinetics.   Values were calculated from equations 1
and 3 in the Appendix.

and in the rate of protein degradation with age can be superimposed.   The areas under the curves will be equivalent to the
total amounts of protein that have been broken down.   It can be
seen that less muscle protein has been catabolized in Fig. 5
(area OXR << area OZT) when the animal is killed at point R as in
Fig. 4.

The animals shown in Figs. 4 and 5 could arise in three ways:
(i) Due to more active or more prolonged replication, twice as
many mononucleate cells are originally incorporated into a normal
number of fibres.   This is followed by a period in which growth
potential is only half expressed i.e. though ribosomes and catabolic enzymes are doubled in number, their individual activities are
temporarily halved.   Growth potential is not fully expressed
until nearer birth or it is perhaps stimulated after birth by
suitable nutrition.   (ii) A normal complement of nuclei is
originally incorporated into a normal number of fibres developing
at their normal rate.   However, at some point between then and
early neonatal life, an equal number of extra myoblasts fuse with
the fibre and begin to accumulate protein at the same rate as the
cells incorporated earlier.   (iii) As situation (ii), except that
the extra myoblasts fuse with themselves to form extra late-
developing fibres.

In real animals, the process by which muscle cells become
specialized in prenatal life is much more asynchronous.   However,
a situation akin to that in Fig. 5 might be possible.   For
example, in pigs, virtually all the muscle fibres are said to have
formed by the 70th day of a gestation period lasting 120 days.
By the 70th day, the number of muscle nuclei that are present will
probably be of the order of 10^9.   An improved pig of the form
illustrated in Fig. 5 could arise if at 70 days of gestation it
possesses an extra 2.5×10^8 mitotically competent precursors of
myoblasts.   These cells would thus account for 25% of the total
nuclear population and they could theoretically result from an
increase of around 1% in the average rate at which individual
cells of previous generations have divided.

This is because the average cell generation time (t) i.e.
the interval between successive divisions of cells, is related to
the number of cells by the formula:

$$t = \frac{\log 2 \times T}{\log n - \log n_0}$$

where T is the overall duration of cell replication.   n and n_0
are the number of cells at time T and at time zero respectively.
If n_0 is taken to be the fertilized egg, it becomes unity and
$\log n_0 = 0$.   Smaller values of t mean faster cell replication.

If, between 70 days and birth, the 2.5×10^8 prospective myoblasts continue to divide rather slowly and yield two generations of daughter cells, these will now number 10^9 i.e. muscle-forming potential at birth has been doubled by mechanism (ii) or (iii).

Alternatively, 2×10^9 cells might be produced at day 70 according to mechanism (i). This will require the average generation time of the precursor cells to decrease in the proportion 9.0/9.3 i.e. by 3%. This figure diminishes to 0.5% if only 10% more cells are required.

The overall number of cells being considered can be reduced drastically without negating the idea that significant variation in the size of cell populations could arise from minor differences in the rate or duration of cell division. For example, to produce 2,000 cells from a single cell where only 1,000 existed before would require that the average cell generation time diminish by only 10%, i.e. in the proportion 3.0/3.3. By the same token, some flexibility is allowed in one's assumption that altered rates of cell division are uniformly spread over the period from the fertilized egg onward. If, for example, the events which caused a doubling of the normal complement of 10^9 nuclei were only to occur after 10^6 cells had been reached, the average cell generation time in this period would still only decrease by 10%.

Even so, the above discussion still highly oversimplified and it merely aims to illustrate principles on which a more efficient meat animal might be contrived. As a factor determining final numbers of cells, the interplay between the rate of cell division and the rate of producing non-dividing specialized cells must be extremely complex. However, Tsanev and Sendov (35) have developed a computer programme for representing the growth of specialized cell populations. In support of my simpler calculations, these authors have found that very significant fluctuations in cell number could theoretically arise from differences of merely 2% in the activities of mechanisms which regulate the rate at which cells divide or are converted to non-dividing, specialized forms.

Underlying my hypothetical animal is the presently untestable assumption that adding extra cells does not impair the efficiency with which muscle protein accumulates per cell. However some other considerations are perhaps encouraging: one need increase neither the number of ribosomes per mononucleate myoblast nor the protein-synthesizing activity of individual ribosomes in the muscle once formed. Nor need the rate of catabolizing unit quantities of muscle protein diminish.

Also, taking a unit of growth potential to be a muscle nucleus surrounded by a constant number of ribosomes programmed to

form muscle protein, doubling the growth potential of its offspring should not require much extra energy expenditure by the
mother.   Thus, a unit of unexpressed growth potential will
obviously weigh less than one whose potential has been expressed
or partially expressed as muscle protein.   How much less is not
clear, but from the rather few published electron micrographs that
are available, I estimate that nuclei and the cytoplasm which
immediately surrounds them occupy only 10% of the transverse area
(and presumably 10% of the mass) of fibres in newborn rats and
late human foetuses.   This figure may be even smaller in animals
such as cattle which are physiologically more mature when they are
born.

        The musculature accounts for around 25% of a newborn mammal's
weight.   If some genetic mechanism simply doubles the total
amount of muscle present at birth, the animal will have to support
about 120% of the weight of normal animals at corresponding stages
of development from birth onwards.   However, if units of genetic
potential were to double in number without synthesizing muscle
protein till just after birth, the birth weight will increase
less, of the order 2.5%.   The animal should be capable of producing twice as much muscle as do normal animals but the time
over which it has to support the extra weight is therefore reduced.

        Summarized, my concept of an improved meat-producing animal
is one which (i) has not unduly increased its birth weight and
gestation period to the point of developing significant problems
of dystocia (ii) which can adequately support its body weight
against gravity, or else presents minimal economic and humanitarian problems if it does not (iii) which if it is lean, will not
be stress-susceptible and (iv) which multiplies its body weight
particularly rapidly and efficiently because of an increased propensity to deposit muscle.

        Such an animal might exhibit these properties by virtue of
possessing a greater number of muscle-forming cells at birth,
expression of the cells' potential to form muscle protein being
somewhat retarded at this stage.   Subsequently, the full
expression of this potential should result either in the particularly rapid development of enlarged fibres which contain an
increased number of nuclei per unit of fibre breadth, or the
appearance of extra late-developing fibres.   The extra precursor
cells should be associated with a population of fibres developing
at a normal rate so that the animal is strong enough to support
itself in the neonatal period.   Given a mechanism of subsequent
muscle growth that is broadly of the form in Fig. 2, the muscle
protein should have been formed more efficiently in the sense that
less energy has been diverted to protein catabolism, to resynthe-
sizing degraded protein and to contractile activity.

7. Some Diagnostic Properties of the Proposed Animal's Musculature

Extra muscle-forming potential in newborn animals could express itself as an excess of prospective myoblasts which then fuse and develop to form additional fibres. The muscle will then possess a normal complement of fibres with a normal genetic potential for growth, plus a population of smaller late-developing fibres. At a given age after birth, therefore, the distribution of fibre breadths will be skewed because of these small fibres. Alternatively, the excess myoblasts could occur randomly among a normal number of fibres and subsequently increase the rate at which the fibres grow.

Published estimates of the number of fibres in individual muscles of conventional meat animals range from 4,000 to 10^7 (36, 37). Though to some extent this might be accomplished by automated techniques, the direct enumeration of fibres in muscles from a large number of animals is rather a daunting task. In principle, one can enumerate fibres indirectly by dividing the total cross-sectional area of a muscle by the average cross-sectional area of individual fibres. However, as with a direct procedure, this requires the excision of at least one complete muscle and the assumption that the number of its fibres is representative of those in the musculature as a whole.

When Stickland and Goldspink (36) examined a prospective test muscle in young pigs, the number of its fibres did correlate with those of other muscles. The authors also confirmed Staun's (38) observation that increased fibre number tends to be accompanied by diminished fibre breadths. However, they did not show that the distribution of fibre breadths was skewed in such cases or obtain evidence that increased fibre numbers are accompanied by an increased rate of growth in the musculature as a whole. Such observations are also incomplete without an estimate of growth potential in individual fibres since different animals can possible distribute the same total potential for muscle growth between different numbers of fibres. A muscle fibre is not a unit of growth potential. It represents growth potential plus the expression, or partial expression of growth potential as muscle protein.

We might now consider the prospective benefits of discovering that the rate and extent to which a muscle fibre grows in breadth is dictated by the number of nuclei occurring along a standard length of the fibre. If such a relationship can be established then one has the prospect of an index of growth potential which is independent of variation in a fibre's dimensions. This is because when fibres are isolated and viewed in longitudinal profile, it is possible to estimate nuclei in fibre segments of

standard length such that the effect of variation in nuclear length
is eliminated (unpublished observation).   When the standard length
is expressed as a number of sarcomeres, the effects of contraction
are also compensated for and by focussing completely through the
fibre the enumeration of nuclei is made independent of fibre breadth.
However, measurements of fibre breadth will still be required to
test the proposition that fibres with an increased number of nuclei
per unit of fibre breadth synthesize protein relatively efficiently.

A procedure of the above sort could in principle also be used
to test whether rapidly growing muscles have acquired a population
of late-developing fibres.   One will again require independent
knowledge that the rate of total muscle growth has accelerated, or
at least that growth in body weight has accelerated in circumstances where fat accumulates slowly.   It is possible, for example,
that muscles possessing enlarged fibres with an increased complement of nuclei, will have fewer fibres.   On the other hand the
above procedure will not require the excision of an entire muscle.
It will suffice to obtain a sample of fibres, perhaps pooled from
more than one muscle in a commercially valuable region of the
animal.

The possible usefulness of measuring total DNA in muscle
should be considered.   On the preceding arguments, one might seek
animals with muscles (a) which are of roughly normal weight at
birth but possess an increased ratio of DNA to protein and (b)
which subsequently grow rapidly while continuing to exhibit this
increased ratio relative to normal muscles.   Once again, an
independent estimate of total muscle mass will be necessary since,
of itself, a low ratio of protein to DNA could simply mean that
the muscle has grown abnormally slowly.

In terms of accuracy, an estimate of gross muscle DNA has the
advantage of being obtained from a sample containing many fibres.
However, there will be errors inherent in the procedures for
extracting the DNA and estimating it.   Moreover, that estimate
will include DNA in the nuclei of capillaries and in cells of
the connective tissue between the muscle fibres.   Such nuclei are
said to account for 15-35% of the total nuclei in a muscles of
chickens and rats (39,40).   Gross estimates of DNA also say
nothing about the interaction between growth in fibre breadth,
length and number;   all three processes being probably some function of nuclear numbers.   This may not matter for the purposes
being discussed.   Finally, though the total amount of DNA in a
growing muscle increases (Table 3), it is not known whether the
additional nuclei which a muscle fibre acquires with age actually
cause its growth in breadth (see below).   If they do not, then
measurements of total DNA are likely to be of little value.

Table 3

Replication of nuclei with age in different muscles.
Estimates where made from measurements of total DNA in
the muscles except for * where microscopy was used.
S = skeletal muscle; H = heart.

Species		Period of growth	−Fold increase in nuclei	Ref.
Mouse	(S)	1-10 weeks from birth	2.4	12
Rat	(S)	1-8 weeks from birth	4.8	"
	(S)	2-12 weeks from birth	1.8-4.2	"
	(H)	0-14 weeks from birth	13	"
Human	(S)	1-15 years from birth	15	"
	(S)	0-65 years from birth	4.0*	50
Pig	(S)	0-32 weeks from birth	22-26	51
Chicken	(S)	0-38 weeks from hatching	18-96	12
Trout	(S)	0.02-3.4 kg body weight	78	"

8.   Unresolved Fundamental Questions

The hypothetical animal of Section 6 presupposes a mechanism
of growth regulation for which there is minimal evidence.   However, many questions are posed that are open to experiment and
some of these will now be summarized.

Essentially one is seeking animals in which a capacity for
rapid and disproportionate muscle growth results from comparatively
active replication of muscle precursor cells in prenatal life.   A
second requirement is that the cells become specialized more
asynchronously.

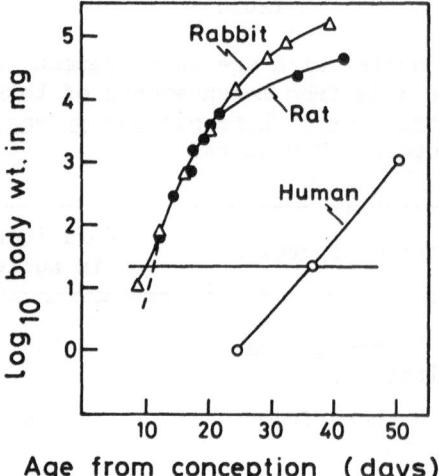

Fig. 6.   Prenatal growth of rats, rabbits and humans.   Taken
from data in references 45-48.

     Fig. 6 illustrates a situation which seems to merit some
preliminary study in a fundamental sense.   This Figure shows that
the rat and the rabbit grow at the same rate in the period when
cells should be multiplying most rapidly before birth.   The
rabbit, however, grows at a fast rate for longer.   It is tempting
to suppose that its cells are dividing actively for longer and to
enquire how the number of nuclei in its muscles and their fibres
differs as a result.   Shown for comparison in Fig. 6 is the
slower growth of a much larger animal, the human.

     Of course, the rabbit weighs more at birth than a rat and
the comparison does not strictly represent the situation one is
seeking.   Thus, it seems worthwhile to supplement a study of the
above sort with an investigation of nuclear numbers in fibres of
different type, since it is known that within individual animals
the breadth of muscle fibres, and hence their rate of growth,
varies with their physiological properties.

     I have previously reviewed three pertinent facts (12):
(a) red, slow-phasic muscles appear to have more nuclei in their

fibres than have fast-phasic muscles, fibres in the red muscles
also being larger, (b) the number of mononucleate cells that fuse
to form a fibre in culture can be increased by altering the conditions of culture, (c) fibres of the slow-phasic type tend to be
somewhat larger in prenatal life.

Collectively, these observations suggest there is some scope
for fibres to differ in growth rate through the number of nuclei
that they have incorporated before birth. Landing, Dixon and
Wells (41) have also found that the breadth of muscle fibres in
humans varies with the number of nuclei occurring along a unit
length of fibre. Similar observations have been made on chickens
(42).

My own unpublished findings on rats and rabbits so far confirm these observations. However, it will not suffice to demonstrate a direct relationship between fibre breadths and nuclear
numbers. One also wishes to know whether each nucleus within a
muscle fibre sponsors the formation of muscle protein. Many, and
hopefully, all the nuclei that initially fuse to form a fibre must
do so. However, elsewhere I have pointed out that although new
nuclei appear to be added to and incorporated within muscle
fibres so they grow with age, no one has yet shown that these
nuclei sponsor the formation of contractile protein (12).

This bears on the second of the mechanisms that I proposed
as an explanation of Fig. 5. It raises the question of whether,
even in embryonic life, a fibre can add new muscle-forming nuclei
along its length. More convincing, perhaps, is the evidence that
the ends of fibres can add such nuclei at their junction with a
tendon (12). Thus, the above mechanism might be operated in this
way or through the addition of nuclei to the tapering ends of
fibres which termina te within the body of a muscle. Swatland
and Cassens (43) have proposed that muscles can grow in breadth
through the elongation of such fibres.

If one aims to increase growth efficiency by manipulating
nuclear numbers in the body of muscle fibres, then one must also
show that it is biologically possible for muscles to grow rapidly
and to possess broader fibres with a greater number of nuclei per
unit of fibre breadth. Such fibres should arise from fibres
whose breadths are only slightly increased at birth, and prefer-
ably not at all, and which either possess a greater number of
internal nuclei per unit of fibre breadth than have normal animals,
or else have more potential myoblasts associated with them.

So far, I have unpublished evidence to indicate that some of
these requirements are met by slow-phasic muscles in comparison
with fast-phasic muscles, but have yet to seek such phenomena
within the fast-phasic category of muscle. As regards genetically-

induced variation in nuclear numbers, my evidence is so far confined to observations of some differences between rabbits and
rats.

Moreover, I have not yet attempted to show that satellite
cells, here taken to be putative myoblasts, actually cause new
myofibrillar protein to be synthesized.

If this cannot be demonstrated then the potential usefulness
of nuclear numbers in muscle fibres as an index of growth potential will principally rely on the existence of mechanism (i) in
Section 6.   It will also depend on one of three conditions being
met: (a) that there is a point in a muscle's development up to
which it can be conveniently sampled without potentially non-
productive nuclei contributing significantly to nuclear numbers,
(b) that the rate at which muscle fibres acquire new nuclei with
age is dictated solely by the number of nuclei that the fibres
initially possess and is not accelerated by mechanisms that dispose
lean animals to stress-susceptibility and (c) that nuclei which
are added later can be identified morphologically.   For example,
I have observed that muscle fibres contain rounded nuclei as well
as the characteristic elongated variety.   It is possible that the
former are acquired later from dividing satellite cells, although
that is speculation at present.

Finally, it remains to emphasize the need for better evidence
regarding the manner in which protein synthesis and degradation in
muscle vary with age.   The limited observations that are so far
available (12) tend to favour a model between Figs. 2B and C, but
that is a tentative conclusion.

Also, I have mentioned the requirement that accelerated muscle
growth should not result in more inefficient growth per unit time.
Thus, it is prudent to note that although red slow-phasic fibres of
the rabbit have particularly abundant nuclei (45; personal observation), there are biochemical observations to indicate that protein
is degraded and replaced comparatively rapidly in slow-phasic or
repetitively contracting fibres.   The evidence depends on measurements of the rate at which the muscle protein incorporates and
subsequently loses radioactive amino acids acquired from the blood
stream.   It is possible that, rather than truly reflecting protein
turnover, these values are simply dictated by the rates at which
the amino acids are gained and lost from red muscles via the
latters' more numerous capillaries.

9.   Harnessing the Answers to Practical Ends

If the proposed mechanism of growth regulation proves to
exist, there will be its practical exploitation to consider.

Fundamental research of the sort I have described should have documented the properties of a range of muscle fibre types from animals of different genotype.   These fibres will hopefully encompass the range of what is biologically possible in terms of their ability to grow rapidly through the number of nuclei that they have incorporated.

The muscle fibres of different species of meat animal might then be compared with this series to see how far their genetic potential for growth by such a mechanism has evolved.   Subsequently, different breeds of the same species might be investigated.

Even if one does find muscle fibres in meat animals to be rather deficient in nuclei, there will still remain the problem of adjusting selection pressures on the animal to favour growth by the mechanism I have indicated.   Naturally, one would hope to employ measurements of nuclear numbers for that purpose, perhaps in association with, rather than as a substitute for estimates of growth rate, efficiency of food conversion and thickness of back-fat.   Accuracy will be a critical factor here.

If this prospect fails to materialize, then one may have to think of meat production in a wider context.   For example, it seems to me that two basic factors on land contend against the more efficient accumulation of cell substance by warm-blooded animals.   One is their necessity to spend metabolic energy in keeping warm, this being most pronounced in small animals with their high surface to body ratio.   The other is the requirement to spend energy in contending against gravity, a factor which becomes most evident in large animals.

One might speculate then that the optimum size for efficient growth and for manipulating units of genetic potential lies some-where in between, at the level of the rabbit and the pig perhaps. Indeed, there is some evidence that the number of nuclei per muscle fibre differs between breeds of rabbit (44).

Casting the net more widely, Table 3 is a compilation of published data on the rate and extent to which nuclei multiply in various muscles as they develop.   Most of the results were obtained from estimates of total DNA and should be treated circum-spectively.   However, it is perhaps more than coincidence that the animals in which nuclei are replicating most extensively, are the chicken and the trout.

The chicken falls in the same range of body size as the rabbit, it is lean and it has proved amenable to selection for increased growth rate.   Fish are also lean and I have already suggested that scope for accelerated cell replication and increased growth efficiency is greatest in water because of the diminished

need for aquatic animals to contend against gravity.   As growth
of the trout was followed from 20 g onwards, we can probably
eliminate the influence of the fact that it hatches at rather an
immature stage of development.   So, a method for estimating the
number of nuclei in muscle or in a portion of a muscle fibre might
be developed to either document the potential of different species
of aquatic and terrestrial animal to form muscle efficiently, or
to indicate which species are most capable of genetic improvement.
Such a method would be free of commercial and subjective bias.

Finally, at the back of our minds should be the idea that
research will eventually prove the impossibility of further
improvements in the efficiency of meat production through genetic
manipulation.   In that case, a rational basis will have been
provided for the redirection of research into other areas of food
production.   However, I think it fair to say that much positive
thinking can still be done and some very challenging experimentation lies ahead before we are ever forced to that conclusion.

ACKNOWLEDGEMENTS

Mr. D. R. Williams kindly drew my attention to the data of
Table 2.   Figs. 2 and 3 are reproduced by permission of the
Cambridge Philosophical Society.   My thanks are also due to
Mr. H. J. H. MacFie for statistical advice, to Miss K. Powis and
Mr. R. Almond for technical assistance, and to my wife for helping
to derive the relationship given in the Appendix.

REFERENCES

1.   Hill, A. V. (1950). Science Progress 38, 209

2.   Schmidt-Nielsen, K. (1970). Fed. Proc. 29, 1524

3.   Walker, E. P. (1968). 'Mammals of the World' Vol. II, p. 545
     (Johns Hopkins Press, Baltimore)

4.   Greenough, P. R., MacCallum, F. J. and Weaver, A. D. (1972).
     'Lameness in Cattle', p. 280   (Oliver and Boyd, Edinburgh)

5.   Oliver, W. M. and Cartwright, T. C. (1968). Texas Agric.
     Expl. Station Technical Report 12

6.   Snell, R. (1971). In 'Breeding for Beef', p. 42 (The Meat
     and Livestock Commission, Bletchley)

7.   Blaxter, K. L. (1971). In 'Breeding for Beef', p. 3 (The
     Meat and Livestock Commission, Bletchley)

8.   Lister, D., Perry, B. n. and Wood, J. D. (1973). Proc. Br.
     Soc. Anim. Prod. 2, Abs. 52

9.   Lister, D.   This vol.

10. Holmes, J. H. G. and Ashmore, C. R. (1972). Growth 36, 351

11. Standal, N., Vold, E., Trygstad, O. and Foss, I. (1973).
     Anim. Prod. 16, 37

12. Burleigh, I. G. (1974). Biol. Rev. 49, 267

13. Ashmore, C. R., Tompkins, G. and Doerr, L. (1972). J. Anim.
     Sci. 34, 37

14. Swatland, H. J. and Cassens, R. G. (1973). J. Anim. Sci.
     36, 343

15. Widdowson, E. M. (1970). Lancet i, 901

16. Kihlström, J. E. (1972). Comp. Biochem. Physiol. 43A, 673

17. Munro, H. N. (1970). In: 'Mammalian Protein Metabolism'
     Vol. III (Ed. H. N. Munro), p. 133 (Academic Press, New York
     and London)

18. Mason, I. L. (1971). Anim. Breed. Abs. 39, 1

19. MacKellar, J. C. (1968). Thesis. Royal College of Veterinary
     Surgeons, London

20. Ouhayon, J. and Beaumont, A. (1968). Ann. Zootech. 17, 213

21. Swatland, H. J. and Cassens, R. G. (1973). J. Anim. Sci.
     37, 885

22. Cooper, C. C., Cassens, R. G. and Briskey, E. J. (1969).
     J. Food Sci. 34, 299

23. Dildey, D. D., Aberle, E. D., Forrest, J. C. and Judge, M. D.
     (1970). J. Anim. Sci. 31, 681

24. Meat and Livestock Commission (1975). Newsletter No. 24
     (Bletchley)

25. Berlin, C. M. and Schimke, R. T. (1965). Molec. Pharmacol.
     1, 149

26. Schimke, R. T. (1973). Adv. Enzymol. 37, 135

27. Fritz, P. J., White, E. L., Vesell, E. S. and Pruitt, K. M.
     (1971). Nature New Biology 230, 119

28. Kleiber, M. (1961). 'The Fire of Life' (Wiley, New York)

29. Lawrie, R. A. (1950). J. agric. Sci. 40, 356

30. Goldspink, G. (1972). In: 'The Structure and Function of
     Muscle' Vol. 1 (Ed. G. H. Bourne), p. 179 (Academic Press,
     New York and London)

31. Boland, R., Martonosi, A. and Tillack, T. W. (1974). J.
     biol. Chem. 249, 612

32. Luff, A. R. and Atwood, H. L. (1971). J. Cell Biol. 51, 369

33. Schiaffino, S. and Margreth, A. (1969). J. Cell Biol. 41,
     855

34. Thurley, D. C., Gilbert, R. F. and Done, J. T. (1967). Vet.
     Rec. 80, 302

35. Tsanev, R. and Sendov, Bl. (1971). J. theoret. Biol. 30, 337

36. Stickland, N. C. and Goldspink, G. (1973). Anim. Prod. 16,
     135

37. Bendall, J. R. and Voyle, C. A. (1967). J. Fd Technol. 2,
     259

38. Staun, H. (1968). 366 Beretn. fra forsøgslab, p. 121
     (Kobenhavn)

39. Enesco, M. and Puddy, D. (1964). Am. J. Anat. 114, 235

40. Marchok, A. C. and Herrmann, H. (1967). Devl. Biol. 15, 129

41. Landing, B. H., Dixon, L. G. and Wells, T. R. (1974). Human
     Pathol. 5, 441

42. Knízetová, H., Kníze, B., Kopecný, V. and Fulka, J. (1972).
     Ann. Biol. Anim. Bioch. Biophys. 12, 321

43. Swatland, H. J. and Cassens, R. G. (1972). J. Anim. Sci.
     35, 336

44. Watzka, M. (1939). Z. mikrosk.-anat. Forsch. 45, 668

45. Altman, P. L. and Dittmer, D. S. (1964). Biology Data Book,
     p. 82 (Fedn Am. Soc. Expl Biol., Washington, D.C.)

46. Barcroft, J. (1946). 'Researches on Pre-Natal Life' Vol. 1, p. 43 (Blackwell, Oxford)

47. Witschi, E. (1956). 'Development of Vertebrates', p. 498 (Saunders, Philadelphia)

48. Winick, M. and Noble, A. (1965). Devl Biol. $\underline{12}$, 451

49. Laws, R. M. and Parker, I. S. C. (1968). Symp. Zool Soc. Lond., Vol. 21 (Academic Press, New York and London)

50. Montgomery, R. D. (1962). Nature, Lond. $\underline{195}$, 194

51. Durand, G., Fauconneau, G. and Penot, E. (1967). C. R. Acad. Sci., Paris $\underline{264}$, 1640

52. Meat and Livestock Commission (1972). Beef Improvement Service-Pedigree Weight Recording Scheme (Bletchley)

APPENDIX (to Section 5)

When growth represents a balance between a constant rate of protein synthesis (s) and a rate of protein degradation ($\frac{dn}{dt}$) that varies directly with the amount of protein (P), then from refs. 25 and 26:

$$\frac{dn}{dt} = kP \dots\dots \text{ eqn. 1, and } \frac{dp}{dt} = s - kP \dots\dots \text{ eqn. 2.}$$

When $t = 0$, $P = 0$ and $\frac{dn}{dt} = 0$.

Integrating eqn. 2, $P = \frac{s}{k}(1 - e^{-kt}) \dots\dots$ eqn. 3

When growth stops, $P = P_{max}$ = a constant from the discussion of Section 5 and $\frac{dP}{dt} = 0 = s - kP_{max}$. Therefore, $P_{max} = \frac{s}{k}$.

P approaches P_{max} with time but never quite reaches it. However, one can calculate the quantity of protein degraded (n_α) in producing a fraction of P_{max} i.e. $P_\alpha = \alpha P_{max}$ where $0 < \alpha < 1$. P_α accumulates in time t .

Integrating eqn. 1: $\alpha n = s(1 - e^{-kt})dt = st + \frac{s}{k}e^{-kt} + c$

since $P = \frac{s}{k}(1 - e^{-kt})$

When $t = 0$, $n = 0$, $e^{-kt} = 1$ and therefore $c = -\frac{s}{k}$

Therefore, $n = st + \frac{s}{k}e^{-kt} - \frac{s}{k} = st - \frac{s}{k}(1 - e^{-kt}) = st - P$

$= kP_{max}.t - P$

Thus, $n_\alpha = P_{max}.kt - P_\alpha \dots\dots$ eqn. 4.

Now, $\frac{P_\alpha}{P_{max}} = \alpha = \frac{\frac{s}{t}}{\frac{s}{t}}(1 - e^{-kt})$, so $e^{-kt} = 1 - \alpha$ and $e^{kt} = \frac{1}{1 - \alpha}$

Therefore, $kt = -\ln(1 - \alpha)$ and from eqn. 4:

$n_\alpha = -P_{max}\ln(1 - \alpha) - \alpha P_{max}$. Thus, since P_{max} is constant, n_α is independent of time for any given value of α. In other words, when muscle fibres have the same growth potential, the amount of protein that is degraded in expressing any fraction of that potential is independent of the time required to do so. This accords with the predictions of the models in Section 5.

FACTORS AFFECTING MUSCLE SIZE AND STRUCTURE

R. McN.Alexander

Department of Pure and Applied Zoology

University of Leeds

This paper is about the structure and arrangement of limb muscles in mammals. It is not specifically about meat animals though the limb muscles of a wide variety of mammals are built to the same basic design.

Most of the musculature of a typical limb consists of extensor muscles, since these are the muscles required to exert downward forces on the ground. Fig. 1a shows how the extensor muscles are arranged in the hind leg of a dog. Though the diagram and the description which follow are based on dogs, the same pattern is found in sheep.

The extensor muscles of the hip are the bulkiest and all the main muscles in this group are parallel-fibred. In a 26 kg Alsatian which I dissected most of their fibres seemed to be 100-150 mm long. The extensor muscles of the knee are pennate with shorter fibres which converge on central tendons. In the same dog most of their fibres were 25-50 mm long. The extensors of the ankle are also pennate, but have even shorter fibres, 10-25 mm long. The shortest fibres are in the plantaris which has a remark- able quadripennate structure.

The extensors of the ankle originate on the femur, so they also act as flexors of the knee. Since their fibres are so short they cannot be stretched much. Consequently the ankle of a dog cannot be flexed (except by very large forces) while the knee is held straight. However, bending the knee allows the ankle to be bent as well. When a dog raises its paw in walking, knee and ankle are bent together. Some of the other leg muscles also cross more than one joint, but this need not concern us here.

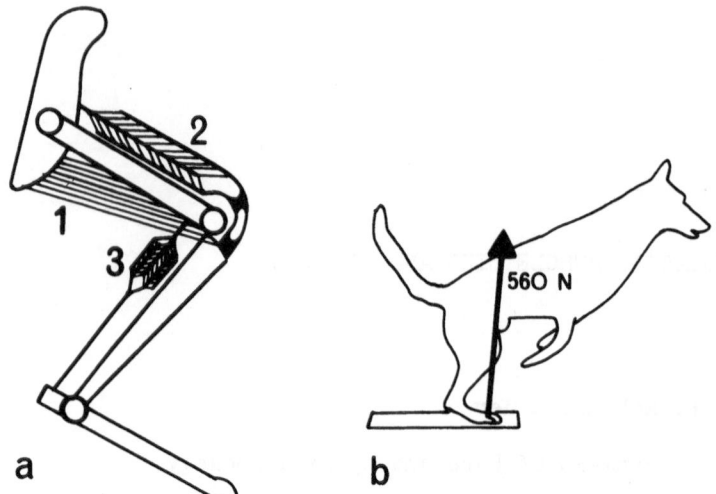

Fig. la.   A diagram of the hind leg of a dog showing 1, the
extensors of the hip; 2, the extensors of the knee and 3, the
extensors of the ankle. Each group of extensor muscles is
represented as a single muscle, and the arrangement of fibres
shown within it is typical for the group.
       b.   An outline traced from a film of a 36 kg dog taking
a running jump from a force platform. The force exerted by the
platform on one foot is also shown.

The main anatomical point which I wish to make is that the
extensor muscles of the hind leg show a gradation, from parallel-
fibred muscles with long fibres extending the hip to pennate
muscles with very short fibres extending the ankle. A rather
similar arrangement is found in the fore leg. The extensors of
the wrist are pennate with very short fibres while the triceps is
partly parallel-fibred and partly pennate, with much longer fibres.
Why should leg muscles be arranged like this?

A possible answer has emerged from a study of jumping by
dogs (1). Jumping was studied because it is a strenuous activity
involving large forces, in which muscles are likely to be used to
the limits of their capability. The subject of the study was an
Alsatian which had been trained, for working trials, to jump a
variety of obstacles on command.

A force platform, set into the floor, was used to record
forces exerted on the ground at take-off. This instrument
produces a record of the three components (vertical, longitudinal
and transverse) of forces exerted on it. Fig. 1b is an outline

traced from a film, of the dog taking off for a running jump. His
hind paws are on the platform and the record shows that at this
instant they were exerting on it a force of 1120 N (3.2 times body
weight) at 84° to the horizontal. Each hind paw was presumably
exerting half this force, so a force of 560 N, at 84°, is shown
acting on the nearer paw. The distance of each joint from the line
of action of this force can be measured, so the moment acting about
each joint can be calculated. The forces exerted by the major
groups of muscles can in turn be calculated, if their moment arms
about the joints are known. The lengths of the muscles can also be
calculated, from the angles of the joints.

Several muscles act at each joint. The calculations show,
for instance, that at the instant shown in Fig. 1b, the extensor
muscles of the ankle must have exerted a total force of 2500 N
(0.25 tonne). Two muscles are involved, and the data do not tell
us how much of this force was exerted by each muscle. In such
cases of doubt, it was assumed that equal stresses acted in all the
muscles of a group. In calculating stresses, account was taken of
the pennate or parallel-fibred structure of the muscles.

Graphs were thus obtained showing the forces calculated for
individual muscles, at successive stages in take-off, and the
changes in length of the muscles. Two of these graphs are shown
in Fig. 2. The biceps femoris (Fig. 2a) is the largest of the
parallel-fibred extensors of the hip. Its fibres shortened by
about 30 mm, or 20% of their initial length, while the force rose
to a maximum and then diminished again. The area under the graph
represents work done by the muscle, in accelerating the dog at
take-off. The gastrocnemius (Fig. 2b) is one of the pennate
extensors of the ankle. It extends by 23 mm as the force increases,
and shortens again as the force diminishes. The graph resembles a
graph of force against length for a spring which is stretched and
then recoils elastically. Work is done on the muscle as it is
forcibly stretched, and by the muscle as it shortens again.

The muscle fibres of the gastrocnemius are initially only
15-25 mm long. It seems inconceivable that they should be
stretched by 23 mm and still be able to exert a substantial force.
However, the gastrocnemius has a long tendon of insertion, about
200 mm including the part to which the muscle fibres attach. The
cross-sectional area of the tendon is 8 mm^2 so the maximum force
of 1000 N implies a stress in the tendon of about 125 MN m^{-2}.
Young's modulus for tendon collagen (2,3) is about 1.2 GN m^{-2} so
the tendon should be stretched by about 10% of its initial length,
or about 20 mm. Thus the muscle fibres may have changed length
very little, while the tendon stretched elastically and recoiled.

The stress which seems to be developed in the gastrocnemius
tendon is remarkably high. Measurements of the tensile strength of

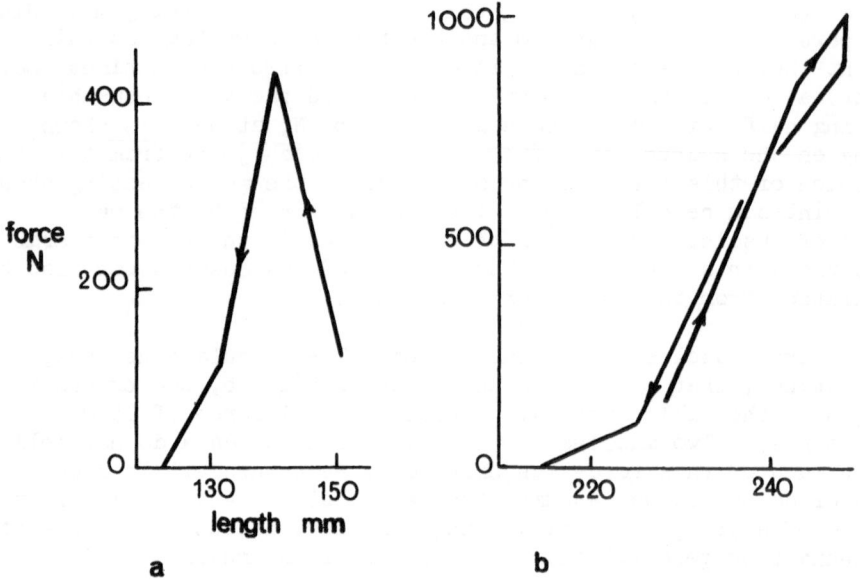

Fig. 2.   Graphs of force against length during take-off for a
running jump for a) the biceps femoris and b) the gastrocnemius
of a 26 kg dog.

tendon (4) have generally yielded values between 50 and 100 MN m^{-2}.
However, stresses are applied slowly in conventional strength tests.
The high stresses which act in jumping act only for a few hundredths
of a second, and might break the tendon if they acted for longer.
Polymeric materials generally can withstand brief stresses better
than maintained ones (5).

     It seems from Fig. 2 (and from similar graphs for the other
muscles) that the large parallel-fibred extensors of the hip and
the much smaller pennate muscles of the ankle have quite different
functions. The parallel-fibred hip muscles are necessary to do the
work required when the animal accelerates or jumps. The pennate
ankle muscles do little net work and are much less important than
their tendons, which act like the spring of a pogo-stick. The
extensor muscles of the ankle probably serve merely to make minor
adjustments to the lengths of their tendons. They do not need to
shorten much so their fibres need not be long, but they must be
able to exert large forces. The hip and ankle extensors must be
appropriately matched in strength, if both are to be used to the
full. They seem to be nicely matched: it was calculated that the

maximum stresses which occurred in jumping were about 270 kN m^{-2} for the hip extensors and 310 kN m^{-2} for the ankle extensors.

Elastic storage of energy in tendons must save energy in locomotion. Animals rise and fall as they run, gaining and losing potential energy. They accelerate and decelerate, gaining and losing kinetic energy. Potential and kinetic energy are lost and regained at each step. If there were no elastic structures, this energy would be absorbed by muscles and degraded to heat, and would have to be wholly replaced by work done by active contraction of the muscles. Elastic tendons can store some of the energy and restore it in an elastic recoil, so that less work has to be done by the muscles. The amounts of energy saved in this way when dogs trot or gallop have not yet been determined, but an investigation of hopping by kangaroos has been completed (6). It appeared that when a wallaby hopped slowly, about 40% of the energy which would otherwise have been required was saved by elastic storage in tendons. There were indications that considerably higher propor- tions of energy might be saved when a kangaroo hopped fast.

An animal with bulky parallel-fibred muscles to work all its joints would have no means of storing elastic energy in locomotion, and the distal parts of its limbs would be unusually heavy. One which had only short-fibred pennate muscles would have light limbs, but little capacity for acceleration. The arrangement shown in Fig. 1a gives the animal the ability to accelerate, and also enables it to save energy by elastic storage. Since the bulky muscles are at the proximal end of the limb the moment of inertia of the limb about the hip or shoulder is reasonably low, so not too much energy is needed to swing the limb forward and back. The muscle fibres of the extensors of the ankle are very short in dogs and sheep. They are longer (relative to the size of the leg) in men and kangaroos, and the extensors of the ankle represent a larger proportion of the total weight of leg muscles.

REFERENCES

1. Alexander, R.McN. (1974)   J. Zool., Lond. 173, 549.

2. Benedict, J. V., Walker, L. B. & Harris, E. H. (1968). J. Biomechan. 1, 53.

3. Matthews, L. S. & Ellis, D. (1968). J. Biomechan. 1, 65.

4. Blanton, P. L. & Bigg, N. L. (1970). J. Biomechan. 3, 181.

5. Ritchie, P. D. (1965) "Physics of plastics".(Iliffe Books, London).

6. Alexander, R. McN. & Vernon, J. A.   (in preparation).

DISCUSSION

Dr. Webster enquired if it were not merely coincidental that the proportions of muscles within the limb of the sheep and the dog were so similar for there were considerable differences in the bone structure of the limb and in the involvement of the spinal column in locomotion. *Prof. Alexander* replied that the difference between the digitigrade and ungulagrade animals lay not in the basic limb proportions, but in the fusion of the metapodials. He agreed that, in some species, there was a contribution of the torso to locomotion, but this was difficult to include in his calculations. *Dr. Widdowson* observed that, in the kangaroo, there was an extraordinary reversal in the relative proportions of the fore and hind limb during growth.

Dr. Burleigh enquired whether the tapered endings of those muscle fibres that did not extend for the full length of the muscle are pulled apart during contraction. *Prof. Alexander* doubted whether such fibres existed, but *Prof. Goss* said that in large muscles there is an inter-digitation of fibres. This was shown in studies of the distribution of motor end-plates, which are usually situated at the mid-point of the fibre. Some muscles have one, others two or three bands of motor end-plates.

Prof. Goss referred to Crawford's experiments with rabbits, in which the tendon of the anterior tibialis was made to run outside the crural ligament. In the treated rabbits, the total length of the muscle plus tendon was the same as in controls, but the muscle now occupied half this length rather than one-third. He enquired what the mechanism for this might be and *Prof. Alexander* suggested that there might be a change in the angle of pinnation of the muscle.

Dr. Widdowson mentioned some of her findings on muscle fibre and nuclear proliferation in newborn and undernourished pigs. Pigs, like humans, have a full complement of muscle fibres at about the time of birth, and this appears to be under genetic control. The runt pig at birth has fewer muscle fibres than its normal littermate. Prolonged undernutrition of a normal individual has little or no effect on the number of muscle fibres, although they contain many more nuclei. *Dr. Turner* wondered to what extent the apparent genetic variation in fibre number at birth was attributable to the uterine environment. *Dr. Widdowson* agreed that undernutrition before birth could reduce fibre number. Her observations on Large White pigs, however, suggested that there was much greater variation in fibre number of well-nourished pigs between than within litters and this she interpreted as being of genetic origin.

Dr. Moody asked what was the significance of the red and white components of a semitendinosus and if this was related to meat quality. He asked whether in double muscling there were more nuclei or more fibres. *Dr. Burleigh* replied that in the semi-tendinosus of the rat there were small red oxidative fibres which could hypertrophy under workload into visibly whiter larger fibres. There appear to be two types of red fibres, the slow contracting type typical of the rabbit semitendinosus and rat soleus, and another kind of fast repetitively contracting fibre with a high ATPase activity found in the diaphragms of small rodents; it may be this which is the red type found in the semitendinosus of larger animals. *Prof. Alexander* speculated that the slow red fibres might be concerned with posture, but there was no quantitative evidence that the numbers were appropriate to the forces required. *Prof. Goss* posed the hypothetical question of whether a very small muscle of an adult rat, transplanted to the site of a potentially larger muscle of an infant rat, would grow larger than nature intended. *Dr. Burleigh* thought that, if subject to the appropriate workload, such a muscle would hypertrophy.

Dr. Fowler asked if new nuclei in hyperplastic muscle fibres arose from cell division at the ends of the fibres. *Dr. Burleigh* replied that nuclei are thought to be added by the fusion to the existing fibre both of myoblasts at the fibre ends, associated with lengthening, and of satellite cells along the fibre length, associated with thickening.

Prof. Cahill remarked that, in man, protein synthesis in muscle is spasmodic; it is stimulated by food and by activity. This works independently in various muscle groups according to their use; in this way the animal is always remodelling itself according to the environment.

The Development of Fatty Tissue

The Development of Fairy Tissue

PHYSIOLOGICAL SIGNIFICANCE OF LIPIDS

G. A. Garton

Rowett Research Institute, Bucksburn
Aberdeen, AB2 9SB

INTRODUCTION

A functional division of the lipids of the animal body was
first made more than half a century ago by Mayer and Schaeffer (1)
who studied the effect of starvation on the lipid content of the
kidneys, liver and muscle tissue of dogs and rabbits and concluded
that there was an 'element constant' (largely phospholipids) and an
'element variable' (triglycerides), though the biological significance of these two major classes of lipid was not to become
evident for many years. The furtherance of knowledge regarding
the part played by phospholipids in the structure and function of
membranes of cells and organelles depended as much on the development of physical methods (subcellular fractionation, electron
microscopy) for their isolation and study as it did on the advent
of chemical techniques (thin-layer chromatography, gas-liquid
chromatography) for the separation and characterization of individual phospholipids. Similarly, it was not until isotopically-
labelled compounds became available that the dynamic metabolic
state of adipose tissue was revealed and not until isolated fat
cells (adipocytes) could be prepared that biochemical events
could be investigated in detail.

Thus lipids, which for so long were considered to be metabolically inert and so did not excite the interest of most physio-
logists and biochemists, began to attract a great deal of research
attention. The first review on the physiology of adipose tissue
was published in 1948 and the subsequent rapid growth of information can be judged from the appearance in 1965 of a Handbook (2)
of the American Physiological Society devoted entirely to adipose
tissue, followed in 1970 by a Supplement (3) describing recent
advances in regulation and metabolic function.

Though certain general functions can be ascribed to lipids (such as thermal insulation and the promotion of intestinal absorption of fat-soluble vitamins), it is proposed in this brief review to concentrate, in the context of the control of growth and productive efficiency of meat-producing animals, on the physiological role of lipids in cell membranes and in white and brown adipose tissue. Reference to the original literature is mainly restricted to selected key papers and to recent reviews.

LIPIDS OF CELL MEMBRANES

Structure of Membranes

Eukaryotic cells (i.e. those of multicellular animals and plants) are characterized by the diversity and complexity of their morphology and intracellular organisation, notably in respect of the nucleus which is surrounded by a membrane, and by the segregation of respiratory enzymes into discrete structures (mitochondria). Each animal cell is bounded by a membrane (plasma membrane) and the organelles are similarly contained. These membranes consist of an ordered juxtaposition of protein and lipid molecules which serve not merely as structural support, but also as a locus for enzymes controlling the passage of metabolites between cells and within cells.

Plasma membranes

On a weight basis plasma membranes from different cell types contain several times more protein than lipid though, because of differences in molecular weight, there are far more lipid molecules than there are protein molecules. Many models have been proposed for the fine structure of plasma membranes (see Vandenheuvel (4), Finean (5)) and there is still much uncertainty and controversy surrounding the topic. Suffice it to note here that, in the electron microscope, most preparations appear to comprise three layers of total thickness 7.5 - 10nm, collectively termed a 'unit membrane'. The structure of this unit membrane can be envisaged (6) as consisting of a central portion containing two monomolecular layers of lipid molecules (mostly phospholipids) such that the hydrophobic 'tails' of the molecules (the fatty-acid moieties) are aligned together, leaving the hydrophilic portions facing outwards to associate with proteins, thus constituting the outer layers of the unit. The unit membrane model was based originally on the dimensions and composition of the nerve myelin sheath which is a specialized, multi-layered membrane system derived from the plasma membranes of cells surrounding the axon (5).

Membranes of organelles

The membranes of mitochondria and the network of membranes which constitutes the endoplasmic reticulum together account for most of the lipids of the organelles. Mitochondria are usually present in large numbers (hundreds) per cell and typically are about 5-10 μm long with a diameter of 0.5 - 1.0 μm. On a dry weight basis, the proportions of lipid and protein are 25-30% and 60-75% respectively and these derive mainly from the outer and inner membranes which are apparently composed of globular proteins in which the hydrocarbon 'tails' of lipid molecules are embedded (7). The endoplasmic reticulum is a structure in which phospholipid-rich lipoprotein membranes form interconnecting channels and tubules to which part of the cell content of ribosomes is attached. Fractionation techniques disrupt the reticulum into fragments ('microsomes') which, as prepared from rat liver cells, contain as much as 55% lipid (dry weight) and can account for about half the total cellular content of phospholipids (8).

Nature of Lipids

Classes of lipid

Whereas there are many published analyses of the phospholipid composition of whole tissues and sub-cellular organelles, very few such analyses have been reported for extra-neural plasma membranes, probably because of difficulties attending their isolation. An exception is the erythrocyte membrane (red cell 'ghost') and the composition of the phospholipids (9) given in Table 1 shows that there is considerable variation between species and, in particular, between ruminants and non-ruminants. In contrast, the constituent phospholipids of different organs and subcellular organelles derived therefrom vary similarly from organ to organ regardless of species (10). (Table 1).

Mitochondria are particularly rich in phospholipids which can account for as much as 90% of the total lipids and, in typical preparations from bovine heart, liver and kidney, the major components were shown to be diphosphatidylglycerol (cardiolipin), choline glycerophospholipids and ethanolamine phospholipids in the molecular proportions of about 1:4:4 respectively (11), though differences exist between the inner and outer membranes (12), (13). The presence of a relatively high concentration of diphosphatidyglycerol is peculiar to mitochondria.

In the phospholipids of the endoplasmic reticulum phosphatidylcholine (lecithin) predominates (13), (14) and no difference in composition was observed between the parts which contained ribosomes and those which did not (15). The Golgi complex of the cell has been shown to have a phospholipid composition

Table 1

Phospholipid composition of erythrocyte plasma membranes
of different species. Values, to nearest whole number,
as % of total phospholipids

Species	Phosphatidyl-choline	Sphingomyelin	Others[*]
Sheep	1	63	36
Ox	7	61	32
Pig	29	36	35
Rabbit	44	29	27
Rat	56	26	18

*Mostly phosphatidylethanolamine, phosphatidylserine
and phosphatidylinositol

intermediate between that of the endoplasmic reticulum and that of
the plasma membrane (16).

Component fatty acids

     Many analyses of the fatty acids of the major and minor
phospholipids of whole cells and organelles have been made (see
Bartley (17) and White (18) from which it is possible to draw
two general conclusions: (i) in a given tissue of a particular
species, the fatty acid patterns tend to be characteristic of the
type of phospholipid, rather than of its cellular location and
(ii) the essential fatty acid linoleic acid (18:2) and its metabolic derivative, arachidonic acid (20:4), are found ubiquitously
and often in relatively high proportions. Examples of the fatty
acid composition of the major phospholipids of the plasma
membrane of rat-liver cells (16) and of ox-heart mitochondria and
microsomes (endoplasmic reticulum) (19) are given in Table 2.
The very high proportion of linoleic acid (84%) in the fatty acids
of the diphosphatidylglycerol of mitochondria is particularly
noteworthy.   (Table 2).

Table 2

Major component fatty acids of the plasma membrane and
of organelles. Values, to nearest whole number, as % by
weight of total fatty acids

Fatty acid	16:0	18:0	18:1	18:2	20:4
Rat-liver plasma membrane					
Phosphatidylcholine	37	31	6	13	11
Phosphatidylethanolamine	25	14	6	6	10
Ox heart mitochondria					
Phosphatidylcholine	23	6	14	37	10
Phosphatidylethanolamine	1	38	4	15	33
Diphosphatidylglycerol	1	trace	9	84	none
Ox heart microsomes					
Phosphatidylcholine	21	10	15	32	9
Phosphatidylethanolamine	5	19	7	22	36

Some Functions of Phospholipids

Whilst other than phospholipids (cholesterol, glycolipids) are
usually present in membranes, it is the inter-relationship between
phospholipids and proteins (as both structural and enzymic components) that is of fundamental significance. Phospholipids function
in membranes in three principal ways, (i) to preserve their
physical integrity, (ii) to promote permeability and transport of
metabolites, and (iii) to participate in enzymic reactions. It is
only possible here to cite a few selected instances of the physiological and metabolic involvement of lipids from the considerable
literature which has accrued in recent years and which has been
extensively reviewed elsewhere (5), (20), (21).

With regard to permeability and transport, it has been demonstrated (see Post (22)) that phospholipid is necessary in plasma membranes for the activity of the 'sodium pump' ATPase which maintains high potassium and low sodium ion concentrations within the cell. Another very different transport process is that of gaseous exchange across the alveolar membranes of the lung and here the presence of phosphatidylcholine with a high content of palmitic acid is required for the correct functioning of these membranes (23).

The dependence of the reactions of the respiratory chain on the presence of mitochondrial phospholipids is well established (24) and it is apparent that they also play a vital role in the structural requirements for coupling oxidative phosphorylation to electron transport (20). Though it has been found that several different phospholipids can restore metabolic activity to mitochondria from which lipids have been extracted, it is significant that diphosphatidylglycerol (characteristic of mitochondrial phospholipids (25) is more effective than others and cytochrome oxidase may, thus, have a specific requirement for diphosphatidylglycerol (26). While the nature of the fatty acids present in diphosphatidylglycerol may not be very important so far as mitochondrial electron transport is concerned, selectivity in favour of linoleic acid is crucial for growth. In the linoleic acid-deficient rat, growth ceases because no new mitochondria can be produced (27).

Another mitochondrial enzyme, 3-hydroxybutyrate dehydrogenase, has been shown to have an absolute requirement for phosphatidylcholine (28) and, in the production of oleic acid from stearic acid, the functioning of the desaturase enzyme system of the endoplasmic reticulum depends on the presence of phospholipid (29).

The foregoing examples of the part played by phospholipids in cell metabolism serve to illustrate the fundamental importance of the so-called 'element constant' in the life of the cell and of the animal as a whole. Though there is a steady synthesis and breakdown ('turnover') of phospholipids of membranes (see Dawson (30), McMurray (20)), there appears to be no requirement for a simultaneous turnover of all types of phospholipid, for most membrane lipids and proteins turn over at the same rate; lipids are usually more quickly replaced, either by the removal and replacement of intact lipid molecules or by the turnover of submolecular components (31). During cell division and growth there is a need for increased synthesis of phospholipids; studies on the hormonal initiation of growth and development have shown that the rates of formation of membrane phospholipids, ribosomes and protein synthetic activity are closely integrated (32).

Special Role of Polyunsaturated Fatty Acids

Only members of three series of polyunsaturated fatty acids

are incorporated into phospholipids (33) and each series is recognised by the location of the double bond nearest to the methyl
group (or ω carbon atom) of the molecule. These polyunsaturated
acids are derived (i) by chain elongation and desaturation of oleic
acid to give the ω9 series, and (ii) from the diet, in the form of
linoleic and linolenic acids (ω6 and ω3 respectively), chain
elongation and desaturation of which gives rise to other members
of these same series, such as arachidonic acid (20:4: ω6). The
term 'essential fatty acids' is usually restricted to the acids
of the ω6 series of which linoleic acid itself and arachidonic
acid are the best-known members. As is indicated below, linoleic
acid gives rise, by successive desaturation and chain elongation,
to γ-linolenic acid, dihomo-γ-linolenic acid and, finally,
arachidonic acid. The 'essential' nature of these acids derives
from the inability of animal tissues to introduce a double bond
into a fatty acid molecule nearer to the methyl group than position
ω9.

$$CH_3(CH_2)_4CH=CHCH_2CH=CH(CH_2)_7COOH$$

linoleic acid (ω 6,9)

$$CH_3(CH_2)_4CH=CHCH_2CH=CHCH_2CH=CH(CH_2)_4COOH^-$$

 γ-linolenic acid (ω 6,9,12)

$$CH_3(CH_2)_4CH=CHCH_2CH=CHCH_2CH=CH(CH_2)_6COOH$$

dihomo-γ-linolenic acid (ω 6,9,12)

$$CH_3(CH_2)_4CH=CHCH_2CH=CHCH_2CH=CHCH_2CH=CH(CH_2)_3COOH$$

arachidonic acid (ω 6,9,12,15)

     The physiological significance of linoleic acid in respect of
mitochondrial diphosphatidylglycerol has already been mentioned
and other specific functions of polyunsaturated fatty acids as
constituents of membrane lipids remain to be discovered, not least
to explain the diverse nature of the effects of feeding animals on
diets deficient in essential fatty acids. These effects vary somewhat between species and, in addition to an overall failure of a
young animal to grow, they can include skin lesions, enlargement
of the heart and kidneys, impaired reproductive ability (irregular
oestrus, degeneration of seminiferous tubules) and changes in the

fatty acid composition of most organs (see reviews of Aaes-Jørgensen (34) and by Alfin-Slater and Aftergood (33)). Retardation of growth was observed in calves fed on a lipid-free diet for about three weeks and, after six weeks, they began to show other deficiency signs, including long dry hair and partial alopoecia; the condition could be prevented or alleviated by giving oils containing essential fatty acids (35).

Additional to the need for polyunsaturated fatty acids for general cellular and metabolic purposes, is their role as precursors of prostaglandins. Although discovered in sheep vesicular glands some forty years ago, it is only within the last decade that the chemistry, biochemistry and physiological activity of prostaglandins have been extensively investigated (for reviews see Horton (36), Hinman (37) and Flowers (38). As they occur naturally prostaglandins are 20-carbon fatty acids containing a cyclopentane ring and many variants of this basic structure occur depending on the number of double bonds, the number of hydroxyl groups and the presence or absence of a keto group. Only two examples will be mentioned here, namely those derived directly from dihomo-γ-linolenic acid and arachidonic acid, i.e. the acids which, as indicated above, are formed successively by chain elongation and desaturation of linoleic acid. By a series of reactions effected by a microsomal multi-enzyme complex, cyclized derivatives possessing functional groups are formed, having the structures indicated below in their 'shorthand' versions.

Prostaglandin E_1 from dihomo-γ-linolenic acid

Prostaglandin E_2 from arachidonic acid

Only when in the free (non-esterified) form are the unsaturated fatty acids substrates for the synthetase (39) and thus a limiting factor in the control of prostaglandin production is phospholipase-induced release of the parent acids from tissue phospholipids (40).

Prostaglandins exert profound physiological activity at concentrations down to 10^{-9} g/g tissue. They are apparently produced

in all animal tissues and it appears that there is a close physiological relationship between prostaglandins and cyclic AMP in the
modulation of hormonal activity. The prostaglandins have many and
diverse effects, for example in reproduction, in nerve transmission,
in muscle contraction and in the regulation of blood supply to
organs (36). In short, they seem to play a fundamental part in the
physiological integrity of the animal body, thereby emphasizing the
essentiality of their fatty acid precursors.

LIPIDS OF ADIPOSE TISSUE

Sites and Structure of Adipose Tissue

It is, of course, well known that adipose tissue is most
extensively present in the subcutaneous, perinephric, omental and
muscular regions of the animal body. This 'visible' fat is termed
white adipose tissue to distinguish it from brown adipose tissue
which is present in new-born mammals of many species and also in
hibernating animals and which serves a special physiological role
(see below). The anatomical distribution of the brown adipose
tissue of neonatal animals varies from species to species (41);
the principal site is usually between the shoulder blades (inter-
scapular) though, in the lamb, most brown adipose tissue surrounds
the kidneys and extends backwards and forwards along the dorsal
wall of the abdomen (42).

The adipocytes of both kinds of adipose tissue contain large
amounts of triglyceride but, whereas in the white adipocyte this
constitutes one large amorphuus globule, the brown adipocyte
(which is smaller than its white counterpart) usually contains a
number of droplets. Brown adipocytes are characterized by their
large and numerous mitochondria, often closely associated with
lipid droplets, and their colour is probably due to their high
content of cytochromes (43), though brown adipose tissue also
derives some of its colour from the erythrocytes of its extensive
vascular network.

Cellularity of Adipose Tissue in Relation to Growth

As long ago as 1909, Bell (44) reported that the diameter of
adipocytes in bovine muscle increased during growth and, many years
later, it was observed (45) that, not only did the size of bovine
adipocytes vary according to the tissue of origin (subcutaneous,
intermuscular, interfascicular), but that the largest average
diameter of adipocytes was associated with the biggest mass of
cells within a particular muscle. Recently, a more extensive
study was made by Hood and Allen (46). Subcutaneous, perinephric
and interfascicular adipose tissue were sampled from the carcases
of cattle of different breeds and ages and it was found that, when
fed on the same ration for the same periods of time, adipose tissue

of (lean) Holstein steers contained fewer and smaller adipocytes
than did corresponding tissues from (fatter) Hereford x Angus
animals. During growth of the steers, increase in mass of adipose
tissue was associated with cellular hypertrophy and hyperplasia.
Whereas hyperplasia was nearly complete in subcutaneous and
perinephric tissue by the time the animals were about eight months
old, hyperplasia in interfascicular adipose tissue was found to be
still active six months later. Thus the 'marbling' of meat (i.e.
visible interfascicular adipose tissue) which is considered
important in relation to palatability and which can be very
variable in amount (47), may depend on a different developmental
pattern to that which obtains in subcutaneous and perinephric
tissues.

In similar studies with growing pigs, Anderson and Kauffman
(48) found that not all adipose-tissue sites develop at the same
rate; they also observed that hyperplasia and hypertrophy of
adipocytes took place simultaneously up to the age of about five
months, after which hypertrophy was primarily responsible for
increase in tissue mass. However, the above studies with cattle
and pigs do not preclude the possibility that continued growth of
the animals could, at some stage, lead to further hyperplasia of
the adipocytes.

Nature of Lipids

White adipose tissue

The lipids consist almost entirely (98-99%) of triglycerides
and countless analyses of their component fatty acids have been
reported for a great many animal species and, in particular, for
the principal meat-producing animals, cattle, sheep and pigs of
different breeds, at different ages and under a variety of
nutritional conditions. These fatty acids represent a mixture
of those synthesized endogenously and those of dietary origin;
the endogenously-produced fatty acids are mainly palmitic and
stearic acids, together with their corresponding unsaturated
counterparts, palmitoleic and oleic acids. This 'basal' composition has been established from analyses of adipose-tissue
triglycerides of animals reared on diets virtually devoid of lipids,
for example, lambs (49) and pigs (50). Whereas in simple-stomached
animals, such as the pig, modifications to the basal fatty-acid
composition of the adipose-tissue triglycerides largely reflect the
nature of the dietary fatty acids (see Garton (51)), in ruminant
animals the rumen micro-organisms modify dietary unsaturated fatty
acids by hydrogenation so that stearic acid and trans isomers of
oleic acid are normally the major fatty acids of exogenous origin
to be incorporated into white adipose tissue (51), (52). In the
context of growth, the depot triglycerides of ruminant animals
show changes in fatty-acid composition which are directly related

to the development of a functional rumen that takes place when roughage forms part of the diet of the young animal. With reference to the growing calf (53), Table 3 illustrates the change which takes place in the weight of the rumen and in the composition of the perinephric triglycerides when, at 50 kg live weight, the animals are given a diet containing concentrates in place of an all-milk diet.

The fatty-acid composition of biopsy samples of subcutaneous adipose tissue of steers and heifers was determined at different stages of growth of the animals and at different times of the year (54). With age the stearic acid content decreased and there was a corresponding increase in unsaturated acids, mostly oleic acid; the relative proportions of myristoleic and palmitoleic acids were higher, and those of palmitic and stearic acids were lower in the (cold) winter months, whilst the reverse obtained in summer. Intramuscular lipids of steers and heifers also show comparable changes associated with age and with season of the year (55). Similar observations have been reported (56) in sheep with respect to ambient temperature and the degree of unsaturation of subcutaneous adipose tissue.

Table 3

Component fatty acids of calf perinephric triglycerides in relation to rumen development. Mean values, for three animals/group, as % by mol. of total fatty acids.

Groups [*]	1	2	3	4
Palmitic acid	34	35	34	21
Palmitoleic acid	8	4	4	3
Stearic acid	8	4	14	25
Oleic acid	47	35	37	25
Trans isomers of oleic acid	None	1	1	15
Mean wt(g) of rumen	210	630	750	3150

[*]
Group 1. Neonatal   Group 3. Given milk to 100 kg
Group 2. Given milk to 50 kg   Group 4. Given milk to 50 kg then
                                                       concentrates to 100 kg

The perinephric triglycerides and those of other internal tissues of ruminants are 'hard' fats (i.e. they have a high content of stearic acid) and they are not infrequently regarded as typical of all the depot fats of ruminants. This is not so and, as Table 4 shows with particular reference to the sheep (57), the fatty-acid composition of the trigylcerides of adipose tissue can vary considerably between anatomical sites, notably between internal and external tissues. It appears that, in ruminants, fatty acids of exogenous origin are preferentially incorporated into the triglycerides of internal adipose tissue (57), (58).

Unusual diets can lead to abnormal fatty-acid composition of the depot fats of ruminants. Thus, when sheep and cattle are given rations containing 'protected' lipids (i.e. coated with formaldehyde-treated casein), the unsaturated fatty acids escape ruminal hydrogenation and are deposited in the tissues (59), as in non-ruminants. Again, when sheep are given diets with a high content of readily-fermentable carbohydrate (barley, maize, wheat), the amount of propionate produced is greater than that which can be metabolized normally. Excess propionate is incorporated into long-chain fatty acids and its primary metabolite, methylmalonate, is similarly utilized, thereby giving rise to numerous branched-chain fatty acids, the presence of which in the triglycerides makes the adipose tissue softer than normal (60), (61).

Table 4

Major Component Fatty Acids in Tissue Triglycerides of a Sheep fed on Hay and Concentrates. Values, to nearest whole number, as % by mol. of total fatty acids.

Fatty acid	16:0	18:0	18:1
Tissue			
Perinephric	24	31	33
Mesenteric	23	26	40
Thoracic	22	17	51
Gluteal	21	10	57
Lower leg	19	5	66
Ear pinnae	14	2	66

Brown adipose tissue

Compared to white adipose tissue, the proportion of triglycerides in the lipids of brown adipose tissue is somewhat lower (75-90%) (62), whilst the amount of phospholipids (particularly diphosphatidylglycerol) is correspondingly higher, reflecting the difference between the adipocytes of the two tissues in the numbers of mitochondria which they contain (60). The triglycerides of the brown and white adipose tissue of several species are similar in fatty-acid composition (see Smith and Horwitz (63)), though there is a tendency for those of brown adipose tissue to contain relatively more stearic acid. The component fatty acids of the perinephric triglycerides of the neonatal lamb include about 60% oleic acid, 20% palmitic acid and 10% stearic acid (64).

Metabolic Functions

The primary function of white adipose tissue is to act as a reservoir of energy for the body and, to this end, it effects the synthesis and storage of triglycerides and releases free fatty acids for oxidation. Whereas the fatty acids released by white adipose tissue are oxidized in other tissues and organs (and, in lactation, utilized for milk-fat synthesis), the fatty acids released in brown adipose tissue are oxidized in situ to provide local heat (so-called 'non-shivering thermogenesis'). It is not possible to give more than a very short description of the complex metabolic activities which combine to make adipose tissue rank as an organ and concerning which a great deal of detailed information is now available (2,3,59,60,65).

White adipose tissue

Four principal and related functions can be ascribed to the adipocytes of this tissue, (i) assimilation through the plasma membrane of pre-formed fatty acids and of metabolites (including glucose and acetate) required for triglyceride synthesis, (ii) conversion of these metabolites into triglycerides, (iii) breakdown (lipolysis) of triglycerides to give fatty acids and glycerol, and (iv) transport of the products of lipolysis out of the cell. In the pig (66) and the sheep (67) (and presumably in other ruminants) adipose tissue is the principal site of fatty-acid synthesis, with the liver participating only to a very limited extent. Though consideration of the control mechanisms governing the functions of adipose tissue is outside the scope of this general account of the physiological role of lipids, it should be mentioned that hormones, particularly insulin, play a key role in modulating assimilative and synthetic activities, whilst other hormones such as adrenaline, glucagon and growth hormone stimulate lipolysis and the discharge of free fatty acids. The relationships between the enzymes which regulate synthesis and dissimilation of adipose-tissue triglycerides is a finely-balanced one which depends on the nutritional state of the animal and which is integrated with the metabolic activity of other tissues.

Following their release from adipose tissue, the free fatty acids are carried in the blood stream, in the form of complexes with plasma albumin, to other tissues and organs. Despite relatively low concentrations in plasma, the half-life of the free fatty acids is many times less than that of plasma glucose. In the tissues the free acids are oxidised (in the mitochondria) to carbon dioxide and water, with the concomitant production of ATP, though some of the acids are converted in the liver to the ketone bodies, β-hydroxybutyrate and acetoacetate which, in turn, can serve as fuel for tissues. The glycerol released from adipose tissue also passes to the liver, where it is utilized for gluconeogenesis.

Brown adipose tissue

As already mentioned, brown adipocytes have the unusual property of producing local heat. Oxidation of fatty acids takes place without, as in other tissues, a 'coupled' phosphorylation taking place resulting in ATP formation. It appears that free fatty acids liberated within the cells are responsible in some way for this 'uncoupling' (68) and so permit the rapid thermogenesis which can be of considerable importance for the survival of some young animals, including lambs, in the first few days of their lives.

REFERENCES

1. Mayer, H. and Schaeffer, G. (1913). J. Physiol. Path. gén. 15, 510.

2. "Handbook of Physiology. Section 5: Adipose tissue". (Eds. A. E. Renold and G.F. Cahill). 1965. (American Physiological Society, Washington).

3. "Adipose Tissue: Regulation and Metabolic Functions". (Eds. B. Jeanrenaud and D. Hepp). 1970. (Academic Press, New York and London).

4. Vandenheuvel, F. A. (1971). In:"Advances in Lipid Research". Vol. 9. (Eds. R. Paoletti and D. Kritchevsky). p. 161. (Academic Press, New York).

5. Finean, J. B. (1973). In:"Form and Function of Phospholipids". (Eds. G. B. Ansell, J. N. Hawthorne and R. M. C. Dawson). p. 171. (Elsevier, Amsterdam).

6. Zahler, P. (1970). In:"Modern Problems of Blood Preservation". (Eds. W. Spielmann and S. Seidl). p.1. (Fischer, Stuttgart).

7. Sjöstrand, F. S. (1968). In:"Regulatory Functions of Biological Membranes". (Ed. J. Järnefelt). p.1. (Elsevier, Amsterdam).

8. Getz, G. S., Bartley, W., Stirpe, F., Notton, B. M. and Renshaw, A. (1962). Biochem. J. 83, 181.

9. de Gier, J. and van Deenen, L. L. M. (1961). Biochim. biophys. Acta, 49, 286.

10. Getz, G. S., Bartley, W., Lurie, D. and Notton, B. M. (1968). Biochim. biophys. Acta, 152, 325.

11. Fleischer, S., Rouser, G., Fleischer, B., Casu, A. and Kritchevsky, G. (1967). J. Lipid Res. 8, 170.

12. Parsons, D. F., Williams, G.R., Thompson, W., Wilson, D. and Chance, B. (1967). In:"Mitochondrial Structure and Compartmentation". (Eds. E. Quagliariello, S. Papa, E. C. Slater and J. M. Tager). p. 29. (Adriatica, Bari).

13. McMurray, W. C. and Dawson, R. M. C. (1969). Biochem. J. 112, 91.

14. Scarpa, A. and Azzone, G. F. (1969). Biochim. biophys. Acta, 173, 78.

15. Glaumann, G. and Dallner, G. (1968). J. Lipid Res. 9, 720.

16. Keenan, T. W. and Morré, D. J. (1970). Biochemistry, 9, 720.

17. Bartley, W. (1964). In:"Metabolism and Physiological Significance of lipids". (Eds. R. M. C. Dawson and D. N. Rhodes). p. 369. (Wiley, London).

18. White, D. A. (1973). In:"Form and Function of Phospholipids". (Eds. G. B. Ansell, J. N. Hawthorne and R. M. C. Dawson). p. 441. (Elsevier, Amsterdam).

19. Wheeldon, L. W., Schumert, Z. and Turner, D. A. (1965). J. Lipid Res. 6, 481.

20. McMurray, W. C. (1973). In:"Form and Function of Phospholipids". (Eds. G. B. Ansell, J. N. Hawthorne and R. M. C. Dawson). p. 205. (Elsevier, Amsterdam).

21. Hawthorne, J. N. (1973). In:"Form and Function of Phospholipids". (Eds. G. B. Ansell, J. N. Hawthorne and R. M. C. Dawson). p. 423. (Elsevier, Amsterdam).

22. Post, R. L. (1968). In:"Regulatory Functions of Biological Membranes". (Ed. J. Järnefelt). p. 163. (Elsevier, Amsterdam).

23. Morgan, T.E., Finley, T. N. and Fialkow, H. (1965). Biochim. biophys. Acta, 106, 403.

24. Green, D. E. and Fleischer, S. (1964). In:"Metabolism and Physiological Significance of Lipids". (Eds. R. M. C. Dawson and D. N. Rhodes). p. 581. (Wiley, London).

25. Fleischer, S., Brierley, G., Klouwen, H. and Slautterback, D. B. (1962). J. biol. Chem. 237, 3264.

26. Awasthi, Y. C., Chuang, T. F., Keenan, T. W. and Crane, F. L. (1970). Biochem. biophys. Res. Comm. 39, 822.

27. Macfarlane, M. G. (1964). In:"Metabolism and Physiological Significance of Lipids". (Eds. R. M. C. Dawson and D. N. Rhodes). p. 631. (Wiley, London).

28. Jurtshuk, P., Sekuzu, I. and Green, D. E. (1963). J. biol. Chem. 238, 3595.

29. Jones, P. D., Holloway, P. W., Peluffo, R. O. and Wakil, S. J. (1969). J. biol. Chem. 244, 744.

30. Dawson, R. M. C. (1966). In:"Essays in Biochemistry". (Eds. P. N. Campbell and G. D. Greville). Vol. 2, p. 69. (Academic Press, London).

31. Omura, T., Siekevitz, P. and Palade, G. E. (1967). J. biol. Chem. 242, 2389.

32. Tata, J. R. (1968). In:"Regulatory Functions of Biological Membranes". (Ed. J. Järnfelt). p.222. (Elsevier, Amsterdam).

33. Alfin-Slater, R. B. and Aftergood, L. (1968). Physiol. Rev. 48, 758.

34. Aes-Jørgensen, E. (1961). Physiol. Rev. 41, 1.

35. Lambert, M. R., Jacobson, N. L., Allen, R. S. and Zaletel, J. H. (1954), J. Nutr. 52, 259.

36. Horton, E. W. (1969). Physiol. Rev. 49, 122.

37. Hinman, J. W. (1972). Ann. Rev. Biochem. 41, 161.

38. Flowers, R. J. (1974). Pharmacol. Rev. 26, 33.

39. Lands, W. E. M. and Samuelsson, B. (1968). Biochim. biophys.
     Acta, 164, 426.

40. Samuelsson, B. (1969). Progr. Biochem. Pharmacol. 5, 109.

41. Afzelius, B. A. (1970). In:"Brown Adipose Tissue".
     (Ed. O. Lindberg). p.1. (American Elsevier, New York).

42. Thompson, G. E. and Jenkinson, D. M. (1969). Can. J. Physiol.
     Pharmacol. 47, 249.

43. Joel, C. D. and Ball, E. G. (1962). Biochemistry, (1), 281.

44. Bell, E. T. (1909).   Am. J. Anat. 9, 401.

45. Moody, W. G. and Cassens, R. G. (1968). J. Food Sci. 33, 47.

46. Hood, R. L. and Allen, C. E. (1973). J. Lipid Res. 14, 605.

47. Blumer, T. N., Craig, H. B., Pierce, E.A., Smart, W. W. G.
     and Wise, M. B. (1962). J. Anim. Sci. 21, 935.

48. Anderson, D. B. and Kauffman, R.G. (1973). J. Lipid Res. 14,
     160.

49. Duncan, W. R. H., Garton, G. A. and Matrone, G. (1971).
     Proc. Nutr. Soc. 30, 48A.

50. Leat, W. M. F., Cuthbertson, A., Howard, A. N. and Fresham,
     G. A. (1964). J. Agric. Sci. Camb. 63, 311.

51. Garton, G. A. (1969). In "International Encyclopaedia of Food
     and Nutrition". Vol. 17. (Ed. D. P. Cuthbertson). p. 335.
     (Pergamon, Oxford).

52. Garton, G. A. (1967). Wld. Rev. Nutr. Diet. 7, 225.

53. Garton, G. A. and Duncan, W. R. H. (1969). Br. J. Nutr. 23,
     421.

54. Link, B. A., Bray, R. W., Cassens, R. G. and Kauffman, R. G.
     (1970). J. Anim. Sci. 30, 722.

55. Link, B. A., Bray, R. W., Cassens, R. G. and Kauffman, R. G.
     (1970). J. Anim. Sci. 30, 726.

56. Cramer, D. A. and Marchello, J. A. (1964). J. Anim. Sci.
     23, 1002.

57. Duncan, W. R. H. and Garton, G. A. (1967). J. Sci. Fd Agric.
     18, 99.

58. Faichney, G. J., Davies, H. L., Scott, T. W. and Cook, L. J.
     (1972). Aust. J. biol. Sci. 25, 205.

59. Scott, T. W., Cook, L. J. and Mills, S. C. (1971). J. Am.
     Oil Chem. Soc. 48, 358.

60. Garton, G. A., Hovell, F. D. DeB. and Duncan, W. R. H. (1972).
     Br. J. Nutr. 28, 409.

61. Duncan, W. R. H., Ørskov, E. R. and Garton, G. A. (1974).
     Proc. Nutr. Soc. 33, 81A.

62. Prusiner, S., Cannon, B. and Lindbert, O. (1970). In:"Brown
     Adipose Tissue". (Ed. O. Lindberg). p. 283. (American
     Elsevier, New York).

63. Smith, R. E. and Horwitz, B. A. (1969). Physiol. Rev. 49,
     330.

64. Garton, G. A. and Duncan, W. R. H. (1969). J. Sci. Fd Agric.
     20, 39.

65. "Adipose Tissue as an Organ". (Ed. L. W. Kinsell). 1962.
     Charles Thomas, Springfield, Illinois, U.S.A.).

66. O'hea, E. K. and Leveille, G. A. (1969). J. Nutr. 99, 338.

67. Ingle, D. L. Bauman, D. E. and Garrigus, U.S. (1972).
     J. Nutr. 102, 617.

68. Hittleman, K. J. and Lindberg, O. (1970). In:"Brown Adipose
     Tissue". (Ed. O. Lindberg). p. 245.
     (American Elsevier, New York).

THE CONTROL OF FAT ABSORPTION, DEPOSITION AND MOBILIZATION

IN FARM ANIMALS

W. M. F. Leat

Biochemistry Department
Agricultural Research Council
Institute of Animal Physiology, Babraham, Cambridge

INTRODUCTION

The lipids of the body comprise those which are an integral part of the cell structure, mainly phospholipids, and those which serve as reserves of energy in the form of depot fat. Triglyceride is the major lipid of this adipose tissue and is a concentrated form of energy containing 40 times the energy of glucose and 280 times that of sodium acetate in isotonic solution (1).

In wild animals adipose tissue represents a valuable reserve of calories and the size of fat depots could decide whether an animal survives during periods of nutritional inadequacy. Even in domesticated animals, for example, hill sheep reared under harsh conditions, survival seems to be related to the reserves of fat in the carcass (2,3). However, in animals reared for meat the deposition of excess fat involves a waste of dietary calories and is economically undesirable.

Consumer preferences for the optimum conformation of a carcass have changed considerably over the years. In the past, when hard physical work was the rule rather than the exception, a fat animal was prized because of the high calorific value of its carcass. Beef animals grew more slowly and were slaughtered at a greater age and weight than they are nowadays. The importance of animal fat has declined with the growth of the vegetable oil industry and the present requirement is for a rapidly growing animal producing a lean carcass.

DEVELOPMENT OF FAT DEPOTS

The newborn farm animal has little reserve of adipose tissue
(1-4%) compared with the human baby (16%) (4,5). During growth
and development more fat is laid down and the carcass of the
adult may contain as much as 30-40% (Fig.1). This is particularly
striking in the pig. The new born pig contains very little fat
(1%), but within a week of birth the percentage of fat in its
body has increased eight-fold (6), and there is a further increase
to bacon weight (7). There is a definite order of development of
tissues with bone preceeding muscle and muscle developing before
fat. Fat depots themselves have a defined order of development
which has been described in several species including wild
as well as domesticated forms, and continues in the order
abdominal, intermuscular, subcutaneous and intramuscular.
The development of the last two depots gives the characteristic plump appearance of the finished animal and the
marbling effect in the meat. As live weight increases there
is a relative rise in the proportions of subcutaneous and intermuscular depots and a decline in those of perinephric and channel
fat. There are also differences between breeds; early maturing
animals (e.g. Hereford cattle) lay down proportionately more fat
at a younger age than the later maturing breeds (e.g. Friesian)
(8,9). The rate of development of the tissues and of the component
fat depots can be modified by diet; a high plane of nutrition
hastening and a low plane of nutrition retarding the rate of
development. However, the order of development of tissues is not
altered, it is merely the time scale which is shortened or
lengthened. One consequence of this is that if a carcass contains
visible intramuscular (marbling) fat it must also contain an
excessive amount of other depot fat, particularly subcutaneous.
This is a very brief summary of the classical work of Hammond and
his colleagues and more detailed information can be obtained
elsewhere (9).

The relative distribution of fat depots varies between
species with the ratio of subcutaneous: intermuscular fat being
2.4 in pigs compared to 1.2 in sheep and 0.6 in cattle (10). The
subcutaneous fat of pigs and cattle has higher lipogenic activity
than the internal depots (11,12). This may be merely a reflection
of the stage of development of fat depots since in young ruminants
internal depot fats have the highest rate of lipogenesis.

The size of a fat depot is dependent on the size and number
of adipocytes it contains. In cattle an increase in the size of
adipose tissue results from both hypertrophy and hyperplasia of
cells, but there are differences between the various depots. In
subcutaneous and perinephric tissues, hyperplasia is virtually
complete at about 8 months of age, and further increase of size

is the result of cell hypertrophy. However, in intramuscular
tissue, hyperplasia is still evident at 14 months of age (13).

In pigs the increase in the amount of adipose tissue is
accounted for by both hyperplasia and hypertrophy up to 5 months
of age, but only hypertrophy after 5 months (14). Restricting
feed intake during the suckling period of the pig has little effect
subsequently on the cell size and number of adipocytes in subcutaneous fat but it reduces both the number and size of cells in
intramuscular fat, (15). However, caloric restriction in early
life does not appear to hold much promise as a method of regulating
fat deposition in the more mature animal.

The fatty acids of the fat depots are formed from lipid
originating in the diet (exogenous) and from lipid synthesised
within the body (endogenous). The site of endogenous synthesis,
and the source of carbon for fatty acid synthesis, varies from one
species to another. In birds the major site of synthesis is the
liver; in the pig and ruminant the adipose tissue is the major
site (16,17). In the pig fatty acids are synthesised from both
acetate and glucose whereas in the ruminant acetate is the primary
source because of the virtual absence of a number of key lipogenic
enzymes.

Although fat depots were originally thought to be rather inert
metabolically,the work of Schoenheimer (18) showed that this was
not so. The fatty acids are in a continual state of flux and
modification; the composition of a fat depot at any one time
representing an equilibrium of synthesis, interconversion,
deposition and mobilization. The amount of depot fat and its
fatty acid composition will, therefore, depend on the relative
proportions of endogenous and exogenous sources which vary from
species to species and are themselves modified by various
physiological parameters such as the type of digestion and stage
of development of the animal.

DIGESTION AND ABSORPTION

Non-ruminants

Dietary lipids are insoluble in water and must be rendered
'soluble' before they can be absorbed from the intestine. The
function of the digestive process is to convert the water insoluble
triglyceride molecule into a form which can dissolve in the
intestinal lumen. Most of the research on non-ruminants has been
carried out on the rat and human but it is unlikely that digestion
of lipids in the pig is any different (19).

The two major secretions required for lipid absorption are
bile and pancreatic juice. Pancreatic juice provides the lipase

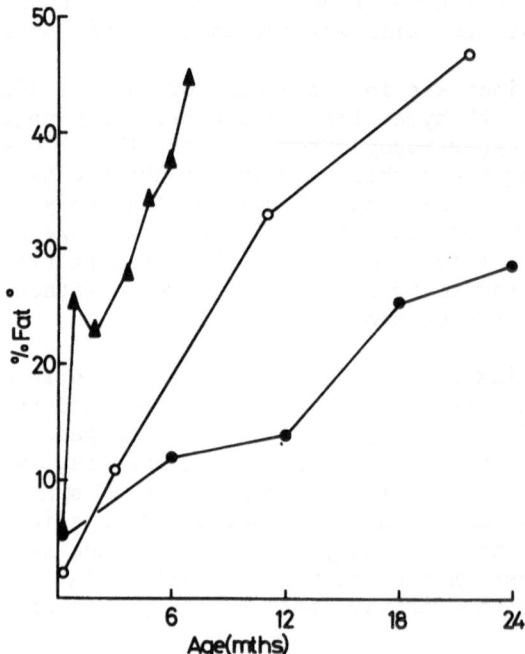

Fig. 1   Changes with age in the percentage fat in the carcass
of pigs (7; ▲), cattle (8; ●) and sheep (9; o).

which hydrolyses the two primary alcoholic groups of triglycerides
to form free fatty acid and monoglycerides which, in contrast to
diglyceride and triglyceride, are soluble in bile salts. Pancreatic lipase does not hydrolyse the triglyceride molecule any
further than is necessary to allow the molecule to dissolve in
bile salts, and in so doing conserves the energy of one of the
three ester bonds. The monoglyceride and free fatty acid form a
mixed micelle with bile salts (20) and pass to the mucosal cell
where the lipids diffuse into the cell and are resynthesised to
triglyceride, mainly via the monoglyceride pathway. The triglyceride is incorporated into chylomicrons which pass into the
lymphatics and thence into the venous circulation. During
digestion and absorption the fatty acid composition of the dietary
lipids is not changed appreciably (c.f. ruminants) and at the
peak of absorption the fatty acid composition of chylomicrons is
very similar to the dietary lipid (Table 1).

Pancreatic juice is necessary for fat absorption in non ruminants, and diversion of this secretion or malfunction of the exocrine pancreas result in steatorrhoea. Bile is necessary for micelle formation and absorption into the lymphatics and depri- vation of bile or disease of the biliary tract does result in lipid malabsorption (21), although some absorption of fatty acids can occur by the portal route. Absorption of lipids can be reduced by feeding bile salt sequestrating agents e.g. cholestyramine.

Ruminants

In ruminant animals dietary lipids consist mainly of triglycerides if the animal is feeding mainly on cereal based con- centrates and mono- and digalactosyl glycerides if it is grazing pasture. The dietary lipids pass first into the rumen where considerable modification occurs before they pass to the site of absorption in the small intestine. The dietary esterified lipids are first hydrolysed by lipases in the rumen and the released free fatty acids, mainly linoleic (18:2) and linolenic (C18:3) acids, are hydrogenated (22,23). The major end product is stearic acid but hydrogeneation is usually incomplete and results in the

Table 1

Major fatty acids (% of weight) of dietary lipids and lymph triglycerides of the pig, cow, goat and sheep

Fatty acid	Pig		Cow		Goat	Sheep		Sheep (Maize oil infused into duodenum)	
	Lymph	Diet	Lymph	Diet (pasture)	Lymph	Lymph	Diet (Hay & Oats)	Lymph	Maize oil
16:0	21.8	28.0	23.8	13.7	26.8	28.1	19.8	16.4	14.0
18:0	8.4	5.9	37.4	1.6	33.3	40.0	2.3	19.6	2.6
18:1	28.2	23.7	23.4	2.6	29.2	14.8	24.0	22.7	28.0
18:2	34.7	40.9	2.4	12.3	2.3	2.8	36.4	37.5	52.6
18:3	1.2	0.9	3.6	66.8	0.9	1.3	16.3	0.9	1.5

formation of appreciable amounts of geometrical and positional
isomers of octadecenoic and octadecadienoic acids. In addition,
the bacteria and protozoa deaminate amino acids and ferment
carbohydrates and the products of these processes are incorporated
into the branched-chain and odd-numbered fatty acids characteristic
of bacterial lipids. Such fatty acids are absorbed from the
intestine and eventually deposited in adipose tissue. As a result,
the fatty acid composition of ruminant adipose tissue is characterized by the high content of stearic acid, the presence of
positional and geometric isomers of unsaturated fatty acids and
appreciable amounts of branched-chain and odd-numbered fatty acids.
A further consequence of this hydrogenation of fatty acids in the
rumen is that the composition of dietary fat will have little
effect on the fatty acid composition of depot fat. In contrast
to the non-ruminant, therefore, the fatty acids absorbed by the
ruminant differ greatly from those in the diet (Table 1).

        As in non-ruminants both bile and pancreatic juice are required
for optimum fat absorption in ruminants. Since hydrolysis has
already occurred in the rumen the function of pancreatic juice must
be other than the hydrolysis of triglycerides. Evidence to date
suggests that pancreatic juice could function through its ability
to hydrolyse biliary lecithin to lysolecithin which may be
involved in the solubilization and/or resynthesis of lipid (24).

DEPOSITION

Fatty acid composition of endogenous depot fat

Pigs. Ever since the classical experiments of Lawes and Gilbert
(25) demonstrated that the carcass of the pig contained more fat
than was ingested in the diet it has been known that other dietary
components, particularly carbohydrate, could be converted into fat.
The pig is enzymatically well endowed for the synthesis and
deposition of fat, and, for example, on diets containing 2-4% fat,
73-82% of the fat deposited is from de novo synthesis (26).
Endogenous fat consists mainly of C16 (30%) and C18 fatty acids
(60-65%) (27,28) with only minor amounts of fatty acids of longer
or shorter chain lengths. The C18 fatty acids consist almost
entirely of stearic acid (11-18%) and oleic acid (48-57%) formed
by desaturation of stearic acid. Negligible amounts of linoleic
(C18:2) and linolenic acid (C18:3) are found since these are not
synthesised by mammalian tissues.

Sheep and cattle. The major characteristic of ruminant depot fat
is its high content of stearic acid. Originally it was thought
that hydrogenation of depot fatty acids in situ after absorption
was the reason but it is now generally agreed that the fatty acids
are hydrogenated in the rumen. However, although an accurate
estimate of the spectrum of fatty acids produced endogenously by

the pig has been made (28) less precise information is available
for the ruminant because it is difficult to assess the contribution
of bacterial fermentation in the rumen. This can be overcome in
part by examining the depot fat from germ-free lambs fed low
fat diets (29), and comparing it with samples from conventional
lambs fed a fat-free diet (30). In both cases, as in the pig,
C16:0, C18:0 and C18:1 acids were the major components with
stearic acid accounting for up to 16% of the fatty acids of the
pig.   This suggests that an appreciable proportion of stearic
acid in ruminant depot fat can be derived endogenously by
synthesis from acetate.

Factors affecting the fatty acid composition
and distribution of depot fat

Diet.   When pigs are fed diets deficient in lipid, the amount of
fat deposited is similar to that in pigs fed 10% maize oil or
beef tallow, although differences in the relative proportion of
the various depots are found (28). Fatty acid synthesis by pig
adipose tissue decreases linearly as the percentage fat in the
diet increases (31). The fatty acids and their CoA derivatives
inhibit acetyl CoA carboxylase which is a rate limiting enzyme in
fatty acid synthesis. Control of the amount of depot fat
deposited in a carcass by reduction in the amount of dietary long
chain fatty acids, therefore, seems to have limited potential.
However, this conclusion might not apply in the case of the
medium-chain (C10 and C12) fatty acids e.g. of coconut oil, which
tend to be oxidised rather than deposited. The addition of
medium chain triglycerides to diets at a suitable stage of devel-
ment might allow a reduced amount of fatty acids to be deposited
in fat depots sufficient to inhibit lipogenesis. Dietary unsaturated fatty acids are deposited in the fat of pigs which becomes
soft and commercially unacceptable as a consequence of its lower
melting point (27). Dietary linoleic acid, in the form of maize
oil, is deposited preferentially in the subcutaneous fat depots,
replacing mainly oleic and palmitic acids, and increases the mass
of this depot relative to other fat depots. It is also of note
that the activity of clearing-factor lipase, which is involved in
the deposition of absorbed fat is higher in the outer subcutaneous
fat of pigs than in the internal depots (32). Saturated dietary
fat as beef tallow for example has little effect on the fatty acid
composition of depot fat in spite of the 24% stearic acid it
contains for the deposition of stearic acid is maintained within
narrow limits compared to sheep and cattle (Fig.2).

Since the fatty acids synthesised from carbohydrate are unsaturated in character, the faster a pig deposits fat the firmer
its depot fat will be. Restricted feeding, on the other hand,
results in the slower deposition of fat which is softer because

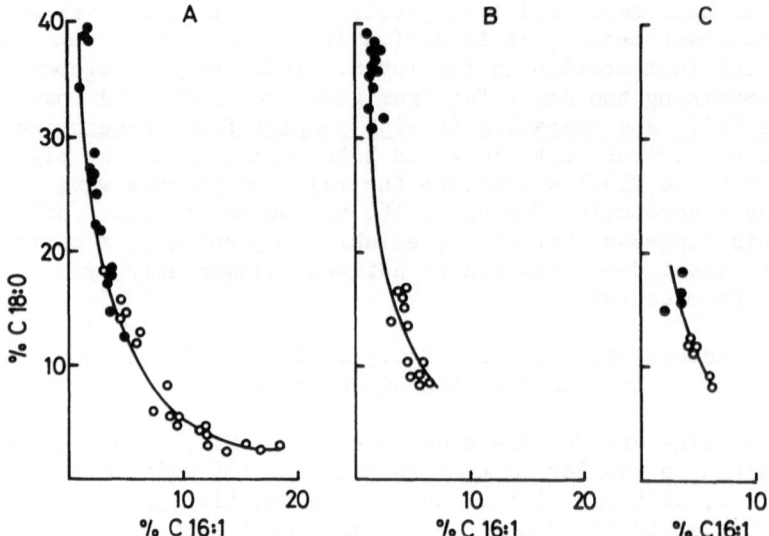

Fig. 2.   Relationship between C18:0 and C16:1 acids in
subcutaneous (o) and perinephric fat (●) of cattle (A),
sheep (B) and pigs (C).

it contains a higher proportion of the dietary unsaturated fatty
acids. In general, however, a useful chemical index of depot fat
softness is the ratio of monoene: saturated fatty acids (33). The
addition of copper to pig diets results in the formation of softer
depot fat because of a change in the ratio of saturated: unsaturated fatty acids, and a redistribution of fatty acids within the
triglyceride molecule (34,35).

     In ruminants feeding unsaturated oils has only a minor effect
on the unsaturation of depot fat because of the hydrogenation
which occurs in the rumen. The major change is in the content of
stearic acid and octadecenoic acid (36,37) but, the fatty acid
composition of ruminant depot fat is not completely refractory to
changes in diet. In sheep (38) and cattle (39) the feeding of diets
rich in grain results in a rather more unsaturated depot fat

(particularly in C18:1) than is found in animals fed roughage
diets. This effect could be due to a more rapid rate of passage
of digesta, incomplete hydrogenation in the rumen or the result
of induced changes in the microbial population of the rumen (40).

Other components of the diet can affect the composition of
ruminant depot fats. The feeding of barley based diets to sheep
results in the formation of softer depot fats (41). This softening
is associated with the enhanced deposition of odd-numbered and
branched-chain acids probably from the increased production of
propionic acid in the rumen. Species differ in this respect for
when cattle are reared on high barley diets the adipose tissue is
firm in texture and there is no increase in the amount of odd
numbered and branched chain fatty acids (W.M.F. Leat, unpublished
observations). This may reflect a difference in the metabolism of
propionic acid in sheep and cattle, or may merely reflect
differences in the ruminal production of propionic acid.

When young ruminants are fed unsaturated oil, particularly if
it is incorporated into milk based diets, the polyunsaturated fatty
acids are incorporated into depot fats (42,43). Dietary fatty
acids are incorporated into depot fats in young ruminants because
either the rumen is not functional, or the rumen is bypassed by
means of the oesophageal groove during suckling. In adult ruminants,
when the rumen is bypassed by infusing oil into the duodenum (44),
or when dietary lipids are protected from ruminal hydrolysis and
hydrogenation (45,46), the depot fats can become highly unsaturated.
However, these dietary polyunsaturated fatty acids are deposited
preferentially in the internal depot fats rather than subcutaneously
as they are in the pig.

Location within the carcase. It has been known for some time that
in sheep, cattle and pigs the subcutaneous fats are softer, i.e.
more unsaturated, than the internal depot fats (27,47,48). There
is also a gradient of unsaturation within the subcutaneous tissue
itself (Table 2). Henriques and Hansen (47) suggested that
temperature differences between the subcutaneous regions and
internal regions could explain the differences in unsaturation.
However, it is now generally accepted that this is an oversimplifi-
cation and it is difficult to formulate a theory applicable to all
animals (49). Callow (50) suggested that for ruminants an adequate
theory must allow for local temperature of tissue, local rate of
fat deposition and local development of fatness. In pigs the concentration of oleic acid is affected by the temperature at which
the animals are kept but the changes are reflected in all fat
depots, suggesting a change in whole body metabolism (51). In
lambs exposure to low temperatures results in the deposition of a
more unsaturated fat (52).

The increased unsaturation of subcutaneous fat relative to

Table 2

Major fatty acids (% wt) of brisket and perinephric
depot fats of a 10 year old Jersey Cow

Depth below skin (mm)	Brisket				Perinephric
	1	50	80	125	
Fatty acid					
16:0	19.4	24.0	21.4	20.3	31.5
16:1	18.1	13.6	11.9	7.6	2.7
18:0	2.9	3.9	4.3	6.4	17.0
18:1	47.8	43.7	48.2	54.4	36.5

internal depots is due mainly to an increase in oleic acid at the
expense of stearic acid. However, in cattle palmitoleic acid
(C16:1) becomes a major component of some subcutaneous samples
(53), but only when stearic acid is less than 10% (Fig. 2). At
levels higher than 10% stearic acid is mainly replaced by oleic
acid.

Sex hormones. Intact male sheep, cattle and pigs are more
efficient than castrate animals in converting feed into live
weight gain, and contain less fat (54,55). At a comparable age or
live weight bulls and rams produce carcasses that are less fat
than those of castrates which in turn are less fat than females.
In pigs, however, the gilt is less fat than the barrow. Castration
retards growth, metabolic activity and utilization of feed and
increases the deposition of fat (56). In sheep and cattle the
female has more unsaturated depot fat than the male (57,58); in
pigs the unsaturation of backfat decreases in the order boars,
gilts, barrows (59). Rams have less stearic acid and more
C18:2 and C18:3 acids than wethers (60), but there is no difference.
between bulls and steers (61). These changes in unsaturation are
usually related to hormonal differences between the sexes which
have differing effects on the desaturase enzymes. However, in
cattle, at least (62), the unsaturation of depot fat appears
related to the degree of fattening of the animal.

The growth rate of steers and wethers is increased by implantation of synthetic oestrogens which alter the pattern of growth,
increasing weight gain, and produce more muscle and less fat (9).

Stage of development. The fat of newborn sheep and cattle contains
less than 10% stearic acid which one expects to rise to a maximum as the
rumen develops and then plateau.   In sheep this appears to be so
but in cattle the prediction is only partially correct for the
content of stearic acids declines after one year of age (62,63,64;
Fig. 3). The decrease in stearic acid is compensated for mainly by
increases in C18:1 acid in perinephric fat and C16:1 acid in
subcutaneous fat. The increase in unsaturation occurs at a time
when cattle enter their fattening phase, and may be related to
changes in the activity of desaturase enzymes at this stage.

Although ruminant depot fats are grouped in the 'stearic rich'
fats (27) there is a considerable variation in the content of
saturated fatty acids (Fig. 2). Cattle seem to have the biggest
variability in fatty acid composition with stearic acid ranging
from 3%-40%. The fats of sheep, goats and pigs are less variable

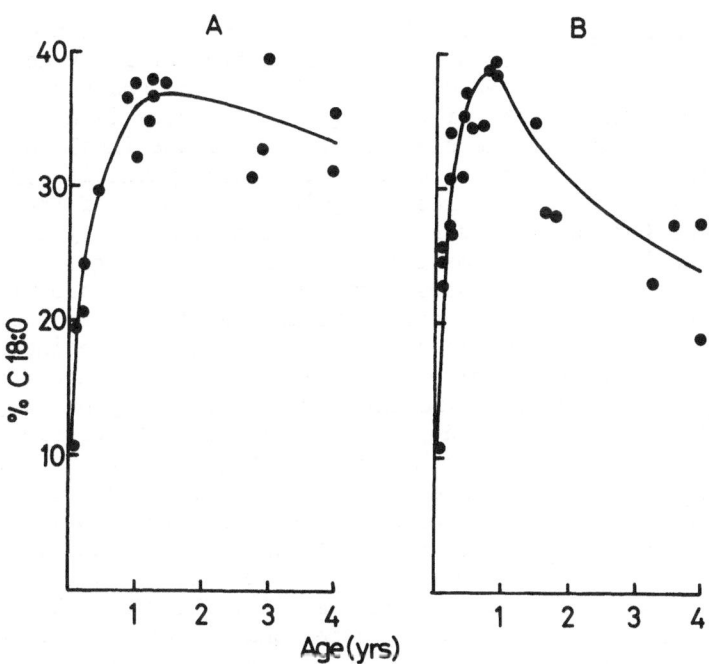

Fig. 3.   Variation with age in the percentage stearic acid
(C18:0) in perinephric fat from sheep (A) and Jersey cattle (B).

with sheep and goats ranging from 10-40% and pigs 10-20%. The relationship between stearic acid and palmitoleic acids (Fig.2)
explains why C16:1 acid is only a minor component of sheep and pig
depot fat whereas it can be a major fatty acid in cattle. The
content of palmitoleic acid only becomes appreciable when
the percentage of stearic acid falls below 8, which occurs
infrequently in sheep and pigs. Although the fatty acid composition of depot fats can be markedly different between cattle,
sheep and pigs, a common relationship appears to exist between
C18.0 and C16:1 acids.

     In pigs there is a selective deposition of saturated fatty
acids in subcutaneous and intramuscular lipid with age (65,66),
mainly an increase in C18.0 acid and a decrease in C18:1, possibly
consequent upon changes in desaturase activity induced by hormonal
change.

Season. In lambs and cattle unsaturation is highest in summer
and lowest in winter (57,67); this may be due to the increased
deposition of unsaturated fatty acids which have escaped hydrogenation.

MOBILIZATION OF DEPOT FATS

     The formation of depot fat is the net result of the processes
of synthesis and mobilization. The mobilization of fatty acids
from the fat depots is under nutritional, hormonal and neural
control. In the depot fats of fasting animals there is a reduction
(32%) in the activity of the clearing-factor lipase involved in
the uptake of lipid, a depression in fatty acid synthesis (99%)
and an increase in lipolysis resulting in the release of free fatty
acids. In the fed animal clearing-factor lipase activity is increased and active lipid deposition and synthesis occurs. Whether
a fat depot deposits or releases fatty acid depends on the tissue
concentrations of cyclic AMP, which are modulated by many factors.
Hormones which increase cyclic AMP concentrations, e.g. epinephrine,
norepinephrine, pituitary polypeptides and glucagon induce lipolysis,
whereas those that reduce cyclic AMP, e.g. insulin, depress lipolysis
and induce fat synthesis and deposition. Pig adipose tissue is
unresponsive to many lipolytic hormones in vitro (68), but effects
can be observed in the living animal (69,70,71). These differences
emphasise the reservations which must apply in extrapolating
responses obtained in vitro to the whole animal. The adipose
tissues from pigs selected for fatness or leanness differ markedly
in their response to lipolytic hormones (72). In pigs selected
for leanness the response is far greater than in pigs selected for
fatness, and the inhibitory effect on lipolysis of serum from lean
pigs is much less than that from fat pigs. These differences
appear to be under genetic control. There is comparatively little
information on lipid mobilization in farm animals but in general

it seems that ruminants are less responsive to lipolytic stimuli
than non-ruminants (73).

GENERAL CONCLUSIONS

There are two major defects in the present day meat carcass;
(a) it can contain too much fat and (b) the fat is mainly saturated
or mono-unsaturated in character, particularly in ruminants.
Excess fat in the carcass is costly, since it represents a wastage
of valuable dietary energy, and may be nutritionally undesirable,
because of the suggested relationship between dietary saturated fats,
raised blood cholesterol and an increased susceptibility to
cardiovascular disease. Although the saturated nature of carcass
fat can be modified by feeding protected unsaturated fats, the
present tenuous relationship between the ingestion of poly-unsat-
urated fatty acids and the reduction in the incidence of heart
disease makes this method of modification premature and possibly
unnecessary in the long term. The more immediate approach to the
problem is to reduce the amount of fat in the carcass. By so doing
the two disadvantages of the meat carcass will be partially
rectified; there will be a more efficient partition of dietary
energy, and the production of saturated fatty acids will be reduced.

The two methods of reducing the fat content of the carcass
would embrace control of deposition and/or lipolysis. The deposition of fat is energetically expensive and up to 20% of the calories
are lost in the formation of triglycerides from glucose and
acetate (74,75). Again, if the depot fats are mobilized the free
fatty acids released would need to be utilized by the body, or the
net result would be the same as trimming the carcase of excess fat.
Control of the deposition of depot fat would therefore seem to be
the most rewarding approach.

Since fat deposition is controlled by the number of fat cells,
which is genetically controlled, the long term approach must be
by breeding for carcasses of low fat content. The selection of
late maturing breeds and a more accurate assessment of caloric
needs during the fattening phase could improve carcass characteristics. On a biochemical level, methods of depressing fat synthesis
at critical phases of development or of partitioning dietary energy
in favour of tissues other than fat would be worth investigation.

REFERENCES

1. Dole, V. P. (1965). In: "Handbook of Physiology. Adipose Tissue."
    Sect. 5, p.13.(Am. Physiol. Soc. Washington D.C.).

2. Gunn, R. G. (1967). Anim. Prod. 9, 263.

3. Russel, A. J. F., Gunn, R. G. and Doney, J. M. (1968). Anim.
    Prod. 10, 434.

4. Body, D. R. and Shorland, F. B. (1964). Nature, Lond. 202, 769.

5. Widdowson, E. M. (1950). Nature, Lond. 166, 626.

6. Manners, M. J. and McCrea, M. R. (1963). Br. J. Nutr. 17, 495.

7. McMeekan, C. P. (1940). J. agric. Sci., Camb. 30, 276.

8. Pomeroy, R. W., Williams, D. R., Owers, A. C. and Scott, B. M.
    (1966). "A Comparison of the Growth of Different Types of
    Cattle for Beef Production." (Royal Smithfield Club, London).

9. Hammond, J. Jr., Mason, I. L. and Robinson, T. J. (1971).
    "Hammond's Farm Animals" 4th Edition (Edward Arnold Ltd.,
    London).

10. Callow, E. H. (1948). J. agric. Sci., Camb. 38, 174.

11. O'Hea, E. K. and Leveille, G. A. (1968). Comp. Biochem.
    Physiol. 26, 1081.

12. Ingle, D. Lt., Bauman, D. E. and Garrigus, U. S. (1972).
    J. Nutr. 102, 609.

13. Hood, R. L. and Allen, C. E. (1973). J. Lipid Res. 14, 605.

14. Anderson, D. B. and Kauffman, R. G. (1973). J. Lipid Res.
    14, 160.

15. Lee, Y. B., Kauffman, R. G. and Grummer, R. H. (1973).
    J. Anim. Sci. 37, 1319.

16. Ballard, F. J., Hanson, R. W. and Kronfeld, D. S. (1969).
    Fedn. Proc. 28, 218.

17. O'Hea, E. K. and Leveille, G. A. (1969). J. Nutr. 99, 338.

18. Schoenheimer, R. (1942). "The Dynamic State of the Body
    Constituents." (Harvard University Press, Cambridge, Mass.).

19. Freeman, C. P., Annison, E. F., Noakes, D. E. and Hill, K. J.
    (1967). Proc. Nutr. Soc. 26, vii.

20. Hofmann, A. F. and Borgstrom, B. (1962). Fedn. Proc. 21, 434.

21. Wiseman, G. (1964). "Absorption from the Intestine" (Academic
    Press, London and New York).

22. Dawson, R. M. C. and Kemp, P. In: "Physiology of Digestion
     and Metabolism in the Ruminant" (Ed. A. T. Phillipson),
     p. 504 (Oriel Press, Newcastle upon Tyne).

23. Garton, G. A. (1967). Wld. Rev. Nutr. Diet 7, 225.

24. Leat, W. M. F. and Harrison, F. A. (1974). In: "Fourth
     International Symposium on Ruminant Physiology," Sydney –
     In Press.

25. Lawes, J. B. and Gilbert, J. H. (1860). Jl. R. Agric. Soc.
     21, 433.

26. Hood, R. L. and Allen, C. E. (1973). J. Nutr. 103, 353.

27. Hilditch, C. P. and Williams, P. N. (1964). "The Chemical
     Constitution of Natural Fats." (Chapman and Hall, London).

28. Leat, W. M. F., Cuthbertson, A., Howard, A. N. and Gresham,
     G. A. (1964). J. agric. Sci., Camb. 63, 311.

29. Leat, W. M. F., Lysons, R. J. and Alexander, T. J. L. (1973).
     Proc. Nutr. Soc. 32, 97A.

30. Duncan, W. R. H., Garton, G. A. and Matrone, G. (1971).
     Proc. Nutr. Soc. 30, 48A.

31. Allee, G. L., Baker, D. H. and Leveille, G. A. (1971). J.
     Anim. Sci. 33, 1248.

32. Enser, M. B. (1973). Biochem. J. 136, 381.

33. Lea, C. H., Swoboda, P. A. T. and Gatherum, D. P. (1970).
     J. agric. Sci., Camb. 74, 279.

34. Taylor, M. and Thomke, S. (1964). Nature, Lond. 201, 1246.

35. Moore, J. H., Christie, W. W., Braude, R. and Mitchell, K. G.
     (1969). Br. J. Nutr. 23, 281.

36. Tove, S. B. and Mochrie, R. D. (1963). J. Dairy Sci. 46, 686.

37. Thomas, B. H., Culbertson, C. C. and Beard, F. (1934).
     Amer. Soc. Animal Production Rec. Proc. 27th Annual Meeting,
     p. 193.

38. Miller, G. J., Varnell, T. R. and Rice, R. W. (1967). J.
     Anim. Sci. 26, 41.

39. Rumsey, T. S., Oltjen, R. R., Bovard, K. P. and Priode, B. M.
     (1972). J. Anim. Sci. 35, 1069.

40. Tove, S. B., and Matrone, G. (1962). J. Nutr. 76, 271.

41. Duncan, W. R. H., Ørskov, E. R. and Garton, G. A. (1972). Proc. Nutr. Soc. 31, 19A.

42. Erwin, E. S. and Sterner, W. (1963). Am. J. Physiol.205, 1151.

43. Stokes, G. B. and Walker, D. M. (1970). Brit. J. Nutr. 24, 435.

44. Ogilvie, B. M., McClymont, G. L. and Shorland, F. B. (1961). Nature, Lond. 190, 725.

45. Scott, T. W., Cook, L. J. and Mills, S. C. (1971). J. Am. Oil Chem. Soc. 48, 358.

46. Faichney, G. J., Lloyd-Davies, H., Scott, T. W. and Cook, L. J. (1972). Aust. J. biol. Sci. 25. 205.

47. Henriques, V. and Hansen, C. (1901). Skand. Arch. Physiol. 11, 151.

48. Moulton, R. and Trowbridge, P. F. (1909). J. Ind. Engng. Chem. 1, 761.

49. Hartman, L. and Shorland, F. B. (1961). N.Z. Jl Sci.Tech. 4, 16.

50. Callow, E. H. (1958). J. Agric. Sci. Camb. 51, 361.

51. Fuller, M. F., Duncan, W. R. H. and Boyne, A. W. (1974). J. Sci. Fd. Agric. 25, 205.

52. Marchello, J. A., Cramer, D. A. and Miller, L. G. (1967). J. Anim. Sci. 26, 294.

53. Dahl, O. (1957). Acta Chem. Scand. 11, 1073.

54. Field, R. A. (1971). J. Anim. Sci. 32, 849.

55. Turton, J. D. (1969). In: "Meat Production from Entire Male Animals." (Ed. D. N. Rhodes). p.1 (Churchill: London).

56. Prescott, J. H. D. and Lamming, G. E. (1964). J. Agric. Sci., Camb. 63, 341.

57. Cramer, D. A. and Marchello, J. A. (1964). J. Anim. Sci. 23, 1002.

58. Terrell, R. N., Suess, G. G. and Bray, R. W. (1969). J. Anim. Sci. 28, 449.

59.   Johns, A. T. (1941). N.Z. Jl Sci. Technol. 22, 248A.

60.   Tichenor, D. A., Kemp, J. D., Fox, J. D., Moody, W. G. and
      DeWeese, W. (1970). J. Anim. Sci. 31, 671.

61.   Clemens, E., Arthaud, V., Mandigo, R. and Woods, W. (1973).
      J. Anim. Sci. 37, 1326.

62.   Embleton, G. A. and Leat, W. M. F. (1971). Proc. Nutr. Soc.
      31, 22A.

63.   Leat, W. M. F. and Embleton, G. A. (1970). Proc. Nutr. Soc.
      29, 48A.

64.   Link, B. A., Bray, R. W., Cassens, R. G. and Kauffman, R. C.
      (1970). J. Anim. Sci. 30, 722.

65.   Sink, J. D., Watkins, J. L., Ziegler, J. H. and Miller, R. C.
      (1964). J. Anim. Sci. 23, 121.

66.   Allen, C. E., Bray, R. W. and Cassens, R. G. (1967). J. Fd
      Sci. 32, 26.

67.   Dahl, O. (1958). Acta. Agric. Scand. Suppl. 3.

68.   Rudman, D., Brown, S. J. and Malkin, M. F. (1963).
      Endocrinology 72, 527.

69.   Rudman, D., Brown, S. J. and Malkin, M. F. (1963).
      Endocrinology 72, 527.

70.   Cunningham, H. M., Friend, D. W. and Nicholson, J. W. G.
      (1963). J. Anim. Sci. 22, 632.

71.   Wood, J. D. (1974). Proc. Nutr. Soc. 33, 61A.

72.   Standal, N., Vold, E., Trygstad, O. and Foss, I. (1973).
      Anim. Prod. 16, 37.

73.   Bauman, D. E. and Davis, C. L. (1974). In: "Fourth Inter-
      national Symposium on Ruminant Physiology", Sydney - In Press.

74.   Milligan, L. P. (1971). Fedn. Proc. 30, 1454.

75.   Baldwin, R. L. (1970). Fedn. Proc. 29; 1277.

Endocrine Regulation

HORMONAL CONTROL OF MUSCLE GROWTH

M. R. Turner and K. A. Munday

Department of Physiology and Biochemistry

The University, Southampton SO9 3TU, U.K.

INTRODUCTION

The endocrine system influences muscle deposition not only by
the direct action of hormones on muscle tissue, but also indirectly,
by regulating voluntary food intake, and the subsequent distribution of nutrients between the tissues of the body in the fed and
fasting state. There are complex interactions between hormones
both in the regulation of hormone secretion, and in the way they
achieve their net effect on the target tissues. It is the importance and nature of some of these interactions which form the
major part of the ensuing discussion.

Endocrine function in the very young is not always the same as
in the post-weaning animal, and it will be shown that the role of
individual hormones in determining growth and nutrient distribution
may vary at different stages of development. Finally comment is
made on changes which can be induced in the development of
endocrine function in the young animal by adverse nutritional
influences during intra-uterine and neonatal life.

NUTRIENT INTAKE AND DISTRIBUTION

Voluntary food intake is regulated by the hypothalamus in
response to sensory information about meal size and composition,
and in the long term, in relation to changes in body weight. The
precision of the regulation of nutrient intake varies between
species being very poor in man, and quite precise in rodents. The
mechanisms involved have not been fully elucidated, but it is
becoming increasingly clear that hormones are an important
component of both the short term regulation of food intake, and

197

the long term regulation of body weight. The secretion of many
hormones is also controlled by the hypothalamus where the releasing
hormones are produced and feedback effects are exerted. Even
hormones which may be secreted independently of hypothalamic
control, e.g. insulin, glucagon and adrenaline, have direct actions
on the hypothalamus, which differs in its metabolic properties from
the brain as a whole. Growth hormone (which stimulates appetite)
and insulin (which inhibits food intake) are of particular importance
in this context. The satiety effect of glucose results from an
increase in the rate of glucose utilisation in the hypothalamus,
and for this to occur insulin is required. Thus endogenous insulin
secretion may be regarded as a more important component of the
glucostat mechanism for appetite control than glucose itself.
Furthermore, in preliminary experiments we have carried out in
rabbits fed amino acid-imbalanced diets the reduction in voluntary
food intake which occurred was associated with changes in both the
basal insulin level, and also the pattern of insulin secretion.
It is possible that the satiety effect of dietary protein also
results, in part, from the stimulation of insulin secretion by
amino acids. Therefore, insulin could be of particular importance
in the normal regulation of food intake.

The regulation of body weight may also be a hormone-mediated
process. It has been postulated that the plasma activity of a
steroid hormone, progesterone has been suggested, will vary
inversely in proportion to the body fat content, and that the
hypothalamus, by monitoring changes in steroid hormone activity,
can adjust the food intake pattern appropriately (1). The known
relationship between body fat and normal sexual function discussed
in a later chapter by Frisch is of interest in this context.

The distribution of nutrients between tissues in the fed and
fasting state is controlled by the integrated action of the
endocrine system. In meal eating, non-ruminant animals, feeding
results in a transient increase in the secretion of insulin, and a
suppression by glucose of the release of growth hormone, cortisol
and adrenaline. Whether or not glucagon secretion is also inhibited
depends on the ratio of glucose to amino acid in the meal (see
below). This insulin dominated situation favours the storage of
dietary carbohydrate and lipid as adipose tissue triglyceride, and
the movement of amino acids into the intracellular compartment of
muscle. As the rate of glucose utilisation in the hypothalamus
begins to fall, growth hormone secretion is resumed, and with both
growth hormone and insulin levels raised, protein synthesis in
muscle is stimulated optimally.

Between meals, plasma insulin falls to its basal level, but
growth hormone secretion continues, and the release of adrenaline,
cortisol and glucagon is stimulated. This endocrine equilibrium
favours the catabolic processes, glycogenolysis stimulated by

adrenaline and glucagon, lipolysis stimulated by growth hormone and adrenaline, proteolysis stimulated by cortisol and gluconeo-genesis stimulated by glucagon, cortisol and adrenaline. All of these catabolic processes would be inhibited by insulin and are 'permitted' between meals by insulin being at its basal level. The flow of energy substrates between tissues during a longer fast, and the way the body minimises the utilisation of body protein as a fuel is discussed later in this volume by Cahill. The situation in ruminant animals, which are not normally subjected to feeding: fasting cycles, is also discussed in an earlier chapter by Armstrong.

GROWTH AND MUSCLE DEPOSITION

For growth to occur, an adequate intake of appropriate nutrients is required. Growth is a composite of hyperplasia (cell division) and hypertrophy (cellular enlargement) which may occur separately or together. Early growth, for example in the fetus, is dominated by cellular hyperplasia and later post-natal growth by cellular hypertrophy. In many tissues, the hyperplastic phase ceases before or fairly soon after birth, and after that time any change in tissue size reflects hypertrophy or atrophy of pre-existing cells. When suboptimal feeding occurs during the hyper-plastic phase, the tissue cell number will be permanently reduced, and subsequent tissue growth may be limited however good the subsequent nutritional environment (2). Conversely, early over-feeding may result in an excessive hyperplasia in the adipose tissue, which results later in life in excessive adiposity (3). So far as muscle is concerned, the number of muscle fibres is determined prior to birth, but it is difficult to identify precisely when the hyperplastic phase ceases in muscle. The concept dis-cussed by Burleigh (this volume p.119) of satellite cell nuclei moving into the intracellular compartment and increasing the potential for the hypertrophy of muscle fibres is an interesting one. However, whatever the fundamental mechanisms involved, their modification by intra-uterine and neonatal nutrition may be as important in determining growth potential in muscle as it has been shown to be in the brain, adipose tissue, and the endocrine system (2).

The hormonal control of foetal growth is poorly understood, although it is known not to be dependent on foetal pituitary hormones. Postnatally, on the other hand, growth hormone is dominant in the stimulation of both cell division and cellular enlargement. The role of hormones, both in the adult and in the immature animal, in stimulating cellular protein synthesis, an essential prerequisite for cell division and cellular hypertrophy, is discussed in the following sections.

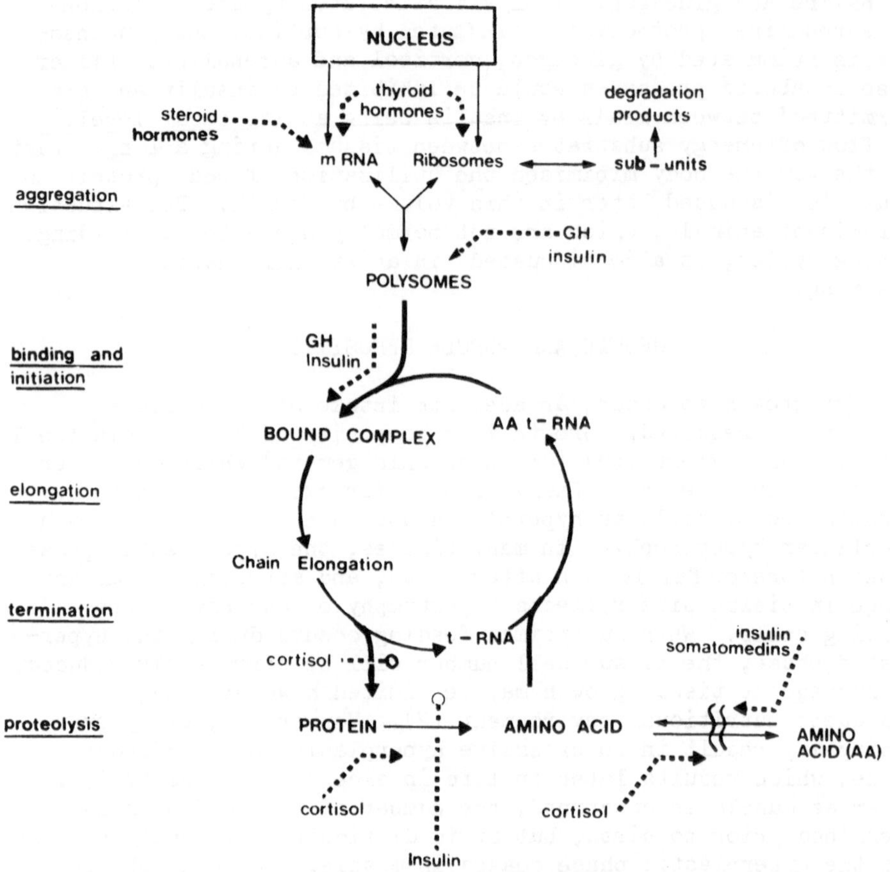

NUCLEUS

steroid .... thyroid .... degradation
hormones hormones products

m RNA   Ribosomes ←——→ sub–units

aggregation

........GH
insulin

POLYSOMES

binding and   GH
initiation   Insulin

BOUND COMPLEX   AA t–RNA

elongation

Chain Elongation

termination   insulin
                                              somatomedins
                    t – RNA
cortisol ....

proteolysis   PROTEIN ——→ AMINO ACID ⇄ AMINO
                                                ACID (AA)

cortisol   cortisol

Insulin

Fig. 1.   Scheme for the control of muscle protein metabolism.

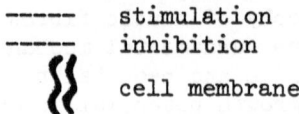

------   stimulation
------   inhibition
          cell membrane

PROTEIN METABOLISM

A simplified scheme for cellular protein metabolism is shown
in Fig. 1. In mammalian systems, the net rate of protein synthesis
is controlled by hormones mainly by variations in the rate of
translation of pre–existing mRNA, and the rate of protein catabolism
(4). The rate of translation is determined by the extent of
aggregation of ribosomes into polysome units, the binding of amino
acids to t–RNA and of amino acyl t–RNA to the polysome units and by

the number of active ribosomes and/or the activity of individual
ribosomes.

Whether or not there is any control exerted at the ribosomal
level over the nature of the protein synthesised is still contro-
versial. On the other hand, hormonal influences on the trans-
cription process may have little effect on the overall rate of
protein synthesis, but by varying the profile of the mRNA released
from the nucleus, specific actions on metabolism may be effected.
Hormones which enter the cell, e.g. steroid and thyroid hormones,
act at the nuclear level and may determine the nature of the
proteins synthesised, whereas hormones which bind to the cell
membrane e.g. growth hormone and insulin, act through their
secondary messenger systems to stimulate the translation process
and hence affect primarily the rate of protein synthesis. A
further point of control is the intracellular concentration of
amino acid, the substrate for the protein synthetic process.
Insulin and probably the somatomedins (see below) stimulate amino
acid uptake in the post-weaning animal by a direct action on the
cell membrane, whereas one consequence of the intracellular action
of cortisol is a net efflux of amino acids from the cell. The
effect of hormones on protein metabolism has been reviewed (5,6).

ACTIONS AND INTERACTIONS OF SOME HORMONES

The endocrine system functions as an integrated unit, the
secretion of most hormones being modulated by other hormones
(Fig. 2). Similarly, the net effect of a hormone on its target
tissues is determined by its interrelationship with other hormones,
and it is a change in the total endocrine equilibrium, not the
effect of any single hormone, which determines the shifts in the
metabolic behaviour of tissues.

Insulin

The effects of insulin on protein synthesis are analogous to
those exerted on triglyceride synthesis in adipose tissue and
muscle, and on glycogen synthesis in muscle, namely, stimulation of
the uptake of substrate into the intracellular compartment,
stimulation of the synthetic pathway, and inhibition of catabolism
(7). The liver differs from muscle and adipose tissue in that
insulin is not required for the uptake of glucose and amino acid.

Insulin stimulates amino acid uptake in muscle consequent
upon an interaction of the hormone with the cell membrane.
Individual amino acids are taken up into the intracellular compartment approximately in proportion to the amino acid composition
of the muscle protein, rather than in proportion to the amino acid
in the extracellular compartment. Even so, in the adult the
uptake effect is not a consequence of de novo protein synthesis in

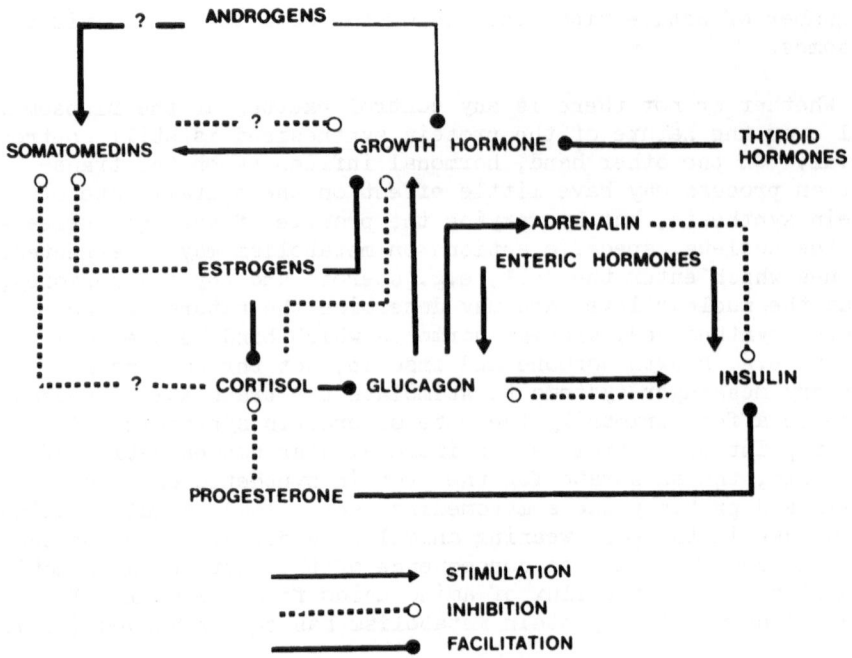

Fig. 2.   Inter-relationships in the secretion of hormones. The
facilitative effect of androgens on growth hormone secretion is
slight compared with that of estrogens and thyroid hormones.
Speculation by the authors is indicated by - ? -

that it is not inhibited by cycloheximide, puromycin or actino-
mycin D. The uptake effects of insulin are reviewed in detail
elsewhere (8,9).

    Insulin also stimulates the translation process, independently
of any effect on amino acid uptake, an action probably mediated by
the inhibition of adenyl cyclase and possibly by the stimulation of
guanosyl cyclase (10). The action of the hormone is to stimulate
ribosomal aggregation, the binding of amino acids to t-RNA, the
binding of amino acyl t-RNA to ribosomes, and an increase in the
number of active ribosomes and/or the activity of individual
ribosomes. This subject has been reviewed recently (11). Although
the action of the hormone on the translation process in muscle is
not dependent on amino acid uptake, blocking the movement of
extracellular amino acid into the cell reduces substantially the
protein synthetic action of the hormone. It has been suggested
that insulin stimulates preferentially the movement of extracellu-

lar, rather than intracellular, amino acid into muscle protein
(12), a view which is in keeping with the fact of insulin being
secreted at times of feeding, but which has not been widely
supported.

In considering the action of any hormone, we must also
consider the circumstances under which its secretion occurs.
Insulin secretion is stimulated by feeding but for a relatively
short period. The major role of the hormone in the body may be
regarded as the storage of ingested nutrients, which is achieved
by stimulating the cellular uptake of glucose, amino acids and
lipids in insulin sensitive tissues and by stimulating lipogenesis,
glycogenesis and protein synthesis. The lipogenic effect of the
hormone is particularly potent.

The secretion of many hormones including insulin is regulated
by neural, hormonal and nutrient stimuli. A scheme for the
control of insulin secretion is shown in Fig. 3. Possible
mechanisms involved are discussed elsewhere (13).

The ingestion of a mixed meal is associated with the release
from the gastro-intestinal tract of the enteric hormones gastrin,
cholecystokinin-pancreozymin (CCK), secretin and enteric glucagon
all of which stimulate insulin secretion. With the absorption of
digested nutrients, insulin secretion is further stimulated by
glucose, and by some amino acids notably leucine, and the basic
amino acids arginine and lysine.

Enteric hormones, particularly CCK and the basic amino acids,
but not leucine, also stimulate the secretion of glucagon, a
hormone which stimulates the release from the liver of glucose,
derived both from the breakdown of glycogen and from the conversion
of amino acid into glucose via the gluconeogenic pathway. Insulin,
in the presence of glucose, is a powerful inhibitor both of
glucagon secretion and of the gluconeogenic pathway. Therefore,
the balance between dietary glucose and amino acids is important
in determining the proportion of insulin to glucagon secreted,
and hence the extent of the 'wastage' of ingested amino acid as a
substrate for glucose biosynthesis (14). The secretion of insulin
is strongly inhibited by adrenaline, so during exercise or stress,
both of which are strong stimuli to adrenaline secretion, insulin
release following a meal would be minimal and dietary amino acid
would be diverted away from protein synthesis.

Growth Hormone

For many years there has been considerable confusion about
the actions of growth hormone (GH). The confusion has arisen
from the extrapolation of data obtained in vivo and in vitro using
hypophysectomised animals, to the intact non-hypophysectomised

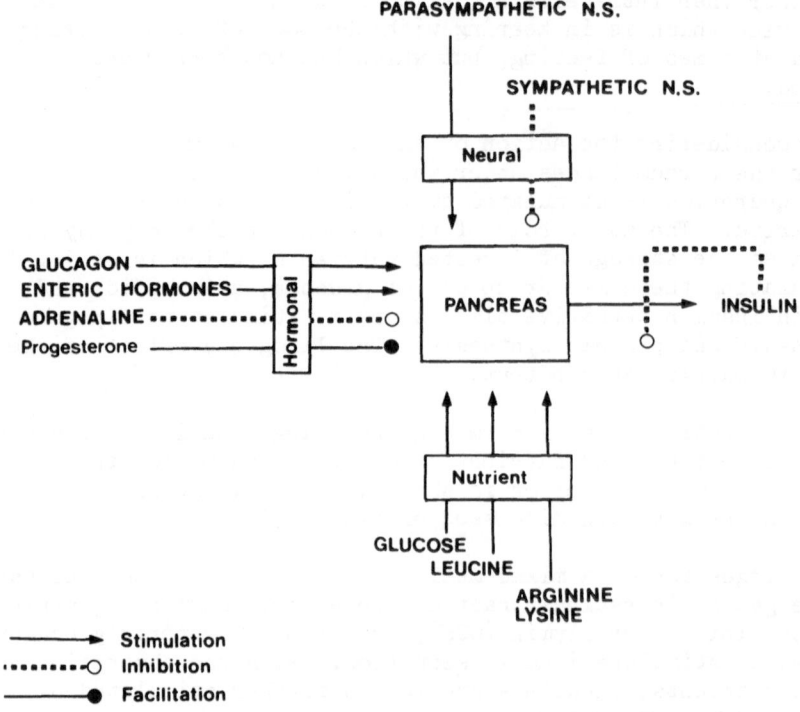

Fig. 3.   Neural, nutrient and hormonal factors which regulate insulin secretion.   Growth hormone may also facilitate insulin release (27).

animal. This classical approach of the endocrinologist has been invaluable in identifying the site of action of many hormones, but gives less information about the effects of hormone secreted in response to NORMAL physiological stimuli. In the case of GH, a false and misleading concept of the action of the hormone has been inadvertently perpetrated in this way. On the basis of in vitro experiments using tissues from hypophysectomised animals, it has become widely accepted that GH stimulates the transport of both amino acids and glucose into cells, as well as stimulating the incorporation of amino acids into protein, and GH has therefore been thought to mimic the action of insulin. This misconception has been compounded further by the insulin-like effects of GH administered to the whole animal, but as we shall see below the in vivo insulin-like actions of GH are mediated by the somatomedins, which are released from the liver under suitable circumstances by the action of GH. On the other hand the direct actions of GH on muscle, liver, and adipose tissue, are not at all insulin-like. We propose that the extent to which GH exerts direct or indirect effects on tissues is modulated by other hormones (Fig. 4), and

that the role of GH in the body varies accordingly from the
stimulation of growth (direct and indirect actions) to the protection of the body protein during a fast, exercise or stress by
stimulating the maximal utilisation of lipid as an energy substrate,
and by maintaining essential protein synthesis in the face of an
ever-decreasing concentration of intracellular amino acid (direct
actions) (15).

We have demonstrated the direct actions of GH in muscle in
vitro from non-hypophysectomised animals using in the incubation
medium concentrations of GH as low as 0.1µg/ml which is only five
to ten times the usual circulating concentration of GH (much of
the in vitro work using tissues from hypophysectomised animals has
been carried out using much higher concentrations of GH). We have
shown that GH stimulates the incorporation into protein of all five
of the amino acids studied, but without having any effect on the
rate of amino acid uptake (16). As a result of the direct action
of GH, the tissue:medium distribution ratio of isotopically
labelled amino acid fell, demonstrating that GH can stimulate
protein synthesis even in the face of a decreasing concentration
of substrate amino acid. Data obtained with GH (Table 1) are
contrasted with those using insulin (Table 2). Similar observations
with GH have been made in liver (17,18).

The uptake effects of GH observed in muscle and adipose
tissue from hypophysectomised rats is transitory, and is prevented
by pre-treatment of the animal with GH, or pre-incubation of the
tissue with GH. Furthermore, the effect is blocked by the
inhibitors of protein synthesis cycloheximide and puromycin and
is dependent therefore on de novo protein synthesis (19). It
seems probable that this uptake effect is a consequence of protein
depletion following hypophysectomy, and that the presence of GH
in the incubation medium initiates protein synthetic activity which
restores the transport functions of the membrane, and permits an
initial movement of extracellular amino acid into the intracellular
compartment, until a normal equilibrium is established. Our view
that this transport effect is a function of protein depletion is
further supported by our observations in vitro, using muscle from
rabbits fed protein deficient diets (20,21). In such tissue GH
as well as stimulating protein synthesis has an uptake effect
which is blocked by cycloheximide (21), just as has been shown in
muscle from hypophysectomised animals.

Therefore in considering the direct actions of GH on muscle
protein synthesis it is necessary to extrapolate to the whole
animal from the data obtained using tissues from non-hypophysectomised animals. These actions clearly do not resemble those of
insulin.

The administration of GH to the whole animal results in both

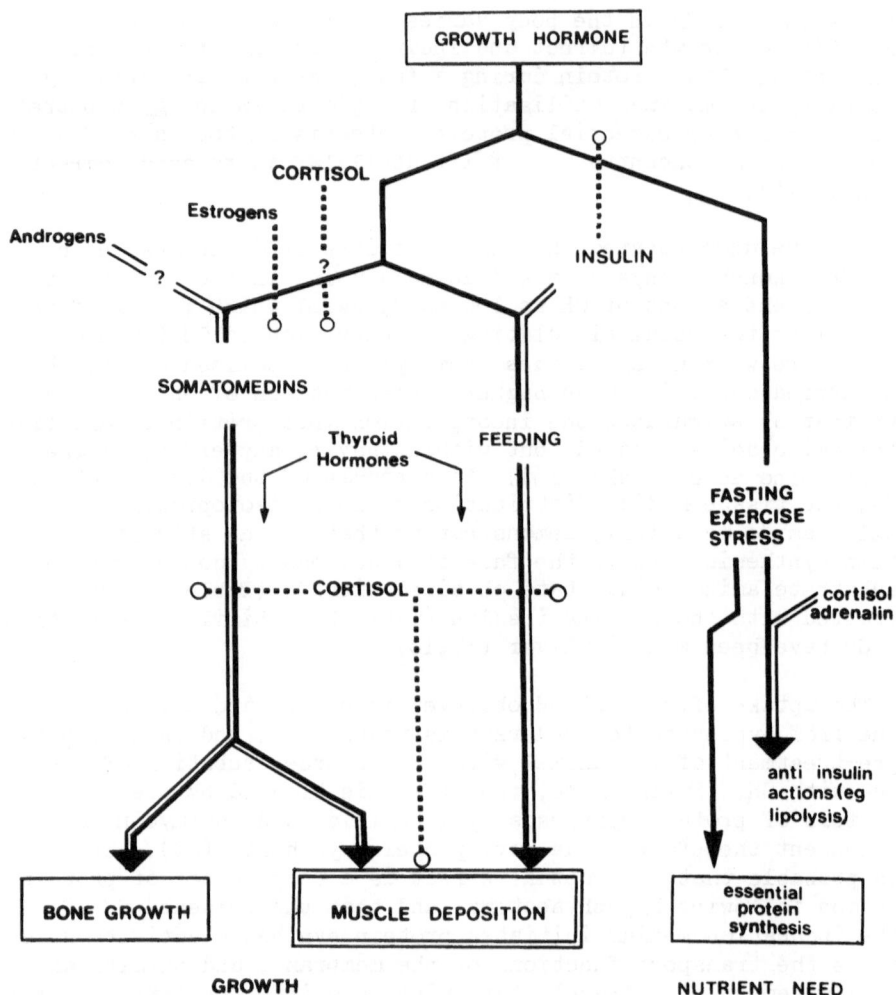

Fig. 4.   Actions and interactions of growth hormone with other hormones, in the control of muscle deposition. Other actions of growth hormone are also indicated. Speculation by the authors is indicated - ? -

protein anabolic and lipolytic effects. In adipose tissue in vitro from hypophysectomised animals it has been shown that, GH in the presence of glucocorticoid hormone, stimulates lipolysis, an effect which is dependent on an initial stimulation of protein synthesis, and may be assumed to result from the synthesis of triglyceride lipase (22). Again this direct action of GH is not insulin-like; it is an effect opposite to that achieved by insulin,

Table 1

Effect of growth hormone (0.5µg/ml) on the net uptake
(n mol/h per g fresh muscle), the incorporation into
protein of amino acids (n mol/h per g fresh muscle),
and on the tissue: medium isotope distribution ratio
(T:M) in muscle, in vitro, from non-hypophysectomised
rabbits. Figures are the mean difference from the
basal value $\pm$ SEM for 5 observations per group. Full
experimental details are published (16).

Significance of differences: [*]P<0.05, [**]P<0.01, [***]P<0.001.

	Net Uptake	Incorporation	T:M
LEUCINE	- 7.2 $\pm$ 9.7	[***] 15.4 $\pm$ 0.9	[**] - 0.71 $\pm$ 0.19
VALINE	- 0.6 $\pm$ 6.7	[**] 17.6 $\pm$ 2.4	[*] - 0.52 $\pm$ 0.15
ARGININE	2.5 $\pm$ 6.4	[**] 11.0 $\pm$ 1.8	[**] - 0.32 $\pm$ 0.15
LYSINE	2.8 $\pm$ 2.3	[**] 12.6 $\pm$ 2.5	[**] - 0.52 $\pm$ 0.10
HISTIDINE	- 0.8 $\pm$ 2.9	[*] 5.0 $\pm$ 1.4	[**] - 0.48 $\pm$ 0.10

a fact which needs to be recognised in seeking to understand the
role of GH in the body.

The direct anti-insulin actions of GH on adipose tissue,
namely lipolysis and inhibition of glucose uptake, and the
direct actions of GH on muscle protein metabolism in the absence
of an increase in insulin secretion (described above), are
appropriate for the protection of body protein during a fast,
exercise and stress, in that they prevent the excessive use of
amino acid as a substrate for the generation of glucose and
metabolic energy. As we shall see below, the major growth
promoting actions of GH are mediated indirectly by the somatomedins
released from the liver under the influence of GH. Nevertheless,
when insulin and GH secretion occur simultaneously after feeding,
the direct action of GH on muscle is truly anabolic, and there is,
therefore, an important interaction between GH and insulin in the
stimulation of muscle protein synthesis.

In pancreatectomised animals, the protein synthetic action
of GH in muscle is minimal (23), possibly because of a breakdown

Table 2

Effect of insulin (0.01 unit/ml) on the net uptake
(n mol/h per g fresh muscle), the incorporation into
protein of amino acids (n mol/h per g fresh muscle),
and on the tissue: medium isotope distribution ratio
(T:M) in muscle, in vitro, from non-hypophysectomised
rabbits. Figures are the mean difference from the
basal value ± SEM for 5 observations per group.
Full experimental details are published (16).

Significance of difference: *P<0.05, **P<0.01.

	Net Uptake	Incorporation	T:M
LEUCINE	*70.4 ± 19.1	*21.6 ± 9.4	*1.90 ± 0.66
VALINE	*31.4 ± 7.8	*16.6 ± 5.2	*0.88 ± 0.20
ARGININE	*45.9 ± 8.0	*21.9 ± 6.7	*0.90 ± 0.27
LYSINE	*20.1 ± 4.9	*8.6 ± 2.3	**0.50 ± 0.10
HISTIDINE	*12.2 ± 2.9	*6.5 ± 1.9	**0.93 ± 0.18

in the protein synthetic machinery, excessive proteolysis, and
a lack of intracellular substrate amino acid, and the effect of
GH in the absence of insulin is mainly the stimulation of
lipolysis (24). The administration of insulin to the animal
restores normal GH function. We have made observations in
vitro which confirm that GH is effective in stimulating protein
synthesis in muscle only when it has been exposed previously to
at least basal amounts of insulin. In muscle in vitro from non-
hypophysectomised alloxan-diabetic rabbits, the stimulation by GH
of the incorporation of 14C-leucine into protein was minimal
(+ 0.7nmol/h per g fresh muscle). However, when the tissue was
incubated with both insulin and GH, protein synthesis was
stimulated to an extent similar to that achieved by GH in control
animals (+ 14.5nmol/h per g fresh muscle), although insulin
alone had only a small stimulatory effect. Therefore, whereas
insulin itself will stimulate muscle protein synthesis during
feeding, this event may occur only once or twice a day in the
non-ruminant animal, so the ability of insulin to stimulate the
movement of extracellular amino acid into the intracellular
compartment, and to sustain the protein synthetic activity of

GH should be regarded as being more important actions of insulin on muscle.

Just as GH requires some insulin for its protein anabolic effect, so insulin is dependent on GH for its protein synthetic action. In hypophysectomised animals, in which the lack of GH and thyroid hormones would result in a deterioration in the protein synthetic machinery, the action of insulin is mainly the stimulation of lipogenesis (25). The administration of GH to hypophysectomised animals restores the protein synthetic action of insulin. The action of insulin on the uptake of glucose, on the other hand, is not impaired by a lack of GH, hypophysectomised animals being particularly sensitive to insulin in this respect, but hypophysectomy does not enhance the stimulation by insulin of amino acid uptake. The concept of GH stimulating the synthesis of a protein which inhibits membrane transport, as part of the normal system for the regulation of glucose uptake, has been proposed (26), and in the hypophysectomised animal the loss of such an inhibitor could facilitate the effect of insulin on glucose uptake.

Therefore, even in the simplest terms, it can be seen that insulin and GH are mutually dependent for their protein synthetic actions in muscle. When the concentration of both hormones is elevated, there is some synergism in their action on protein synthesis. This is true not only in the whole animal, but also in muscle in vitro, as we have shown using tissue from normal rabbits. In these experiments the protein synthetic action of insulin together with GH tended to be greater than the sum of the effects of insulin and GH separately (Table 3).

Growth hormone secretion, like insulin secretion, is controlled by neural, hormonal and nutrient factors. Whereas insulin is regulated mainly by nutrient and hormonal stimuli, GH release is influenced most strongly by neural stimulation (27) during exercise, stress and non-rapid eye movement (non-REM) sleep. In addition, glucose in the presence of insulin, is a powerful inhibitor of exercise and stress induced secretion, and is important in modulating the pattern of GH secretion during feeding, the effect being mediated by an increase in the glucose utilisation rate in the hypothalamus. As glucose utilisation falls from the elevated rate achieved during absorption, the inhibition of GH secretion is released, and the stimulatory actions of some amino acids such as arginine, lysine and histidine, and of glucagon are able to be expressed. A scheme showing the main factors regulating GH secretion is given in Fig. 5.

In addition to the regulatory influences depicted, there is a diuranal rhythm in the release of GH, secretion during the night being substantially more than that which occurs during the day.

Table 3

Effect of growth hormone (0.1μg/ml), insulin (0.01 unit/
ml) and of growth hormone plus insulin on the net uptake
(n mol/h per g fresh muscle), and the incorporation into
protein of valine (n mol/h per g fresh muscle), and on
the tissue:medium isotope distribution ratio (T:M) in
muscle, in vitro, from non-hypophysectomised rabbits.
Figures are the mean differences from the basal value ±
SEM for 5 observations per group.
Full experimental details are published (16).

Significance of differences: * P < 0.05, ** P < 0.01, *** P < 0.001.

	Net Uptake	Incorporation	T:M
Growth hormone	− 1.4 ± 1.2	**5.7 ± 0.9	− 0.23 ± 0.10
Insulin	**19.6 ± 2.6	*5.0 ± 1.7	*0.29 ± 0.08
Growth hormone and insulin	*22.6 ± 6.0	***14.0 ± 1.0	**0.22 ± 0.05

Children secrete more GH than adults, especially at night, and
it is thought that growth hormone secretion occurring during
sleep may be of particular importance in the stimulation of
growth (28). During the day, GH also has a protein anabolic
effect when insulin secretion occurs simultaneously, but
exercise and fasting are more potent stimuli to GH secretion
than food, and the main effect of the hormone during the waking
hours will be the stimulation of lipolysis.

The control of the secretion of GH is complex, and the
action of the hormone is directed to growth, including muscle
growth, or to lipolysis, by the influence of other hormones.

Somatomedins

The insulin-like activity, and much of the growth promoting
effect of growth hormone is mediated by the somatomedins, a
family of hormones produced in the liver under the influence of
GH. The properties of GH now attributed to somatomedins were
described previously in terms of observed actions of GH in the
hypophysectomised animal, or in isolated tissues from hypophysectomised animals, as 'thymidine factor activity' 'sulphation

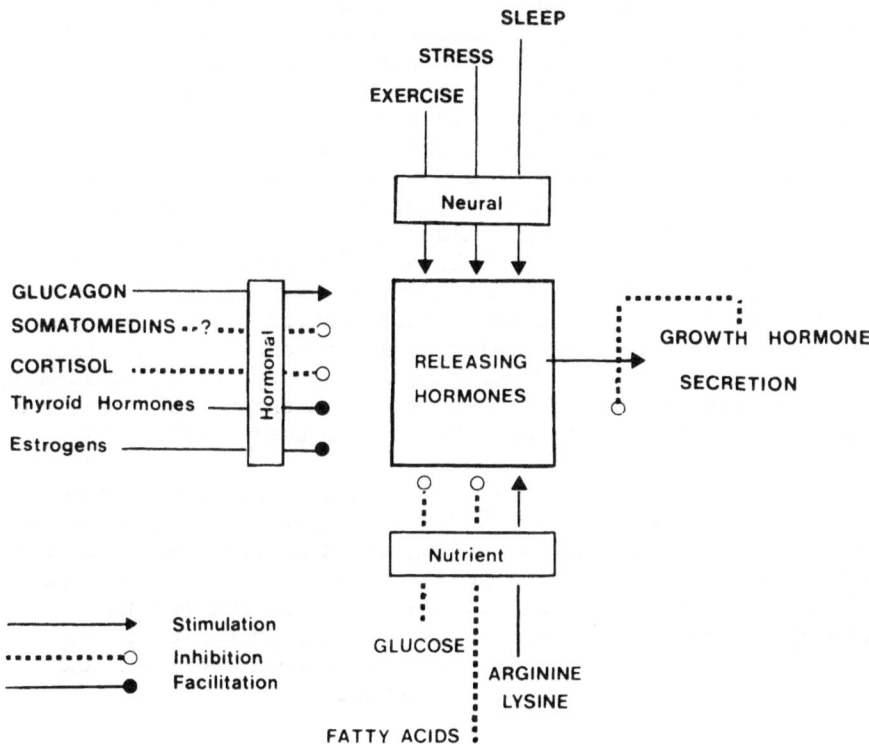

Fig. 5.   Neural, nutrient and hormonal factors which regulate
growth hormone secretion. Speculation by the authors is
indicated by - ? -

factor activity' and 'non-suppressible insulin-like activity'.
The history and current state of knowledge of the somatomedins is
well reviewed (29,30,31).

     Three somatomedins have been identified so far, but until
they have been isolated, it will not be possible to define precisely their individual metabolic actions. Nevertheless it appears
at this stage, that one of the somatomedins mediates the
important mitogenic action of GH whilst the other two have
insulin-like properties. The somatomedins, which in the ensuing
discussions will be regarded as a single entity, have a potent
effect on bones in which they stimulate all aspects of growth.
Investigations in muscle, in vitro, unfortunately from hypophysectomised rats, have indicated that somatomedins are more potent in
stimulating protein synthesis in muscle than GH itself (32).
However, until investigations have been made in tissues from non-
hypophsectomised animals the role of somatomedins in muscle and
adipose tissue metabolism cannot be evaluated.

The release of somatomedins by the liver is not dependent
solely on the circulating GH concentration. It has been shown
that estrogens inhibit somatomedin release (33), and if somatomedins
rather than GH itself are responsible for the feedback inhibition
of GH release, the co-existence in the female of enhanced GH
secretion and slower bone growth than in the male, would be
explained. This observation also stresses the importance of
somatomedins in growth. This is further emphasised by an
observation in protein-energy malnourished infants who do not
grow despite the circulating GH level being increased approximately three-fold. It has been shown recently in these infants
that the circulating somatomedin concentration is low, but that
on re-feeding the GH level falls, the somatomedin level rises and
growth recommences (34). Nevertheless, by suppressing somatomedin
production with estrogens, it has also been shown that not all of
the growth effect of GH is mediated by the somatomedins (33).

In view of the known action of somatomedins on bone growth,
the effect of androgens pre-puberty being predominantly on bone
growth, and the dependence of androgens on GH for their anabolic
effects, it is tempting to propose that androgens exert an action
on the liver opposite to that of the estrogens and stimulate
somatomedin production. Whether or not this proves to be the
case it is likely that some interaction between androgens and
somatomedins occurs during normal growth.

The interaction between insulin and GH in the fed state is
particularly favourable for protein synthesis and it is possible
that after the transitory rise in plasma insulin concentration
resulting from feeding, the somatomedins could substitute for
insulin in promoting the movement of extracellular amino acid into
the intracellular compartment. In the longer fast on the other
hand, the insulin-like properties of somatomedins would conflict
with the action of GH in stimulating lipolysis, and it is probable
that somatomedin release is inhibited in the fasted state by yet
another hormone. In view of the implications for growth of the
diurnal rhythm in cortisol secretion (discussed below), the fact
that the liver is a known target tissue for cortisol, and that the
only hormone known to modulate GH-stimulated somatomedin release
is a steroid, we feel that cortisol is a likely candidate. However, this is speculative and a thorough investigation of the
physiology of somatomedin release is clearly necessary.

Cortisol

The stimulation of muscle deposition by the endocrine system
is dominated by growth hormone, but its effects are achieved only
in the presence of at least basal circulating amounts of insulin
and thyroid hormones and in the absence of excessive cortisol

activity. Cortisol has a catabolic effect on muscle protein
metabolism, probably by inhibiting protein anabolism, but possibly
by stimulating the catabolic process as well. Whatever the
mechanisms involved, the consequence of the action of cortisol is
a breakdown of muscle protein, a net efflux of amino acid from
muscle tissue, an increased rate of utilisation of amino acid as
an energy substrate in muscle, and an increased rate of conversion
of amino acid to glucose in liver. These effects are all rational
when considered in the light of the stimulation of cortisol
secretion, which occurs in response to fasting, exercise and
stress, all three situations requiring an increase in the availability of energy substrates. However, the effect of cortisol is
even more subtle. It appears to divert the action of GH away from
muscle protein synthesis and yet favour the protein synthetic
action of GH in adipose tissue, the result of which is an increased
rate of lipolysis. Thus the action of insulin, even when the
concentration is only minimally increased above the basal value,
favours a protein synthetic action of GH in muscle, and the action
of cortisol favours the lipolytic action of GH in adipose tissue,
a process which would be inhibited by insulin. The balance
between insulin and cortisol may therefore be critical in determining the effectiveness of GH in promoting muscle growth. The
circumstances under which cortisol is secreted, namely fasting,
exercise and stress, would all result in an inhibition of insulin
secretion, and this is a further refinement in the interaction
between these hormones.

    The view has been expressed that growth occurs mainly during
sleep (28). To discuss this view, it is necessary to examine the
endocrine equilibrium during sleep. The secretion of GH is
stimulated during non-REM sleep, the amount of GH released at
night being substantially more than that secreted during the
waking hours. Insulin secretion, on the other hand, is minimal
during the long overnight fast and insulin is not able to interact
with GH at this time to favour the stimulation of muscle protein
synthesis by GH. Under circumstances of increased GH secretion
and decreased insulin secretion, the possibility of GH being
lipolytic rather than protein anabolic is open. However, during
sleep this appears not to be the case (28), possibly because the
diurnal rhythm in the secretion of cortisol is such that the
circulating concentration at night is minimal at the time when GH
levels are elevated, and just as a lack of insulin may permit the
lipolytic actions of GH and adrenaline, the lack of cortisol may
permit the protein anabolic action of GH in muscle, despite the
lack of insulin. It would be interesting to know about the
synthesis of somatomedins during sleep especially if, as we have
speculated, cortisol is one of the hormones which modulates
somatomedin production. The possibility has still to be tested
that during sleep, the growth promoting action of GH is largely
mediated by the somatomedins.

DIETARY EFFECTS ON THE SECRETION AND ACTION OF HORMONES

In addition to the specific effects on hormone secretion of some nutrients, notably glucose, leucine, basic amino acids, and alanine, the composition of the diet in the longer term alters both the secretory capacity of some endocrine glands, and the hormone:tissue interaction. For example low carbohydrate diets reduce the insulin secretory capacity of the β cells to a glucose challenge (35) and reduce the rate of disappearance of glucose from plasma following insulin injection (36). Similarly, eating low protein diets results in a reduction in insulin secretion (37,38) and in changes in tissue responses to insulin (39-41). Food restriction affects insulin function in a similar way.

Both food restriction and low protein diets reduce the ability of the pituitary to secrete GH in response to nutrient stimuli (15,37,42) and decrease the responsiveness of muscle to the protein synthetic action of the hormone (21,39). From a practical point of view, it is the permanent changes in endocrine function produced by small aberrations in maternal and neonatal nutrition which need emphasising. Thus feeding rabbits during pregnancy on diets containing 10% of soya bean protein has been shown by us to cause drastic changes in the secretion and actions of insulin and GH in the newborn offspring, despite there being no reduction in total food intake during gestation (42,43). Maternal food restriction has a similar effect. The extent to which these changes in endocrine function are reversible in the neonatal period remains to be established, but the well-known studies of Widdowson and McCance (44) have demonstrated the principle that when sub-optimal nutrition occurs sufficiently early in life, the changes induced are irreversible. This effect of sub-optimal nutrition has been demonstrated in terms of growth, food conversion efficiency, growth of adipose tissue, development of the brain, and development of pituitary function (2) and it is an urgent requirement that the effect of maternal and neonatal nutrition on endocrine and metabolic development be investigated fully, not only from the point of view of growth potential and food conversion efficiency, but also in terms of the subsequent development of normal reproductive ability, a physiological process dependent on the interaction of many hormones, and known to be affected by diet.

ENDOCRINE FUNCTION IN THE IMMATURE ANIMAL

The hormonal actions and interactions described so far occur in the post-weaning animal, but may not do so in the fetus and neonate. There is limited data on the very young, but what there is indicates important functional differences both in hormone secretion, and in hormone:tissue interactions. The differences in

the very young could be regarded as immaturity in endocrine
function appropriate to each stage of development of the progeny.
Thus, for example, the endocrine and metabolic equilibrium in the
fetus may be more appropriate to the intra-uterine environment
than would an adult endocrine status. To highlight some of these
differences, the secretion and actions of insulin and growth
hormone in the very young will be described and compared with the
adult.

The isolated pancreas from the fetus of both man and the
rabbit is barely responsive to glucose, but both leucine and
glucagon will elicit substantial responses (45,46). These
observations are compatible with the view that insulin in the
fetus relates more to growth and the metabolism of amino acid,
than to glucose, and this is in contrast to the situation in the
adult, in which the main function of insulin is the stimulation
of lipogenesis. However, the metabolic disturbances in pre-
diabetic and poorly controlled diabetic mothers result in hyperplasia of islet tissue in the offspring, which then shows
substantial insulin secretory responses to glucose. At birth such
infants are overweight, excess fat being the major component, but
are also long for their gestational age, and are liable to suffer
from neonatal hypoglycemia consequent upon exaggerated insulin
secretory responses to glucose. So when metabolism is disturbed
in this way, insulin function in the newborn more nearly resembles
that of the adult, yet this creates health problems for the offspring.

We have studied endocrine function in newborn rabbits. As in
fetuses, glucose does not stimulate insulin secretion, but leucine
and glucagon do (42). The basic amino acids, which we have shown
to substitute for glucose in the stimulation of insulin secretion
in the adult, are also ineffective in the newborn (42). We have
shown also that the action of insulin in the newborn differs from
that in older animals. In muscle from newborn rabbits in vitro,
insulin does not stimulate amino acid uptake, but does stimulate
protein synthesis, even at the expense of the intra-cellular amino
acid (39). In terms of adult endocrine function this could be
described as a growth hormone-like action of insulin. Even at 6
weeks of age when the rabbits were weaned, the ability of insulin
to stimulate amino acid uptake in muscle had not developed. On the
other hand, insulin does stimulate glucose uptake in the newborn
as we have shown by injecting insulin and measuring glucose disappearance from the plasma, but by comparison with adults the
newborn may be regarded as being insulin resistant in these terms.
Thus the poor glucose tolerance observed in the very young of
several species results both from a reduced insulin secretion in
response to glucose, and from a reduced effectiveness of insulin
in stimulating the movement of glucose into target tissues.

During postnatal development, the insulin secretory response to glucose develops, and the tissue sensitivity to the uptake effects of insulin increases (40,41). The different nature of the action of insulin on glucose uptake in the very young may be regarded as a protective adaptation in endocrine function in that any tendency to hypoglycemia would be likely to damage severely the developing brain.

In the fetus, GH appears to be unnecessary for growth. Post-natally on the other hand, GH is a dominant factor in controlling both hyperplastic and hypertrophic growth in many tissues, including the skeleton and muscle, but it has not been established how soon after birth the offspring becomes sensitive to the growth promoting actions of GH. In the neonate, GH levels are high, but there is no negative feedback on the hypothalamus (47). It is not known whether somatomedin release occurs in the newborn, but as liver function continues to develop after birth, it is conceivable that the lack of feedback by GH in the newborn is an index of a lack of synthesis by the liver of somatomedins. As we have discussed above, the somatomedins could be an important component of the feedback regulation of GH secretion, and are an important component of the growth promoting action of GH. Our own studies in newborn rabbits show that there is a normal secretion of GH in response to an arginine challenge (42) and that in muscle, in vitro, GH is a potent stimulus to protein synthesis (43). The action of GH expressed as a percentage of the basal value is the same in the newborn as in the adult, but because the basal rate of protein synthesis is so much higher in the newborn than in the adult this represents, in absolute terms, a greater stimulation by GH.

In our studies so far in the newborn animal, therefore, there has been no indication of differences, such as were observed for insulin, in the nature of the secretion of GH, or the direct action of GH on protein metabolism, when compared with the post-weaning animal. The direct lipolytic effect of GH in the newborn has not been studied. We feel that endocrine function in the very young needs to be defined fully, and the physiological significance of differences from the adult evaluated.

REFERENCES

1.   Hervey, G. R. (1971). Proc. Nutr. Soc. 30, 109.

2.   Albanese, A. A. (Ed.) (1973). Nutr. Rep. Int. 7, No. 5.

3.   Lemonnier, D. and Alexiu, A. (1974). In: "The Regulation of Adipose Tissue Mass" (Eds. J. Vague and J. Boyer) p. 158 (Excerpta Medica Foundation, Amsterdam).

4.   Young, V. R. (1974). J. Anim. Sci. $\underline{38}$, 1054.

5.   Trenkle, A. (1974). J. Anim. Sci. $\underline{38}$, 1142.

6.   Manchester, K. L. (1970). In: "Mammalian Protein Metabolism"
      Vol. 4 (Ed. H. N. Munro) p. 229 (Academic Press, New York
      and London).

7.   Cahill, G. F. Jr., Aoki, T. T. and Marliss, E. B. (1972)
      In: "Endocrine Pancreas" (Ed. R. O. Greep and E. B. Astwood)
      p. 563 (American Physiological Association, Washington, D.C.)

8.   Riggs, T. R. (1970). In: "Biochemical Actions of Hormones"
      Vol. 1 (Ed. G. Litwack), p.166 (Academic Press, New York and
      London).

9.   Morgan, H. E. and Neely, J. R. (1972). In: "Endocrine
      Pancreas" (Ed. R. O. Greep and E. G. Astwood), p. 323
      (American Physiological Association, Washington, D.C.

10.   Cuatrecasas, P. (1974). Ann. Rev. Biochem. $\underline{43}$, 169.

11.   Manchester, K. L. (1975). In: "Protein Metabolism and
      Nutrition" - in the press (Butterworth, London).

12.   London, D. R. (1972). Proc. Nutr. Soc. $\underline{31}$, 193.

13.   Lacy, P. E. and Malaisse, W. J. (1973). Rec. Progr. Horm.
      Res. $\underline{29}$, 199.

14.   Le Febre, P. J. and Unger, R. H. (Ed.) (1972). "Glucagon"
      (Pergammon, Oxford).

15.   Turner, M. R. (1972). Proc. Nutr. Soc. $\underline{31}$, 205.

16.   Reeds, P. J., Munday, K. A. and Turner, M. R. (1971).
      Biochem. J. $\underline{125}$, 515.

17.   Clemens, M. J. and Korner, A. (1970). Biochem. J. $\underline{119}$, 629.

18.   Liberti, J., Wood, D. M. and DuVall, C. H. (1972).
      Endocrinol. $\underline{90}$, 311.

19.   Pecile, A. and Müller, E. E. (Ed.) (1968). "Growth Hormone"
      (Excerpta Medica Foundation, Amsterdam).

20.   Turner, M. R., Reeds, P. J. and Munday, K. A. (1971).
      Excerpta Medica Foundation Int. Congr. Ser.

21. Turner, M. R., Reeds, P. J. and Munday, K. A. (1975).
     Brit. J. Nutr. - in the press.

22. Fain, J. N., Dodd, A. and Novak, L. (1971). Metabolism 20,
     109.

23. Milman, A. E., DeMoor, P. and Lukens, F. D. W. (1951).
     Amer. J. Physiol. 166, 354.

24. Sirek, O. V., Hotta, N. and Sirek, A. (1971). In: "Diebetes"
     (Ed. R. R. Rodriguez and J. Vallance - Owen), p. 175
     (Excerpta Medica Foundation, Amsterdam)

25. Wagner, E. M. and Scow, R. O. (1975). Endocrinol. 61, 419.

26. Young, F. G. (1968). In: "Growth Hormone" (Ed. A. Pecile and
     E. E. Muller), p. 139 (Excerpta Medica Foundation, Amsterdam).

27. Merimee, T. J. and Rabin, D. (1973). Metabolism 22, 1235.

28. Hunter, W. M. (1972). Proc. Nutr. Soc. 31, 199.

29. Van Wyk, J. J., Underwood, L. E., Hintz, R. L.,
     Clemmons, D. R., Voina, S. J. and Weaver, R.P. (1974).
     Rec. Progr. Horm. Res. 30, 259.

30. Uthne, K. (1973). Acta Endocr. (Copen.) Suppl. 175.

31. Hall, K. and Luft, R. (1974). In: "Advances in Metabolic
     Regulation" Vol. 7, (Ed. R. Levine and R. Luft), p. 1
     (Academic Press, New York and London).

32. Salmon, W. D. Jr. and DuVall, M. R. (1970). Endocrinol. 87,
     1168.

33. Widemann, E. and Schwartz, E. (1972). Endocrinol. 34, 51.

34. Pimstone, B. L., Becker, D. J. and Hansen, I. D. L. (1973).
     In: "Endocrine Aspects of Malnutrition" (Ed. L. I. Gardner
     and P. Amacher) p. 73 (Kroc Foundation, Santa Ynez,
     California).

35. Hales, C. N. and Randle, P. J. (1973). Lancet 1, 790.

36. Himsworth, H. P. (1935). Clin. Sci. 2, 67.

37. Turner, M. R., Allen, K. A. and Munday, K. A. (1974).
     Proc. Nutr. Soc. 33, 56A.

38. Allen, K. A., Ayres, C. E., Munday, K. A. and Turner, M. R.

(1975)
Proc. Nutr. Soc. - in the press.

39. Allen, K. A., Munday, K. A. and Turner, M. R. (1974).
    Proc. Nutr. Soc. 33, 113A.

40. Heard, C. R. C. and Turner, M. R. (1967). Diabetes 16, 96.

41. Heard, C. R. C. and Henry, P. A. J. (1969). Clin. Sci. 37, 37.

42. Turner, M. R., Allen, K. A. and Munday, K. A. (1974).
    Proc. Nutr. Soc. 33, 38A.

43. Allen, K. A., Munday, K. A. and Turner, M. R. (1974).
    Proc. Nutr. Soc. 33, 112A.

44. Widdowson, E. M. and McCance, R. A. (1960).
    Proc. Roy. Soc. B. 152, 88.

45. Milner, R. D. G. (1969). J. Endocrinol. 44, 267.

46. Milner, R. D. G., Ashworth, M. A. and Barson, A. J. (1972).
    J. Endocrinol. 52, 497.

47. Kaplan, S. L. and Grumbach, M. M. (1972). In: "Growth and
    Growth Hormone" (Ed. A. Pecile and E. E. Muller), p. 382
    (Excerpta Medica Foundation, Amsterdam).

PROTEIN-FAT INTERACTIONS

G. F. Cahill, Jr. and T. T. Aoki

Joslin Research Laboratory, Department of Medicine
Harvard Medical School and the Peter Bent Brigham
Hospital, Boston, Mass. 02215

Most knowledge of the major interrelationships between fat
and protein is derived, unfortunately for those attending this
conference, from experiments in man, dog, and rat (in that order).
The pre-eminence of human experimentation as the major source of
this knowledge is due, apart from the great need for clinically
relevant information, to several factors: man's large blood
volume, the ready availability of advanced techniques for regional
vessel catheterization, subject cooperation, and finally, the
relative uniformity of Homo sapiens, obviating major differences
in size, strain, the lesser effect of many environmental phenomena
such as diet, known to alter animal responses. This brief paper
will therefore be concerned primarily with data derived from man
with reference to experiments which might be relevant to animals
of agricultural value, and which, in turn, might lead to further
insights into the factors leading to variations in adipose tissue
and muscle growth in such animals.

One must always keep in mind the energy requirement of both
terrestrial and aerial fauna and the obvious need for an energy-
dense fuel, particularly where mobility is crucial. Unfortunately,
the accumulation of 1 gram of either glycogen or protein in tissue
necessitates the accumulation of 2-4 grams of water, resulting in
only 1 kcal or less per gram of stored tissue. Parenthetically,
virtually every molecule of protein in man is committed to a
specific and important role, e.g. as an enzyme, as contractile
protein in muscle or as plasma albumin, etc. At present, there is
no known form of storage of nitrogen for storage sake alone. In
contrast to glycogen and protein, triglyceride in adipose tissue
comprises 70-95% of total tissue weight, and therefore yields close
to the theoretical 9 kcal/gram of tissue. Finally, the

inability of animals to convert acetate to pyruvate, or in overall
terms, fat to carbohydrate, due to the irreversible action of
pyruvate dehydrogenase, has important implications. Once a
molecule of acetate is formed, be it from carbohydrate or from
protein, it can be oxidized or incorporated into long chain fatty
acids, but cannot contribute to net carbohydrate or protein
synthesis (Fig. 1). Consequently, long chain fatty acids can only
be stored (as triglyceride) or oxidized to yield energy. Thus an
animal can use protein and its constituent amino acids or carbohydrate and its fundamental unit, glucose, to fill body protein and
carbohydrate stores to an appropriate extent. The excess is converted into long chain fatty acids, esterified with glycerol and
stored in adipose tissue. In other words, adipose tissue is the
caloric buffer (or capacitor) between the animal and its environment, and man seems to have developed this caloric bank extremely
successfully.

A few simple physiologic facts and schemes can be used to
emphasize the aforementioned. In a meal eater who is a non-ruminant,
fuels enter in a pulsatile fashion. In Fig. 2 ingestion of either
a pure carbohydrate or a standard mixed meal is illustrated. The
first priority of the glucose in man is for brain fuel, and the
second priority is for replenishment of glycogen in liver and muscle.
In addition, in the presence of this glucose load, muscle preferentially oxidizes glucose as fuel. Indeed, in the postprandial
state, the heart meets over 80% of its energy needs by oxidizing

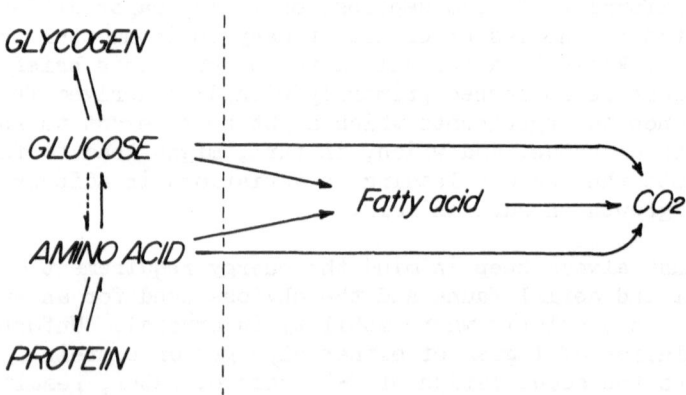

Fig. 1.   Relative ease of interchange between glucose and amino
acids (the non-essentials); however, once acetate or long chain
fatty acids are made (to the right of the broken line), the
calories are irreversibly committed and the fatty acids can only
be directly oxidized or stored for later oxidation.

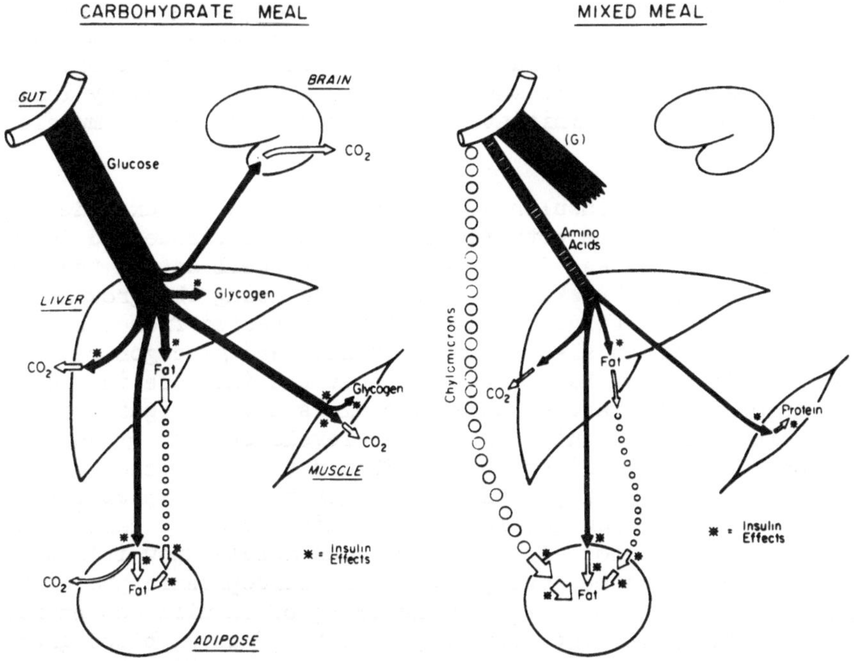

CARBOHYDRATE MEAL   MIXED MEAL

Fig. 2.   A pure carbohydrate meal (on the left) and a mixed meal
(on the right) showing the relative flow of fuels from the gut to
the various organs. Asterisks mark some of the sites of insulin
action. In the mixed meal, the carbohydrate (glucose) arrow has
not been shown to simplify the figure and only the amino acid and
triglyceride fluxes are shown.

glucose to CO_2. The surplus glucose from the meal can also be
converted into fatty acids either by adipose tissue or by the liver,
in which case they are incorporated into very low density lipoprotein triglyceride and exported via the blood for later incorporation into adipose tissue. All of these processes result from an
increase in insulin concentration (1), the increase serving as the
signal to these tissues that there is adequate glucose entering the
system. A most important unknown, however, is the processes which
determine how much glucose is to be stored as glycogen and how much
used for lipogenesis. Some data suggest that as glycogen stores
become larger, the crowding of the outer tiers render them less
accessible to enzymes for addition of further glucose units, but at
present this concept remains a speculation. In any case, it is
easy to see that the pattern of meal eating, the differential
sensitivity of various tissues to insulin and many other factors,
may all participate in the disposition of the various foodstuffs.

    In between meals,(Fig. 3) liver glycogen maintains blood
glucose mainly for brain, and the other tissues switch from glucose
to the use of free fatty acids. Glucose utilization is thereby
diminished, particularly in muscle, and this is achieved by insulin
levels being too low to initiate glucose transport across muscle
cell membranes.

    The low level of insulin serves as the signal to exclude
glucose utilization from both muscle and adipose tissue and for
adipose tissue to release free fatty acids as well as for the liver
to initiate glycogen breakdown. Recent data using the growth
hormone inhibiting peptide (somatostatin) extracted from the
hypothalamus and which also inhibits insulin and glucagon release
from pancreatic islets have suggested that the increase in glucagon
concentration in the fasted state also plays a very significant
role in maintaining circulating glucose levels by both glycogenolysis
and gluconeogenesis in liver (2,3). Should the abstinence persist
for over 12 or more hours, hepatic glycogen becomes depleted (4),
and gluconeogenesis from muscle-derived amino acids is called upon
to maintain blood glucose levels. How does muscle know it's time
to divest itself of some actin, myosin and sarcoplasmic protein?
The signal appears to be a further lowering of insulin concentration.
Should the fasting persist even longer, for 1-2 days, blood glucose
is supported almost completely by gluconeogenesis (Fig. 4 and 5).

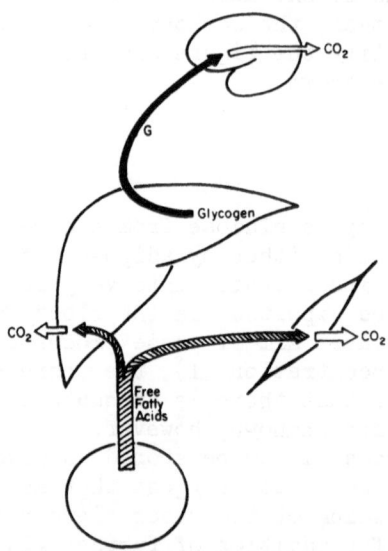

Fig. 3.   Between meals (interprandial), hepatic glycogenolysis
maintains blood glucose levels; other tissues use mainly free
fatty acids released from adipose tissue.

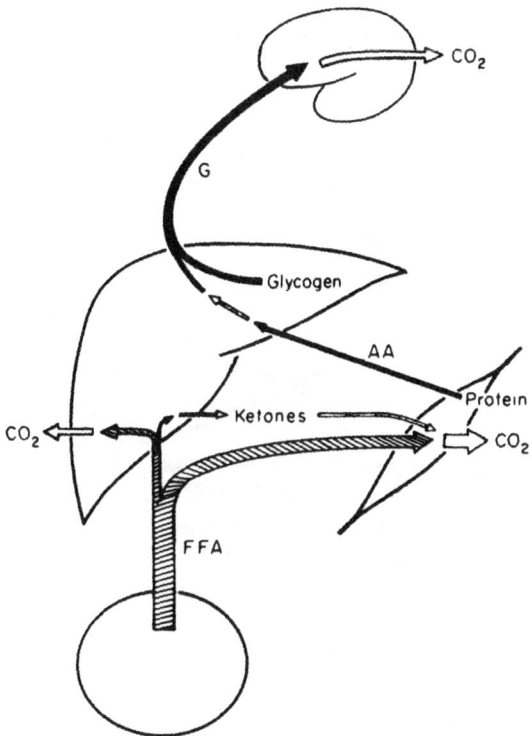

Fig. 4.   With more prolonged deprivation (overnight fast), liver
glycogen becomes depleted and gluconeogenesis and ketogenesis begin.

    If one administers glucose in small quantities (Fig. 6),
sufficient to provide brain with most or all of its glucose,
gluconeogenesis ceases and muscle amino acid breakdown becomes markedly attenuated. This process, the classical "nitrogen-
sparing" effect of carbohydrate, appears to be mediated by a slight
increase in insulin concentration. How efficient is this process?
In adult man,150g of glucose/day can diminish urinary nitrogen loss
from a level of about 7 g nitrogen/m^2 body surface to about 3 g. If
one gives 600 g/day of glucose, an amount supplying more than the
total caloric need of the entire body, nitrogen excretion falls
only 1 g further to 2g/day(5). Thus in terms of its nitrogen balance,
muscle appears to be exquisitely sensitive to insulin.

    If, in the next phase of fasting, the pattern in Fig. 6 were
to continue, such prolonged starvation would rapidly deplete
muscle nitrogen to such a degree that viability would be in jeopardy
after 2-3 weeks. What occurs, however, is a series of metabolic
adaptations whereby the utilization of ketoacids by muscle diminishes; their level in blood rises to 6-8 mM, a concentration

EARLY STARVATION

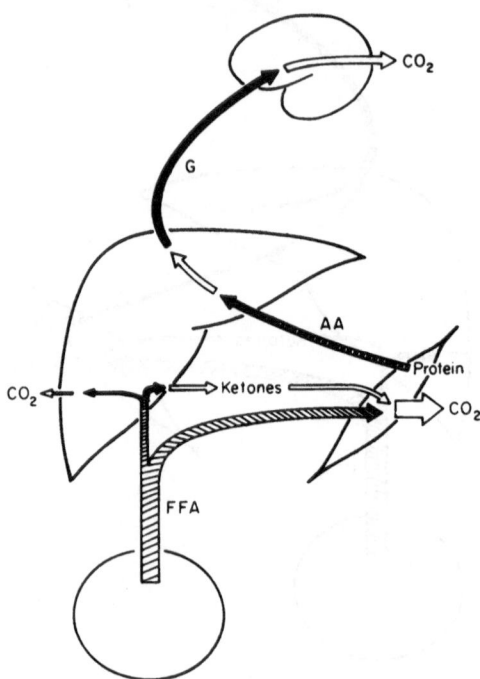

Fig. 5.   After 2-3 days of starvation, muscle-derived amino acids provide gluconeogenic substrate for liver. Ketoacid production in the liver is fully operative. There is probably ketoacid uptake by the brain already at this stage as circulating levels of ketoacids begin to increase, but this is not shown in the figure for simplicity.

sufficient to permit diffusion into the central nervous system adequate to displace glucose as fuel (1). Thus even the brain begins to use fat, but fat in a modified water-soluble form; e.g. acetoacetate and β-hydroxybutyrate (Fig. 7). The importance of all of this, however, is that muscle proteolysis becomes markedly attenuated now that gluconeogenic amino acids are no longer necessary for liver to make glucose for brain. Paradoxically, insulin levels are even lower, which would be expected to result in the reverse, an even further proteolysis instead of nitrogen conservation, and this novel phenomenon needs further explanation. First, however, intermediary metabolism of muscle protein and its control should be clarified (Fig. 8).

Insulin has been shown by many to stimulate muscle uptake of certain amino acids and to enhance the protein-synthetic machinery

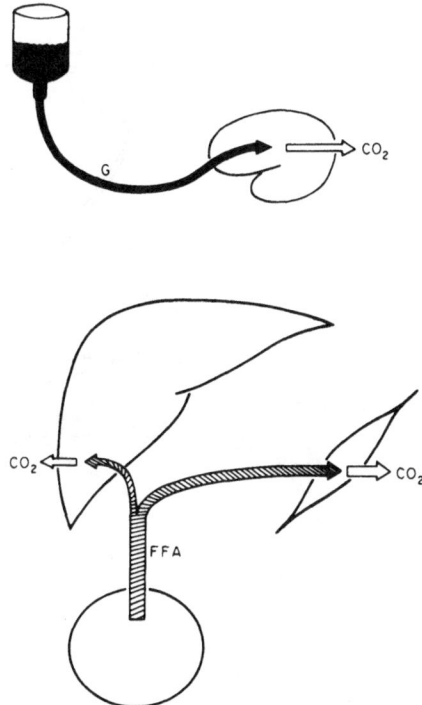

Fig. 6.   Small amounts (100 gm/day) of glucose (intravenous
"D & W"), by increasing circulating insulin levels, suppress
muscle protein breakdown as well as hepatic glucose production.

inside the muscle cell. Recent data have also suggested another
effect in inhibiting proteolysis, the two effects therefore
supplementing each other (6). The amino acids inside the muscle
cell which are produced by proteolysis can either be used for
resynthesis of the protein or can be metabolized or released. But
here is where some important metabolic events may take place. If
one examines the pattern of release of amino acids from muscle
either in the absorption or prolonged fasted state (Fig. 9),
alanine and glutamine are released far out of proportion to their
content in total muscle, (7,8), or in any known muscle protein.
Leucine, isoleucine and valine, which together comprise 15-20% of
muscle protein are barely released at all due to transamination
and oxidation of the ketoacid remnant inside the cell as their
$-NH_2$ groups contribute to formation of the excess alanine or
glutamine.

      Glutamate, on the other hand, is taken up into the muscle. Te-
leologically, muscle appears to be redirecting its released nitrogen
into the two ideal glucogenic fuels for liver, alanine and gluta-

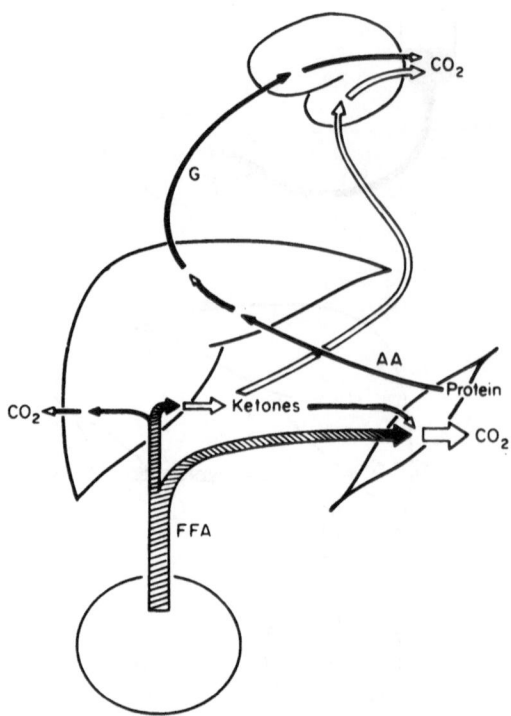

Fig. 7.   With prolonged starvation, as ketoacid levels increase,
brain markedly decreases glucose oxidation. Muscle proteolysis
is also attenuated due to the elevated levels of free fatty acids
probably sparing the irreversible oxidation of the branched-chain
amino acid derivatives, as discussed in the text.

mine, the latter also serving as the ideal ammoniagenic substrate
for kidney.

     The branched-chain amino acids, leucine, isoleucine and
valine, are transaminated with α-ketoglutarate in the muscle to
form glutamate, which can then form alanine or glutamine. The α-
ketoanalogue residue of the branched-chain amino acid is oxidized
in situ. Of extreme importance is the peculiar metabolic pathway
of the branched-chain amino acids, whose ketoanalogues, like
pyruvate, require NAD+ and free Co-enzyme A for the next and
thermodynamically irreversible oxidative step. Fatty acids have
been shown to be most important in inhibiting oxidation of the
branched-chain amino acids. In so doing, they prevent the draining
off of three of the essential materials without which protein
synthesis cannot occur. Thus one can think of the Randle cycle as
a process whereby fat spares not only carbohydrate oxidation, but

BRANCHED CHAIN AMINO ACID METABOLISM IN MUSCLE

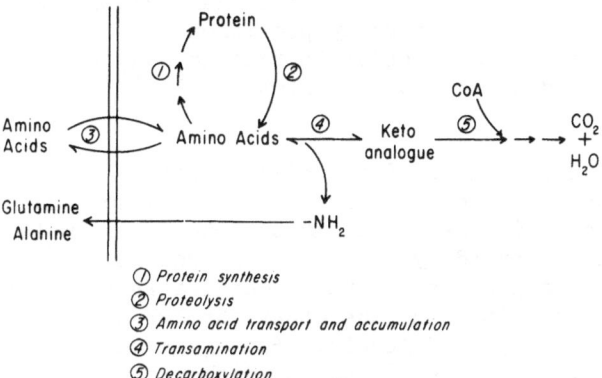

① *Protein synthesis*
② *Proteolysis*
③ *Amino acid transport and accumulation*
④ *Transamination*
⑤ *Decarboxylation*

Fig. 8.   Scheme of muscle amino acid and protein interrelationships. Insulin stimulates at sites (1) and (3) and inhibits at
site (2). Elevated free fatty acids appear to inhibit at site (5).

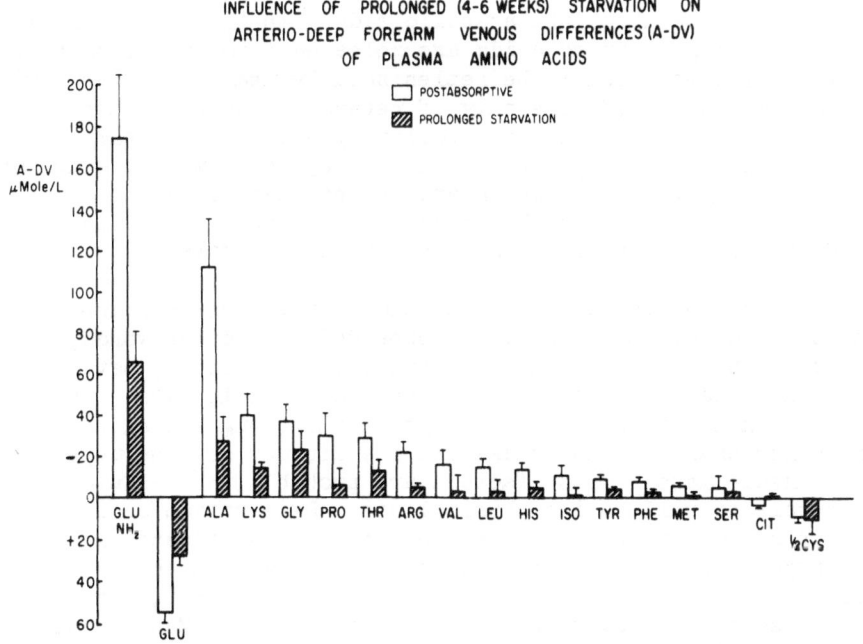

Fig. 9.   Pattern of amino acid release from human forearm muscle
both in the postabsorptive state and after prolonged starvation,
showing the predominant release of glutamine and alanine and the
uptake of glutamate, as well as the decrease in overall amino acid
release after prolonged starvation.

also nitrogen mobilization by conserving the branched-chain amino acids in muscle. Recent experiments by Sapir et al (9) have shown that the previously known "minimum" nitrogen loss during starvation in man can be lowered even more by provision of the ketoanalogues of the branched-chain amino acids to fasting man, lending further support to the evidence for their pivotal role in nitrogen conservation.

We have described these interrelationships in very gross physiological terms not only for brevity, but also because their biochemical elucidation is as yet incomplete. In simple terms, nitrogen can be conserved by two metabolic routes, one involving insulin, and this appears to be the mechanism whereby exogenous carbohydrate spares nitrogen, the other through an elevation of circulating free fatty acid levels. But again we have a problem, since these fatty acids are highest in diabetic ketoacidosis, a state associated with a large nitrogen loss. Apparently a small amount of insulin is also needed for the free fatty acids to exert these effects on nitrogen conservation.

So far we have focused on the control of muscle protein catabolism, emphasizing insulin and free fatty acids; now to say what little we know about the anabolic side. In the postabsorptive state shown in Fig. 4 amino acids are released from muscle protein, and obviously, these need to be replenished by meals. In man (10), insulin appears to be capable only of retarding their release. A net uptake has been achieved only following a protein-containing meal, and even then the uptake by the resting forearm muscle is not very dramatic (11). Thus insulin and an increment in amino acid concentration appear both to be important, and as will be discussed later, exercise may play an even more significant role.

Normal man, unless on an amino acid or nitrogen-deficient diet, maintains his muscle mass maximally expanded. In other words, any daily increment in protein intake results in a precisely matched increment in urea excretion, not in increased muscle mass. This occurs even in individuals with excess insulin, suggesting that factors other than protein intake and insulin may set the upper limit of muscle mass. These factors are probably the cellularity of the muscle as well as its physical use. Denervation or inactivity results in rapid atrophy and increased use in significant although limited hypertrophy. Goldberg has shown that even in the face of overall body depletion of nitrogen, as in a fasting rat, increased use of a muscle can result in net hypertrophy (12). Man can also expand his muscle protein, but only by stepping up his daily usage of that muscle, not by increasing protein intake. It should be emphasized, however, that the minute to minute effect of exercise on muscle protein synthesis or catabolism in man has not yet been carefully studied, particularly in relation to meals. For example, can muscle hypertrophy occur more easily per unit exercise in the wave of hyperamino acidemia following a protein-containing meal?

All of the aforementioned, however, is applicable to adult man, a plateaued animal, and may not be directly pertinent to an animal capable of continued growth, such as pre-adult man or many animals. More important, is any of this information applicable in the ruminant, which is, in effect, continuously fed? Until amino acid uptake and incorporation into muscle is determined in ruminants on different feeding regimens, an answer cannot be given. It is still possible that muscle protein synthesis and breakdown may be pulsatile as it is in man, and as techniques for the determination of protein turnover become more sophisticated, it is highly probable that the ruminant may be found to have these intermittent waves of synthesis and proteolysis over which hormonal and physical controls can be exerted.

Finally, in those countries in which the ruminant is an inter- mittent feeder and is faced by starvation during famine or dry seasons, it is obviously crucial that its lipid reserves be capable of conserving its protein reserves for both its own survival as well as that of the human population dependent on it in times of deprivation.

REFERENCES

1.   Cahill, G. F., Jr. (1971). Diabetes 20, 785.

2.   Koerker, D. J., Ruch, W., Chideckel, E., Palmer, J., Goodner, C. J., Ensinck, J. and Gale, C. C. (1974). Science 184, 482.

3.   Alford, F. P., Bloom, S. R., Nabarro, J. D. N., Hall, R., Besser, G. M., Coy, D. H., Kastin, A. J. and Schally, A. V. (1974). Lancet ii, 974.

4.   Hultman, E. and Nilsson, L. H. (1971). In: "Muscle Metabolism in Exercise", (Ed. B. Pernow and B. Saltin). p. 143. (Plenum Press, New York).

5.   O'Connell, R. C., Morgan, A. P., Aoki, T. T., Ball, M. R. and Moore, F. D. (1974). J. Clin. End. Met. 39, 555.

6.   Cahill, G. F., Jr., Aoki, T. T. and Marliss, E. G. (1972). In: "Handbook of Physiology", Sect. 7, Vol. 1, "Endocrine Pancreas". p. 563. (Williams and Wilkins, Baltimore).

7.   Felig, P., Marliss, E. B., Pozefsky, T. and Cahill, G. F., Jr. (1970). Science 167, 1003.

8.   Marliss, E. B., Aoki, T. T., Pozefsky, T. and Cahill, G. F., Jr. (1971). J. Clin. Invest. 50, 814.

9. Sapir, D. G., Owen, O. E., Pozefsky, T. and Walser, M. (1974).
    J. Clin. Invest. 54, 974.

10. Pozefsky, T., Felig, P., Tobin, J. D., Soeldner, J. S. and
    Cahill, G. F., Jr. (1969). J. Clin. Invest. 48, 2273.

11. Aoki, T. T., Müller, W. A., Brennan, M. F. and Cahill, G. F.,
    Jr. (1973). Diabetes 22, 768.

12. Goldberg, A. L. (1971). In: "Cardiac Hypertrophy",
    (Ed. N. Alpert). p. 39. (Academic Press, New York).

DISCUSSION

   Dr. King enquired about the role of lipid-mobilizing factor
in the control of lipid metabolism. Prof. Cahill thought that,
in general, lipid-mobilizing factors were artefacts. A number of
factors had been isolated from both the hypothalamus and the
pituitary which induced lipolysis in vitro, but there is little
evidence that they have a similar rôle in the living animal. For
example, hypophysectomized animals have normal lipid mobilization
under a wide range of conditions of stress, exercise or starvation.
There is, however, at least one piece of contrary evidence. In a
rare syndrome affecting children, there is a total loss of adipose
mass. Factors isolated from their blood resemble releasing
factors from the hypothalamus. Dr. Webster asked what part
catecholamines play in insulin release: he pointed out that β-
agonists, such as isoprenaline in pharmacological quantities,
stimualte insulin release; was there a comparable physiological
effect of catecholamines? Dr. Turner thought it likely that in
living animals insulin release was inhibited by adrenalin, whether
of peripheral or adrenal medullary origin.

   Dr. Braude suggested that efforts to reduce the backfat thickness of growing pigs might, in some cases, have led only to a redistribution of the same total amount of lipid. Prof. Cahill
doubted whether fat distribution could be altered by nutrition,
since differentiation confers a certain cellularity to each site.
Fat distribution can, however, be affected by the sex hormones.
Dr. Turner added that, with reduced insulin and thyroid hormones,
but without a reduction in cortisol, lipid could be mobilized
preferentially from subcutaneous depots, whilst sparing internal
deposits.

   Dr. Fuller noted that, whereas a growing animal typically
retained about one-third of its dietary nitrogen, Prof. Cahill
had mentioned that the black bear in hibernation completely prevented nitrogen loss. Did the mechanism lie in the inhibition of
amino acid degrading enzymes? Prof. Cahill emphasized the
distinction between the conservation of nitrogen in starvation
and the accretion of nitrogen by growing animals; the highest
efficiency of accretion that he knew of was 60% in rehabilitating
children. Prof. Kielanowski remarked that sucking pigs may
retain up to 90% of their nitrogen intake.

Overall Control of Growth

THE RIGHT SIZE

Richard J. Goss

Division of Biological and Medical Sciences
Brown University, Providence
Rhode Island, U.S.A. 02912

The twentieth century has witnessed the accumulation of an
impressive backlog of data on the details of growth mechanisms.
As a result, we now have a wealth of information about how cells
divide and enlarge, synthesizing DNA, RNA and proteins as they do
so. Hardly a tissue has escaped the inquisitive explorations of
students of regeneration and compensatory growth. From time to
time, someone has ventured a hypothesis to explain how growth
may be controlled, and every combination of inhibitor and stimu-
lator seems to have been suggested (1). Out of this bewildering
profusion of facts and theories, one would hope that a unifying
hypothesis might emerge, one which will embrace all forms of
growth from embryogenesis to the pathology of ageing, and all
levels of organization from the subcellular to the organismal.
It would be presumptuous to pretend that in our present state of
ignorance the time has come for such a breakthrough. It would be
inexcusable, however, if we did not continue our efforts to fathom
the implications of growth. The unabashed speculation which
follows is an attempt to paint a picture of growth not in the
usual minute scientific detail but with broad strokes of the brush.
In view of the generalizations inevitably to be included, however,
the reader is reminded that "all generalizations are wrong,
including this one."

BODY SIZE VS. ORGAN SIZE

The body is the sum of its parts. Accordingly, its size may
be determined by innate limitations in the growth of one or more
of its component organs. Yet each part of the body is a predictable proportion of the whole, and their sizes may therefore be a
function of the overall body mass. The central problem of growth

237

regulation is to find the limiting factor responsible for fixing
not only the absolute size of the body but also the relative
sizes of its constituent parts. The question is raised: Is body
mass the point of reference for organ growth or is there a built-
in limit to the growth of each organ which determines the size
beyond which the body as a whole cannot grow?

The allometric equation (2) testifies to the dependence of
organ size on body mass. Phylogenetically, the relative sizes of
organs have evolved as genetic adaptations to physiological needs.
In the ontogenetic sense, changing allometric relationships in the
growing animal reflect alterations in the physiological conditions
at different stages of maturation. Thus, in both the evolutionary
and the developmental sense, it is true, as Brody (3) has pointed
out, that "the organism changes geometrically so as to remain the
same physiologically."

ORGAN AND TISSUE GROWTH

The potential for enlargement of an organ can be divided into
four increments (fig. 1). There can be little doubt that the
growth of organs is at least in part a function of the physiological
demands impinging on them. It is equally obvious, however, that
such influences cannot account for the entire development of an
organ. Disuse almost always leads to atrophy, but almost never
brings about the total disappearance of an organ or tissue. The
sizes of many endocrine glands, for example, decrease markedly
if their hormones are exogenously administered or, in the case of
pituitary target organs, after hypophysectomy. The kidney
atrophies if its arterial blood pressure is reduced to the point
where glomerular filtration becomes impossible (4). The urinary
bladder loses more than half its weight if deprived of the hydro-
static pressure of urine inflow (5). Unused bones become
osteoporotic (6), and their associated skeletal muscles undergo
atrophy after denervation, tenotomy or immobilization (7). In
these and other instances, atrophy is a relative phenomenon, for
although unused adult organs lose weight on an absolute scale,
their counterparts in young animals may continue to gain weight
but at a much reduced rate. Physiological deprivation reduces
organ mass only to a basal level perhaps comparable to that which
develops in the early prefunctional stages of maturation. The
persistence of atrophic organs may be looked upon as the nucleus
out of which subsequent regrowth may be generated should the need
arise.

Over and above the basic minimal mass of an organ, much of
its normal size is represented by that growth which has been
stimulated by ordinary physiological activities. Indeed, it
would seem that this severalfold enlargement of a normally
functioning organ over its basic minimal mass may be correlated

with the well-known redundancy which characterizes virtually all organs and tissues. It is this built-in superabundance of organ mass which provides the comfortable margin of safety in living systems which permits their survival despite the loss or incapacitation of over two-thirds of such vital parts as the liver, kidneys and lungs.

There are few organs of the body which will not enlarge above their normal dimensions when overworked. The administration of appropriate trophic hormones is well known to stimulate the growth of target organs while promoting increased functional activities. The salivary glands will grow in response to such diverse experimental interventions as the repeated amputations of the lower incisors of rats, the addition of proteolytic enzymes to the diet or the injection of isoproterenol (8). The exocrine pancreas enlarges if an animal is fed a diet of raw soybean containing a trypsin inhibitor (9). The size of the remaining kidney increases in compensation for its missing partner after unilateral nephrectomy (10). The liver likewise regenerates after partial hepatectomy, but can also be made to grow by the administration of phenobarbital (11). The heart is famous for its capacity for hypertrophy in the face of hypertension, the two ventricles responding differentially to systemic versus pulmonary hypertension (12). In the case of skeletal muscle, mounting evidence suggests that exercise and tension may pay a fundamental role in maintaining their structure and promoting their hypertrophy (13), even to the extent of overriding the otherwise atrophic effects of denervation or immobilization. All such cases of compensatory hypertrophy have much in common with that fraction of normal organ growth which is triggered by ordinary functional activities. Time and again, it has been shown that compensatory growth is an adaptive reaction to overwork, and like its normal counterpart, it is reversible.

A final stage in the enlargement of an organ may be referred to as pathological growth. This is an extension of compensatory hypertrophy brought on by such heightened levels of functional demand that the tissues, however much they may try to adapt, are not capable of coping with the relentless pressure for more work. Irreversible damage coupled with cellular exhaustion lead to such pathological consequences as nephroschlerosis, emphysema and heart failure.

The so-called "normal" size to which an organ grows is a relative thing. It is relative to the size of the body as well as to the prevailing physiological conditions to which it responds. For example, a normal gonad in the breeding season is considerably larger than a normal one at other times of the year. Similarly, the normal red blood cell count is defined for conditions at sea level, but this would be anemic at higher altitudes. For many organs, therefore, it is not the absolute size to which they grow

that is genetically determined. Rather, it is their relationship
to appropriate physiological conditions that has been selected for
in the course of evolution thus insuring that each organ will
adjust its size to the work to be done. Some organs, however, can
adjust better than others, depending on their proficiency for
increasing the numbers of functional units they contain.

A functional unit may be defined as the smallest irreducible
sub-division that is still capable of carrying out the specific
functions for which the organ has been differentiated (14,15). In
some tissues, the functional unit is equivalent to the cell, as in
the case of erythrocytes or leukocytes. In others, it may be a
subcellular entity, such as myofibrils or platelets (both of which
depend upon cells for their very existence and origin). In most
cases, however, cells must be organized into histologically complex
units in order to do their jobs. The various kinds of acini and
follicles found in many glands are cases in point, as are intestinal
villi, pulmonary alveoli, renal nephrons, seminiferous tubules, the
cords of hepatic lobules and the osteons of bones.

INDETERMINATE ORGANS

Certain organs have the capacity for unlimited adaptive growth,
not so much because their cells may never lose their proliferative
potential, but because these cells also retain the capacity to
arrange themselves into new functional units at the histological
level of organization. If part of the liver is removed, for
example, the remaining portions soon grow back to restore the
original mass of the organ. If this operation is repeated, the
same thing happens again, and persistent experimenters have forced
rat livers to regenerate each month for a year without depleting
their resources (16). The conclusion is inescapable that the
liver has the capacity for unlimited growth under conditions of
chronic stimulation. Other organs are likewise capable of remarkable feats of growth. The pancreas can be almost completely destroyed
by daily injections of ethionine, yet restore itself to normal as
soon as the treatment is discontinued (17). The adrenal cortex
can regenerate from such few residual cells as may adhere to the
inner lining of its empty capsule (18). A fragment of ovary is all
that is needed to regenerate a functional organ (19). A single
subcutaneous injection of $CdCl_2$ leads to ischemia and nearly complete degeneration of the testes. In due course, however, the
interstitial tissues (but not the seminiferous tubules) grow back
from what must have been a very few surviving cells beneath the
tunica albuginia (20). Thus, many of the body's glands, both
exocrine and endocrine, appear capable of reconstituting their
original mass after severe depletions. Presumably, only the lack
of sufficient stimulation prevents them from growing to excessive

Fig. 1.   Increments of organ growth. The basic size of an organ
represents that which forms in prefunctional stages of development
and persists after disuse atrophy. The normal dimensions of an
organ equal the sum of its basic size plus such additional growth
as may take place in response to ordinary physiological activities.
Compensatory hypertrophy is reversible adaptive growth caused by
functional overload. Pathological growth includes overgrowth so
excessive as to be irreversible.

dimensions. Because they have the potential for far more growth
than they may ever have a chance to express, they may be classified
as indeterminate organs.

DETERMINATE ORGANS

Not all parts of the body fall into this category. There are
many organs and tissues which cannot grow beyond certain prescribed
size limits. They are accordingly referred to as determinate
organs. Their restricted growth potential is attributable to the
fact that their functional subunits are unable to increase in
number beyond early developmental stages when there occurs a
switch from hyperplasia to hypertrophy. Growth may continue
beyond this point, but it is achieved solely by the enlargement of
pre-existing units until adult dimensions have been attained. Not
unexpectedly, these determinate organs have limited capacities for
compensatory growth compared with the seemingly limitless potentials

of indeterminate organs. This is not to say that they are without
the capacity for growth beyond normal dimensions, however. After
unilateral nephrectomy, the remaining kidney increases its size,
but it does so by multiplying and enlarging its cells without
forming new nephrons. Likewise, the remaining lung can compensate
for its missing partner by doubling its size, but this is unfortunately not accompanied by the production of new alveoli. The
heart is able to enlarge its ventricles when overworked, but this
is achieved by hypertrophy of its myocardial fibres rather than
the production of new ones. The central nervous system, which
ceases producing new functional units earlier than any of the
other organs, is least endowed with the capacity for regeneration.
These and other organs are examples of body parts which are not
without limited capacities for adaptive growth but which react to
deficiencies in a surprisingly inefficient manner, namely, the
enlargement of functional units rather than their proliferation.

      What is the rationale for the existence of determinate and
indeterminate organs? It would seem that mammals might be more
efficient in their competition for survival if all of their tissues
possessed unlimited regenerative potential. Even more puzzling is
the fact that the very organs that are least capable of regeneration
are among the most vitally essential parts of the body. The kidneys,
lungs, heart and brain are seriously handicapped in making up for
deficiencies in their mass or adapting to increased workloads. On
the other hand, we find that many glands in the body, organs which
are convenient to possess but many of which are not essential for
survival, have extraordinary powers of repair and regeneration.
Clearly, there must be some reason why, during the course of evolution, less important organs acquired greater powers of growth
than the more important ones. One can only conclude that it is
not to the advantage of the species as a whole for its individual
members to be potentially immortal, as presumably would be the
case should all tissues of the body be capable of unlimited repair
and regeneration.

INDETERMINATE BODY SIZE

      The foregoing remarks apply to mammals. By comparison with
lower vertebrates, and even some of the invertebrates, mammals are
seen to be a special case of how organisms grow, and by no means
are they to be considered typical. Indeed, along with their warm-
blooded avian cousins, mammals are in many respects very unique
creatures, and from the point of view of growth their efficiency
does not always compare favourably with that encountered in lower
forms.

      The distinction between determinate and indeterminate organs
in mammals does not necessarily apply to cold-blooded vertebrates.
The reason for this is twofold. In the first place, many of the

cold-blooded vertebrates are endowed with remarkable powers of
regenerating lost appendages, from fish fins to lizard tails (21).
Perhaps the most extraordinary capacity for regeneration is found
in the salamander limb, which if amputated forms a bud, or
blastema, on the stump which then differentiates into an exact
replica of the missing appendage. Whole new skeletal elements and
muscles are formed de novo during the course of regeneration. Yet
it is a curious thing that if a bone or a muscle is excised from
an otherwise intact limb, these same animals are no more able to
replace them by in situ regeneration than are mammals.

The other attribute of lower vertebrates which distinguishes
them from the warm-blooded ones is the potential they possess for
unlimited body growth (22). Birds and mammals mature to a pre-
ordained adult size which is species-specific. Although the same
may be true of some lower vertebrates, there are many other species
for which no typical body size can be predicted. This is especially true of those fishes which continue to enlarge throughout
their life spans and so far as we know may enjoy an indefinite
longevity.

It is in this perspective that the limitations of mammalian
organ growth are particularly interesting. With the exception of
the lung, fishes possess the same organs which in mammals are
classified as determinate. How, then, can these organs in lower
vertebrates keep pace with the ever expanding dimensions of the
body as a whole? In the case of the kidney, there appears to be
no prescribed number of nephrons that can develop as is the case
in mammals. If glomeruli are counted in fishes representing a
range in body size, there is found to be a linear increase in
nephron number throughout life (23). Indeed, fish living in fresh
water develop more glomeruli in their kidneys than do the same
species inhabiting marine environments. In frogs and iguanas,
larger specimens are found to have more glomeruli than smaller
ones (24). How, where, and from what cells these new nephrons are
formed is not known. The problem of similitude is even more
interesting in the case of the heart. Regenerative and proliferative capacities of mammalian hearts are conspicuous by their
absence. In fishes, however, it is necessary for the heart to
keep pace with body growth over a wide range of sizes. Studies
have shown, however, that the dimensions of individual myocardial
fibres are unchanged from small fish to large ones. Therefore,
we must conclude that new ones are produced from time to time as
the mass of heart muscle increases with body size, yet it has
never been established whether such cells are formed by division
of differentiated muscle fibres or by the differentiation of new
ones from a pool of reserve cells. The same problem exists with
reference to the central nervous system of fishes, but in the near
absence of factual evidence one can only speculate that here too
there must be a lifelong source of cells from which neuroblasts

can be recruited as the brain and spinal cord expand with the body
mass. In this regard, it is interesting to note that the growing
neural retinae of fishes can augment their populations of photo-
receptors by the differentiation of new ones around the margins (25).

Nearly everything about the body of a fish is endowed with the
potential for indefinite growth. The fins, for example, grow not
so much by internal expansion as by terminal addition. It is
probably no coincidence that their normal growth takes place in
the same way that they regenerate after amputation. Similarly,
scales enlarge by adding new growth rings in the manner in which
they regenerate.

Gills pose a special problem. Each gill arch supports a
number of filaments from which a series of paired lamellae branch
off at right angles. As the fish grows, new lamellae are added to
the ends of each filament, and the number of filaments per gill
arch also increases. The mechanism and control of these events
have never been seriously explored.

The teeth of fishes are especially interesting. As in so many
other lower vertebrates, they are replaced in continuous succession
throughout life. Moreover, as a fish grows its total number of
teeth also increases, a phenomenon also reported in amphibians.

These and other aspects of morphology bear witness to the
fact that fishes are designed for unlimited growth. Inevitably,
they represent a compromise between increases in size versus
numbers of body parts, but in contrast to mammals they benefit far
more from the advantages of hyperplasia over hypertrophy.

EVOLUTION OF BODY SIZE

The body sizes of animals have become increasingly determinate in the course of evolution. Some invertebrates, most
notably the crustaceans, are famous for their apparent inability
to limit body size. Lobsters, for example, are not known to lose
their capacity for molting and growing, as the record-breaking
sizes of occasional specimens bear witness. Nevertheless, many
other invertebrates are capable of limiting their body sizes by
ceasing to grow with the approach of sexual maturity.

The evolution of body size regulation among vertebrates
appears to be related to aquatic versus terrestrial habits. As
they abandoned the buoyancy of their aquatic environment, vertebrates were forced to invent some mechanism to arrest their body
growth lest their bulk become too much for them to support on
land. This led to the evolution of the cartilaginous plate in the
long bones, a structure well adapted to insure the cessation of
growth by its programmed disappearance at a predetermined time in
the life cycle.

The bones of fishes are not only solid, but they tend to enlarge by appositional growth. In amphibians and many reptiles, the bones have acquired a marrow cavity and their epiphyses remain cartilaginous. Hence, their appendicular skeleton can undergo indefinite elongation owing to the persistence of cartilage on their ends which retains the capacity for growth. It was among some of the reptiles that the first cartilaginous plate interposed between the epiphysis and diaphysis made its appearance, and it is this remarkable structure which is responsible in large measure for the arrest of growth among birds and mammals when they reach maturity. Parenthetically, it is worth noting that some of the cetaceans may have regained secondarily the capacity for unlimited body growth. Recent studies have shown that in seals and whales not only do the teeth continue to grow by the annual deposition of increments of dentine, but the mandibles also continue to enlarge by the laying down of bony lamellae on the surface of the bone (26). Hence, it is no coincidence that these represent the only mammals whose age can be determined by counting growth rings in various structures of the body.

Although epiphyseal closure is clearly a major factor in stopping body growth, it may not be the only one. Just as the elongation of long bones in birds and mammals is predetermined, so also is the ultimate size to which determinate organs can grow. In view of the fixed number of functional units which develop in determinate organs, it is entirely possible that they may serve as limiting factors to the overall enlargement of the body. Indeed, the organism cannot increase its size beyond the species-specific limit without running the risk of outgrowing such vitally essential organs as its lungs, kidneys, heart and brain. The existence of determinate organs in warm-blooded mammals may therefore hold the key to body growth.

HOW FIXED IS THE NUMBER OF FUNCTIONAL UNITS?

It is by no means certain that the species-specific number of functional units in determinate organs is necessarily predetermined. Although partial ablation or functional overload in adults is known not to bring about the production of extra functional units, there is evidence that if the deficiency is created at an early stage of development when the organ is still normally producing such functional units, it may be induced to make extra ones.

Unilateral pneumonectomy in puppies (27) and kittens (28), for example, has been reported to cause the remaining lung to produce more pulmonary alveoli than it would normally make, although the same operation in adults leads solely to the enlarge-

ment of pre-existing alveoli. Much the same situation obtains in
the case of the kidney. Recent studies on the effects of unilateral nephrectomy in infant rats have shown that the remaining
kidney which has still not acquired its full complement of nephrons,
can be induced to form (29,30) more than it would normally. Indeed,
if one kidney is removed from newborn rats, the opposite one
produces an average of 63% more nephrons than normal (31). Such
evidence suggests that what has been regarded as the presumably
immutable normal number of functional units may in fact be a
physiological adaptation to the functional needs of the growing
organism. Unhappily, this does not explain why such organs
eventually lose the capacity for manufacturing more functional
units as they approach maturity. One can only assume that they
may exhaust their supply of unspecialized cells as the process
of differentiation overtakes proliferation in the developing organ.

ORGAN TRANSPLANTATION

Another way to approach this problem is to transplant organs
between small and large animals to find out if the sizes of such
grafts will adjust to the dimensions of the new host, and also if
the population of functional units will likewise adapt. There are
two ways to do this. One is to exchange parts between two species
of different sizes, the other is to transplant organs from younger
animals to older ones. Interspecific grafts have been made in
amphibian embryos, in which various organs and appendages have
been transplanted between species of different intrinsic sizes.
In general, when eyes (32), limbs (33), or even the entire halves
of the bodies (34) are grafted from embryos of small species to
those of large ones, or vice versa, they tend to grow to their
own species-specific sizes irrespective of how mismatched they
may become. These are all cases in which the autonomy of an
organ's growth is correlated with the lack of functional regulation by the body as a whole (35). However, when physiologically
interdependent parts from animals of different sizes are combined
in their embryonic stages, as the lens and eye cup, then each
adjusts its growth to the other to give rise to a harmonious
structure of intermediate dimensions (36).

Heterochronic transplantations can be carried out between
individuals of different ages but of the same species. By
grafting an organ from an infant animal into an adult it is
possible to learn if the potential for growth that would have been
expressed had the organ never been removed from the donor will
still take place in an adult environment. In the case of the
kidney, transplants have been performed between puppies and adult
dogs, but always after having first removed the host's own
kidneys. Under these circumstances the grafted infant kidneys
continue to grow (37,38), but there is no way to determine whether
this is an expression of an innate potential for growth or is in

compensation for the renal deficiencies of the host. Organ
transplants into intact hosts are urgently needed to resolve this
fundamental problem of growth determination.

Comparable experiments have been done with a few other organs.
For example, neonatal mouse spleens grafted subcutaneously into
adults fail to grow unless the host has been splenectomized (39).
Furthermore, the size to which they grow is inversely proportional
to the number of infant spleens grafted to any given host (40).
In the case of adipose tissue, grafts of mouse fat depots proved
to be more successful in hosts deprived of fat than in intact
mice (41). Hearts from infant rats anastomosed by means of
microvascular surgery to the abdominal aorta and vena cava of the
host continue to grow to normal adult dimensions despite the fact
that they are auxiliary organs presumably pumping no more blood
than needed for their own nourishment (42). Finally, grafts of
young bones to nonfunctional sites e.g., subcutaneous (43),
intrasplenic (44) or beneath the kidney capsule (45), grow to
nearly normal lengths.

THE PROBLEM OF ASYMMETRY

The precision with which the growth of living systems is
regulated is no more clearly demonstrated that in the bilateral
symmetry which is the basis for the body plans of most animals.
So it is that when a natural imbalance occurs in a paired organ,
there is need to understand how, if growth is genetically
controlled, the expression of the genes can be so different on
the two sides of the body. Therefore, one is tempted to look for
nongenetic explanations for the occurrence of asymmetrical
structure.

Perhaps the most well-known case in point is the lopsided
structure of the cardiac ventricles. The fact that the left
ventricle is normally about twice the size of the right one is
often assumed to be a genetic adaptation to the different workloads of these two sides of the heart, particularly when it is
noted that in the giraffe heart, which must pump against an
unusually high head of pressure, the left ventricle is disproportionately large (46). Nevertheless, it is instructive to note
that prior to birth the two ventricles of mammalian hearts are the
same size, in keeping with their equal prenatal workloads. The
inequity that is established after birth, therefore, is clearly
a physiological adaptation to the closure of the ductus arteriosus
and the establishment of two separate circulatory circuits of
unequal size.

Mammalian lungs are also asymmetrical, but in this case their
inequities are apparent from the very beginning. It is tempting
to conclude that the smaller size of the left lung is correlated

with the more sinistral position of the heart in the chest cavity,
but for obvious reasons it is impossible to prove experimentally
whether the inequity between the left and the right lung is
hereditary or just another physiological adaptation. An extreme
instance of pulmonary asymmetry is found in snakes. Apparently
as an adaptation to their elongated bodies, the right lung is
highly developed while the left one is small or rudimentary. The
failure of the left lung to develop raises the question of whether
or not it is held in abeyance by the functional right lung.
Preliminary experiments in which most of the right lung has been
tied off indicate that the rudimentary left one can now be
stimulated to undergo compensatory growth, albeit to a very limited
extent.

In snakes it is the right ovary which tends to be longer than
the left one, but in birds it is the left ovary that is functional,
the right one remaining rudimentary. In bats, it is usually the
right ovary which is functional, a situation paralleled in the
mountain viscacha, an inhabitant of the Andes. Pregnancy in these
forms tends to occur only in the right horn of the uterus.

Numerous cases of asymmetry are found among the invertebrates,
not the least of which is the imbalance between the claws of
crustaceans. Some species are all either right-handed or left-
handed, as in the fiddler crabs, while in others (e.g., lobsters)
the asymmetry is distributed 50-50. The fact that in some (but
not all) crustaceans the asymmetry is reversible by amputating the
dominant appendage and allowing the opposite one to become the
larger while the original large one is replaced by a small claw,
indicates that this asymmetry is not genetically determined (21).
A similar situation occurs in the regenerative reversal of
asymmetry in the operculum of Hydroides, a tube dwelling annelid
(47). Apropos of this, it would be interesting to find out if the
hectocotyl arm, which is the third tentacle on the right side of
the male octopus and is modified as an intromittent organ, could
be made to differentiate on the opposite side following amputation
of the original one.

When organs capable of regeneration are also asymmetrical,
they provide an opportunity to test the possible genetic control
of such asymmetries. The antlers of reindeer and caribou exhibit
an interesting instance of asymmetry in that the brow tine, the
most proximal branch which grows down over the snout, is usually
larger on one side than the other (48). In most specimens the
left brow tine is developed as a branching or palmate outgrowth
while the right one is a shorter unbranched spike. In other cases
it is the right brow tine which is dominant, while sometimes both
of them are fully developed and in rare cases neither may be.
Observations on individual animals over two or more years have
shown that the pattern of brow tine asymmetry is not necessarily

the same in successive sets of antlers regenerated from one year
to the next. This may be taken to indicate that the asymmetry of
these structures is not an inherited character, but leaves
unanswered the provocative question of how the uneven development
of brow tines is controlled.

Perhaps the most perplexing instance of asymmetry is the
clinical condition known as hemihypertrophy (49). In these
fortunately rare cases, a patient may grow larger on one side of
his body than the other. This unlikely condition excites one's
interest as much as it defies explanation, but warns us against
accepting the normal bilateral symmetry of our bodies as proof that
the two halves are not separately controlled.

CAN BODY SIZE BE MANIPULATED?

Aside from the well-known modulating influences of temperature,
nutrition and surface-volume relationships, factors which affect
an animal's size by altering its metabolic rate, the determination
of the basis body size for each species remains a matter for
conjecture (50). It is well established that the sizes of cells
are approximately the same in animals of widely divergent dimensions
(51). The chief exceptions to this rule are the nerve and muscle
cells which tend to be larger in bigger animals. This is undoubtedly
correlated with their inabilities to divide. Nevertheless, they
are also more numerous in larger animals, owing to the relative
prolongation of their hyperplastic phases in embryonic development.
Yet even in animals whose cell sizes are experimentally altered,
as in the polyploid salamanders created by Fankhauser (52), the
animal itself grows to normal dimensions although it contains
larger but fewer cells. The control of body size therefore would
appear to transcend whatever factors may influence the sizes of
individual cells.

Evidence suggests that body size is hereditary and that
although all mammalian eggs are approximately the same size the
rates at which they develop forecast the ultimate dimensions to
which the body will grow. The sizes of embryos and fetuses are,
however, not unaffected by their prenatal environments. Indeed,
their growth may be profoundly influenced by such factors as their
position in the uterus, litter size, placental dimensions, maternal
hormones and degree of hybridization. Perhaps it is a combination
of these factors that accounts for the interesting effects of
maternal body size on the growth of offspring in experiments on
horses (53), sheep (54) and cattle (55) in which large and small
strains have been bred to mismatch the fetal and maternal sizes.
These studies show that genetically small fetuses grow larger than
normal when carried by large mothers, and vice versa. Moreover,
such differences are not to be explained entirely by changes in the
lengths of gestation, and it is particularly interesting that the

size differences of the offspring persist through much of their
postnatal development. Clearly, while the maternal environment
cannot override the influences of genetics, it can exert some
long-lasting effects on the subsequent growth and development of
the progeny.

If one's objective were to promote the growth of outsized
animals, there are some intriguing and apparently unexploited
prospects for manipulating maternal-fetal relationships. It is
not outside the realm of possibility that we might someday learn
how to foster-mother the embryos of one species in the uterus of
another. To be sure, it is a common practice to transport eggs
from one place to another in the uterus of a rabbit, but in such
cases the rabbit is only a vehicle. Presumably the foreign eggs,
if not removed from the rabbit, would fail to implant and would
eventually perish, but it is by no means certain why this happens.
Very few studies of interspecific egg transfers have been published,
but available evidence suggests that some exchanges work better
than others. Eggs from sheep and goats can be transferred to each
others' uteri, whereupon some embryos may implant and continue
their development for limited periods of time seldom exceeding
50 days (56,58). Rat and mouse ova have also been exchanged, and
develop as far as blastocysts but fail to implant (59). Presumably
there is some kind of incompatability in such combinations, but
its basis e.g., immunological or physiological, is not known.

If this potentially important field of investigation were to
receive the attention it deserves, then it might be possible to
discover the nature of the barrier that prevents such interspecific
pregnancies. If such a breakthrough were to permit us to grow the
eggs of one species in the mothers of another, the results would
be of considerably more than hypothetical interest. It is
conceivable that if the ova from a small species were transferred
to a large one the sizes of the offspring might exceed the normal
species-specific range heretofore assumed to be genetically
controlled. The agricultural implications of such a tour de force
would far outweigh the risks of failure. The academic spin-off
in terms of advancing our basic knowledge of growth regulation
should be an even greater incentive to explore these compelling
possibilities.

ACKNOWLEDGEMENTS

Original experimental results or ideas by the author herein
reported have arisen during the course of projects supported by
research grants GM-18805 and HL-16336 from the United Stated
National Institutes of Health.

REFERENCES

1. Goss. R. J. (1972). In: "Regulation of Organ and Tissue
    Growth". (Ed. R. J. Goss). p. 1. (Academic Press, New York).

2. Huxley, J. (1932). "Problems of Relative Growth". (Methuen,
    London).

3. Brody, S. (1945). "Bioenergetics and Growth". (Van Nostrand-
    Reinhold, Princeton, New Jersey).

4. Masson, G. M. C. and J. Hirano (1969). In: "Compensatory
    Renal Hypertrophy". (Eds. W. W. Nowinski and R. J. Goss).
    p. 235. (Academic Press, New York).

5. Goss, R. J. and S. D. Singleton (1971). Proc. Soc. exp.
    Biol. Med. 138, 861.

6. Sevastikoglou, J. A., U. Erikson and S. E. Larsson (1969).
    Acta Orthopaed. Scand. 40, 624.

7. Gutmann, E. and P. Hnik (Eds.) (1963). "The Effect of Use
    and Disuse on Neuromuscular Functions". (Elsevier Pub. Co.,
    Amsterdam).

8. Schneyer, C. A. (1972). In: "Regulation of Organ and Tissue
    Growth". (Ed. R. J. Goss). p. 211. (Academic Press, New York).

9. Lepkovsky, S., F. Furuta, T. Koike, N. Hasegawa, M. K. Dimick,
    K. Krause and R. J. Barnes (1965). Brit. J. Nutr. 19, 41.

10. Nowinski, W. W. and R. J. Goss (Eds.) (1969). "Compensatory
    Renal Hypertrophy". (Academic Press, New York).

11. Argyris, T. S. (1971). Devl. Biol. 25, 293.

12. Goss, R. J. (1971). In: "Cardiac Hypertrophy". (Ed. N. R.
    Alpert). p. 1. (Academic Press, New York).

13. Steward, D. M. (1972). In: "Regulation of Organ and Tissue
    Growth". (Ed. R. J. Goss). p.77. (Academic Press, New York).

14. Goss, R. J. (1966). Science 153, 1615.

15. Goss, R. J. (1967). In: "Control of Cellular Growth in Adult
    Organisms". (Eds. H. Teir and T. Rytomaa). p. 3. (Academic
    Press, New York).

16. Ingle, D. J. and B. L. Baker (1957). Proc. Soc. exp. Biol.
    Med. 95, 813.

17. Fitzgerald, P. J. (1960). Lab. Invest. $\underline{9}$, 67.

18. deGroot, J. and C. Fortier (1959). Anat. Rec. $\underline{133}$, 565.

19. Pencharz, R. I. (1929). J. exp. Zool. $\underline{54}$, 319.

20. Gunn, S. A., T. C. Gould and W. A. D. Anderson (1963).
     J. Nat. Cancer Inst. $\underline{31}$, 745.

21. Goss, R. J. (1969). "Principles of Regeneration". (Academic
     Press, New York).

22. Goss, R. J. (1974). Perspect. Biol. Med. $\underline{17}$, 485.

23. Nash, J. (1931). Amer. J. Anat. $\underline{47}$, 425.

24. Marshall, E. K., Jr. (1934). Physiol. Rev. $\underline{14}$, 133.

25. Lyall, A. H. (1957). Quart. J. Microsc. Sci. $\underline{98}$, 101.

26. Brodie, P. F. (1969). Nature $\underline{221}$, 956.

27. Longacre, J. J. and R. Johansmann (1940). J. Thoracic Surg.
     $\underline{10}$, 131.

28. Bremer, J. L. (1937). J. Thoracic Surg. $\underline{6}$, 336.

29. Bonvalet, J.-P., M. Champion, F. Wanstok and G. Berjal (1972).
     Kid. Internat. $\underline{1}$, 391.

30. Imbert, M. J., G. Berjal, N. Moss, C. de Rouffignac, and
     J. P. Bonvalet (1974). Pflügers Arch. $\underline{346}$, 279.

31. Canter, C. E. and R. J. Goss (1975). Proc. Soc. Exp. Biol.
     Med. $\underline{148}$, (in press).

32. Twitty, V. C. (1930). J. Exp. Zool. $\underline{55}$, 43.

33. Harrison, R. G. (1924). Proc. Nat. Acad. Sci. $\underline{10}$, 69.

34. Church, G. (1953). Proc. Nat. Acad. Sci. $\underline{39}$, 877.

35. Twitty, V. C. (1940). Growth, Suppl. 109.

36. Harrison, R. G. (1929). Arch. Entw.-Mech. Org. $\underline{120}$, 1.

37. Ross, G., Jr., M. D. Cosgrove, P. Dragan, P. Mowat,
     J. Battenberg and W. E. Goodwin (1970). Surg. Forum $\underline{21}$, 526.

38. Baden, J. P., G. M. Wolf and R. D. Sellers (1973). J. Surg. Res. 14, 213.

39. Metcalf, D. (1963). Austral. J. Exp. Biol. 41, 51.

40. Metcalf, D. (1964). Transplantation 2, 387.

41. Liebelt, R. A., L. Vismara and A. G. Liebelt (1968). Proc. Soc. Exp. Biol. Med. 127, 458.

42. Dittmer, J. E., R. J. Goss and C. E. Dinsmore (1974) Amer. J. Anat. 141, 155.

43. Felts, W. J. L. (1959). Amer. J. Phys. Anthrop. 17, 201.

44. Chalmers, J. (1965). In: "Calcified Tissues". (Eds. L. J. Richelle and M. J. Dallemagne). p. 177. (Univ. of Liege).

45. Noel, J. F. and E. A. Wright (1972). J. Embryol. Exp. Morphol. 28, 633.

46. Goetz, R. H. and E. N. Keen (1957). Angiology 8, 542.

47. Schochet, J. (1973). J. Exp. Zool. 184, 259.

48. Davis, T. A. (1973). Forma et Functio 6, 373.

49. MacEwen, G. D. and J. L. Case (1967). Clin. Orthop. Rel. Res. 50, 147.

50. Cloudsley-Thompson, J. L. (1970). Sci. Jour. 6, 24.

51. Conklin, E. G. (1912). J. Morph. 23, 159.

52. Frankhauser, G. (1955). In: "Analysis of Development". (Eds. B. H. Willier, P. A. Weiss and V. Hamburger). p. 126. (W. B. Saunders Co., Philadelphia).

53. Walton, A. and J. Hammond (1938). Proc. Roy. Soc. B, 125, 311.

54. Hunter, G. L. (1956). J. agric. Sci. Camb. 48, 36.

55. Joubert, D. M. and J. Hammond (1958). J. agric. Sci. Camb. 51, 325.

56. Warwick, B. L. and R. O. Berry (1949). J. Hered. 40, 297.

57. Bowerman, H. R. L. and J. L. Hancock (1963). J. Reprod. Fert. 6. 326.

58. Hancock, J. L., P. T. McGovern and J. T. Stamp (1968).
     J. Reprod. Fert. Suppl. 3, 29.

59. Tarkowski, A. K. (1962). J. Embryol. Exp. Morph. 10, 476.

THE CENTRAL CONTROL OF GROWTH: ITS CONNECTION

WITH AGE-DEPENDENT DISEASE

P. R. J. Burch

Department of Medical Physics, University of Leeds

The General Infirmary, Leeds LS1 3EX

INTRODUCTION

Humans and farm animals alike begin life as a single cell, the zygote. The human organism grows for a period of about 20 years after fertilization and at maturity comprises 10^{14} cells. By any reckoning this constitutes a prodigious increase in cell number. But even more impressive are the phenomena of cytodifferentiation and the assembly of differentiated cells into those characteristic morphological structures that comprise the organs and vasculature of the body. In turn, these multiple components are anatomically and functionally integrated to serve the requirements of the whole organism.

The instructions for this dynamic and almost inconceivably elaborate process of growth and development reside, largely or wholly, in the genome. The final outcome – the realization of the *growth-potential* – is remarkably impervious to various temporary delays imposed by adverse environments. Thus, even when monozygotic twins are delivered with strikingly discordant birth weights, they become virtually identical, both morphologically and psychologically, by the age of 10. We are all familiar with the analogous phenomenon of 'catch-up' growth.

The growth of an organism from the zygote entails, therefore, much more than a net increase in size. During development, cells acquire, at the appropriate time and location, specialized bio-chemical, physiological, morphological and anatomical functions. The overall form of the organism is determined by the shape and contact relations of individual cells and extra-cellular tissues.

Although different organs follow distinctive growth curves, a
balance and harmony is preserved throughout the entire development
process. Not that the balance is a simple one. For example, the
human body shows an approximate left-right mirror symmetry, but
the departures from strict symmetry, and the rare but regular
occurence of viable anomalies such as *situs inversus totalis* and
dextrocardia are , perhaps, even more intriguing.

This enigma - one is tempted to say miracle - of growth
constitutes one of the most outstanding of the unsolved problems of
biology. How can so vast and intricate a process be accomplished
with so few errors? To unravel the secrets we shall require, not
only a detailed description of the growth of visible structures,
but an understanding of the genetic, molecular and cellular
mechanisms and interactions that guide and coordinate the intricate
steps of development.

THEORIES OF GROWTH

In view of the daunting complexity it is scarcely surprising
that no comprehensive theory of growth, cytodifferentiation and
development has yet emerged. Several authors have proposed that
cells in a given tissue synthesize specific inhibitors, that, when
present at a high enough local concentration, prevent that tissue
from growing (1,2,3,4,5,6,7).

Tyler believed that complementary pairs of substances,
resembling antigen and antibody, are present in each cell: they
are supposed to constitute a system that forms the basis of cell
structure (5). Differences in the rates of production of different
'antigens' give rise to cytodifferentiation. However, Tyler did
not attempt to describe the mechanisms that control the 'rates of
production' of the antigens and the size of differentiated tissues.

In Weiss's theory, 'templates', specific to the cell type, are
synthesized and confined within the cell where they regulate
growth in proportion to their concentration: each cell also produces
antagonistic complementary molecules - the 'antitemplates' - that
can block 'templates' (6, 7). Antitemplates diffuse freely out of
cells into the extracellular space and circulation, but also return
back to their cells of origin. They are continually catabolized
and excreted, but their loss is compensated by continuous production. When a stationary equilibrium between intra- and extracellular concentrations of antitemplates is reached, growth ceases.
This model invokes a form of negative-feedback - without, however,
a specified central comparator - but it fails to account for various
findings from experiments (8,9,10) on liver regeneration in
partially-hepatectomized rats and mice, and on compensatory kidney
hypertrophy in unilaterally-nephrectomized animals of several
species.

Rose's theory (4) is formally analogous to Weiss's (6,7). It postulates that specific inhibition of normal growth results when a critical concentration of inhibitor(s) is attained. 'Like inhibits like'.

Druckrey (3) introduced the idea of a two-stage system of feedback. The first stage resembles Weiss's system of 'templates' and 'antitemplates', but in the second stage, afferent signals flow from target cells to a 'higher' regulating centre. This centre exports mitotic stimulators (effectors) when the concentration of afferent signals declines.

Bullough (1,2) believes that the ultimate control of the mitotic rate in a tissue resides within that tissue and is exercised by 'chalones' which are tissue-specific, but neither species nor class-specific inhibitors. In my view, much of the experimental evidence for 'chalones' is consistent with the view that they are involved in the regulation of *asymmetrical* but not *symmetrical* mitotis - to borrow Osgood's (11) terms. Postnatal growth, which entails a net increase in tissue size, depends on symmetrical rather than asymmetrical mitosis.

Regenerative growth of the liver requires symmetrical mitosis of parenchymal cells in the liver remnant: the experiments of Fisher *et al.* (12) cannot be reconciled with the idea that liver regeneration is effected through a decrease in concentration of liver 'chalones'. Their experiments, utilizing ingenious surgery with two portions of liver in series, implicate an increase in the level of blood-borne positive mitogenic factors in regenerative growth, as opposed to a decrease in level of mitogenic inhibitors.

Control by the lymphoid system

In 1963, Burwell (13) stated: 'The unity of the integrity of the body depends not only upon the interdependence of *different* tissues and organs but also, and perhaps equally importantly, upon the interrelationship between the cells of any one differentiated tissue and cells of a *similar* differentiation wherever they may lie in the body'. He proposed ...'that in the mammal an essential function of lymphoid tissue is to establish and maintain *morphostasis* for many of its differentiated tissues'. (Weisz (14) defined *morphostasis* as the steady state condition that maintains a particular pattern.)

Burwell's thesis arose from reflections on: (i) the biological significance of species-specific, strain-specific, individual-specific and tissue-specific antigens; (ii) the inhibitory effects of tissue extracts on normal and regenerative growth; (iii) the anatomical distribution of lymphatics, lymph nodes, spleen and thymus; (iv) the pathways taken by transplantation antigens - and,

by inference, of normal tissue factors – within lymph nodes; and
(v) the biological significance of allograft and xenograft rejection.
Burwell believed that all these features could be understood and
explained if the physiological function of the lymphoid system is
the establishment and maintainance of morphostasis (13).

    Central homoeostatic control of growth. Relation between
    the mechanisms of growth control and age-dependent disease.

    Burwell's original ideas (13) led directly to the theory of
growth, cytodifferentiation and disease that we developed together
(15). We hold that diseases whose age patterns satisfy certain
mathematical criteria arise through a specific type and number of
somatic gene mutations in cells of the central, homoeostatic system,
that normally regulates the growth of differentiated target tissues
throughout the body (9,16,17,18,19).

    Burwell first acquainted me with his views in early 1963 when
I was studying the age patterns of diseases in man that are widely
regarded as 'autoimmune'. My main objective was to try to distinguish between the 'disturbed-antigen' and 'disturbed-tolerance'
theories (20) of autoimmunity. The 'disturbed-antigen' theory
holds that damage to a target tissue (for example, thyroid)
inflicted by trauma, viral infection, etc., releases previously
sequestered antigens towards which there is no immune tolerance:
as a result, an autoimmune response against the released antigenic
material is elicited. The autoantibodies attack the damaged
tissue; they exacerbate the damage and a vicious spiral is created.
(20).

    It had become obvious that the reproducibility and the detailed
mathematical form of the age patterns of autoimmune diseases were
incompatible with this version of the 'disturbed-antigen' theory.
On the other hand, these same properties were strikingly consistent
with the main tenets of Burnet's 'forbidden clone' theory (20) of
'disturbed-tolerance' autoimmunity.

    As part of his general theory of acquired immunity, Burnet had
argued that the genes that code for humoral and cellular antibodies
undergo spontaneous somatic mutation (20). Occasionally, and by
chance, a self-reacting or auto-antibody will be coded by a mutant
gene. According to Burnet, a physiological monitor should normally
detect and then eliminate any lympoid cell synthesizing an auto-
antibody. But for one reason or another, the monitor may be
expected to fail from time to time, thus allowing the mutant
lymphoid stem cell to propagate a (normally) 'forbidden clone' of
similarly-mutant descendent cells. Such lymphoid cells, or their
secreted humoral auto-antibodies, will then attack those target
cells in the body that carry complementary antigenic determinants.
Although I now believe that the details of Burnet's theory (20) are

incorrect, his basic concept of *forbidden clones* seems to me to contain a truth of far-reaching consequences.

Since 1963, I have studied an enormous volume of evidence, not only for the so-called 'autoimmune' diseases, but for many other well-defined natural diseases, malignant and non-malignant, infectious and non-infectious, of early and late onset. In every instance – given reliable diagnosis – the data agree with the idea that age-dependent disease is initiated spontaneously by a small number of specific somatic mutations. A very simple biological model (17) – a direct descendent of Burnet's 'forbidden clone' theory fits and explains all the reliable age patterns I have examined so far. Moreover, the same model can explain how a few random somatic mutations – or only one – can give rise to simultaneous *multi-focal* or diffuse lesions (17,18). (Burnet's 'forbidden clone' type of theory avoids the impossibility of having to explain how rare and *random* events can occur simultaneously at multiple sites.) The forbidden clone can 'amplify' the effect of even a single random event, in a single cell, so that it can involve numerous target cells simultaneously at one or multiple foci.

Although the range of agreement between theory and observation was wide I encountered some remarkable and unexpected features that could not be convincingly explained by Burnet's theory: (i) the average rate of a given initiating somatic mutation appears to be effectively constant from around birth, throughout postnatal growth, to the end of the lifespan; and (ii) when the same disease affects both males and females, the average rate of a given initiating somatic mutation is either the same in both sexes, or it is twice as high in females as in males. Another property (see below) gave me somewhat more trouble, although ultimately it converted me to the essence of Burwell's theory.

When the overall evidence relating to (i) and (ii) was examined in detail only one plausible inference could be drawn: the number of cells at somatic mutational risk with respect to the initiation of a given age-dependent disease remains effectively constant from around birth, throughout post-natal growth to the end of the lifespan (17). I had expected the number of cells 'at risk' to increase roughly in proportion to body size and that the number in men would, on the average, exceed the corresponding number in women. Neither expectation was confirmed.

Of course, the simple connexions observed between the average rates of initiation of various diseases in man and the complement (a) of X-chromosomes (16,17) and (b) of chromosome 21 (17,19) are readily explained if somatic mutation of genes on (a) the X chromosome and (b) autosome 21 can contribute to the formation of forbidden clones. However, there is no indication from immunogenetics that X-linked genes code for polypeptide chains in immunoglo-

bulins and hence, where the somatic mutation of X-linked genes is
implicated, it is most unlikely that forbidden clones of immuno-
globulin autoantibodies are the primary pathogenic agents.

    But what is the *biological* significance of the constancy of
the number of cells at somatic mutational risk?

    Suppose Burwell is essentially right? Suppose the lymphoid
system, or a part of it, regulates the growth of target tissues?
In that case a negative-feedback control arrangement would be
needed, with efferent and afferent pathways and a comparator in
the central control apparatus (Fig. 1). Biology, in common with
the electronics engineer, cannot dispense with these elementary
requirements of negative-feedback control systems. In particular,
a *comparator* of some form is needed to furnish a fixed yardstick,
or datum, against which the size of the target tissue can be
'measured'. It occurred to me that a *fixed number* of *stem cells*
in the central control apparatus would be admirably suited to
function as a comparator. Indeed, it is not too easy to imagine a
more plausible alternative. Although not convinced of his case
at that stage I took the precaution of moderating my objections to
Burwell's arguments.

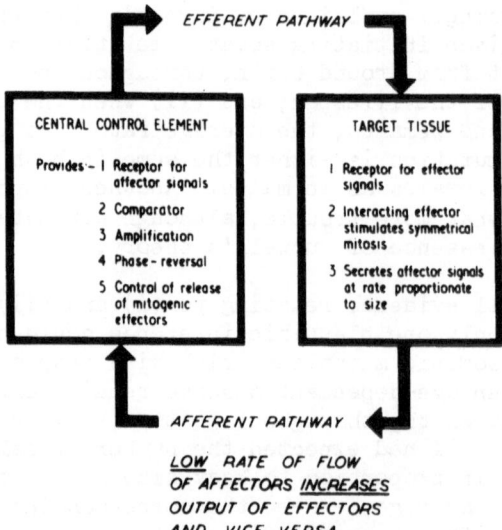

Fig. 1.   Outline of the basic requirements for the negative-
feedback control of the growth of a target tissue by a central
control element (17). Mitogenic effectors, exported from a
central control element along the efferent pathway, have to be
able to recognize their cognate target tissue. Similarly,
affectors, secreted from the target tissue along the afferent pathway, need to recognize their cognate control element in the central
system.

To my mind, the outstanding challenge of Burwell's scheme (13) was the problem of tissue recognition. He believed that this depended on a complementary, or antibody-antigen, type of relation between lymphoid control and target recognition factors. The high specificity of such relations is fully recognized and, as such, it would be admirably suited to growth control as many investigators have appreciated. Nevertheless, I contended (16) that such relations are characteristic of pathogenic, autoimmune-type inter-actions, and that Burwell's scheme (13) offered no scope for autoimmunity. I must also point out that, in a negative-feedback control of growth, the problem of recognition arises at both ends of the feedback loop. Effectors have to be able to recognize the cells of their target tissue, but also, affectors have to be able to recognize their specific control elements in the central apparatus. Although I had an intuition that the evidence for so-called autoimmune diseases contained the solution to this problem of recognition it was some months before I discovered the answer.

The details of the evidence and argument have been given elsewhere (15,17) and only the conclusions will be described here. These relate to the role of the genes that predispose to age-dependent, that is *autoaggressive,* disease. I inferred that pre-disposing genes perform a *dual* function. They code for complex recognition macromolecules, both in 'central' and in 'target' cells. In other words, a specific autoaggressive disease involved a system in which, prior to somatic mutation, parts of recognition molecules in central cells are identical to corresponding parts of cognate molecules in target cells. Identity between complex molecules such as polypeptide chains frequently gives rise to specific self-association interactions, based, presumably, on London-van der Waal's self-recognition interactions (21,22,23,24,25).

Identity between central and target recognition molecules provides, therefore, an admirable basis for growth control (Fig.2). A specific target tissue differs from every other tissue but its recognition molecules (*tissue coding factors* or TCFs) bear an *identity* relation to those recognition molecules (*mitotic coding proteins* or MCPs) that are synthesized by its controlling element in the central system of growth control. No greater economy in the use of recognition genes could have evolved. It is also of interest that another argument (17), unconnected with the evidence for age-dependent disease, shows that if growth in complex organisms is centrally regulated, no genetic basis for recognition tissue other than the identity one described here could have evolved. Any system based on one set of genes in the central control and a *different* set in target tissues would rapidly break down due to gene mutation in germ cells alone. A species relying on such a system could not survive: neither could it evolve. This argument can be regarded as a 'proof' - or at least corroboration - of the one derived from the evidence for age-dependent disease.

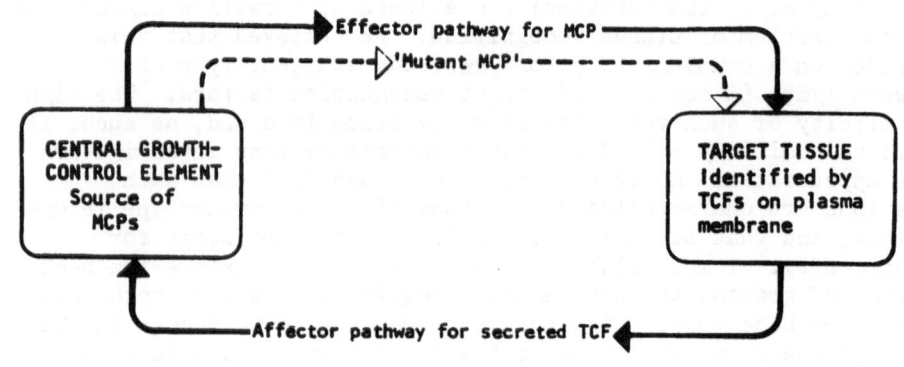

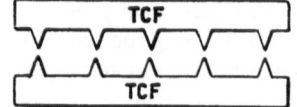

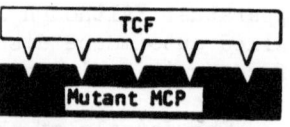

TCF - TCF interaction between   MCP - TCF interaction promotes   Complementary interaction
similar contiguous cells in-   symmetrical mitosis in target   between 'mutant MCP' and
hibits mitosis   cell   TCF leads to disease

Fig. 2.   Inferred recognition relations in the negative-feedback
control of growth (17).   Each distinctive target tissue is characterized by a *tissue coding factor* (TCF) that is a complex of major
and minor histocompatibility antigens, a classical 'tissue-
specific' antigen and subtler determinants that distinguish one
mosaic element of a classical 'tissue' from other such mosaic
elements.   TCFs are carried on the plasma membrane of target cells
and secreted in humoral form, perhaps as high density α-lipoproteins,
to act as affector signals.   Cell-contained TCFs, or their component parts, determine cytodifferentiation.   *Mitotic control proteins*
(MCPs) from the central system effect symmetrical mitosis in cognate
target cells.   The recognition protein components of MCPs, and those
of cognate TCFs, are coded by the same genes and therefore have the
same amino acid sequences.   In the control system that uses cellular
effectors, MCPs are carried on the plasma membrane of T-lymphocytes.
Humoral MCPs are, perhaps, to be found in the α_2-macroglubulin
fraction.   They are distinguished from humoral TCFs by the 'non-
recognition' protein and/or non-protein components.

     Normally, contiguity between neighbouring cells in an organized
tissue inhibits symmetrical mitosis through TCF-TCF interactions.
Specific MCP-TCF interactions are needed to overcome this inhibition.
In age-dependent, or *autoaggressive* disease, the identity relation
between a normal MCP and its TCF is replaced by a specific complementary (and pathogenic) relation between the mutant MCP and the
target TCF.

I have estimated (17) that the total number of distinctive tissue elements in the body might be as high as $\sim 10^{10}$. Because the total amount of DNA in the haploid human genome suffices to specify only $\sim 6 \times 10^6$ genes – assuming each gene codes for a polypeptide chain of 20 000 daltons – it is clear that each recognition molecule (mitotic control protein, tissue coding factor) must be made up of a *combination* of distinctive polypeptides (15,17). We deduce that MCP-TCF recognition molecules are constructed on a hierarchical principle, each being assembled from the following components: major histocompatibility antigens forming part of the TCFs of most cells in the body, and, at successively higher levels of specificity, minor histocompatibility antigens; classical tissue-specific antigens; and, finally, the subtlest determinants that distinguish one mosaic element of a classical tissue from other such elements.

Furthermore, I conclude that TCFs act not only as recognition molecules in growth control but also as the primary determinants of cytodifferentiation (17). I infer that TCFs, or their component parts, determine cytodifferentiation partly through their intrinsic properties and partly by derepressing specific sets of genes (17).

In complex organisms such as mammals, a substantial proportion of the genetic material must be devoted to specifying tissue recognition molecules for use in growth control and cytodifferentiation. These requirements are frequently overlooked, or even unrecognized, and distinguished genetic authorities maintain that the human genome has only about 30 000 or 50 000 functional structural genes. Davidson and Britten (25) have recently shown, however, that the sea urchin embryo, at the 600-cell stage, has about 14 000 functioning genes and they presume that, throughout the lifetime of the organism, the total number of such genes will be many times greater than 14 000. It is surely not too presumptuous to suppose that the genetic complement of mammals greatly exceeds that of sea urchins.

EXPERIMENTAL EVIDENCE FOR THE CENTRAL CONTROL OF GROWTH

The evidence is too voluminous to review in detail here but brief reference will be made to pertinent experiments on liver regeneration and compensatory growth of the kidney.

An early suggestion that circulating humoral factors promote liver regeneration came from experiments by Bucher (8) on pairs of rats in parabiotic union. Cells in the liver of the 'normal' partner undergo a marked increase in DNA synthesis in response to the ablation of the liver in its parabiotic partner (8, 27). More elaborate experiments along these lines by Sakai (10) have confirmed the existence of a short-lived humoral factor that promotes liver growth. The experiments of Fisher *et al.* (12) referred to above are important in showing that regeneration depends on an increase in concentration of a mitogenic agent and not a decrease in concentration of a mitotic inhibitor.

Some early experiments of Czeizel, Vaczo and Kertai (9) showed that 500R of x-rays delivered to the liver remnant of a two-thirds partially-hepatectomized rat, with the remainder of the animal shielded from radiation, had a negligible effect on regeneration after one week. However, the same dose delivered to the whole body of the rat, with the liver remnant shielded, reduced regeneration to 69 *per cent* at one week. The inhibitory effect of whole-body irradiation (500R) was almost completely reversed when 2×10^7 syngeneic, unirradiated bone marrow cells, were injected within two hours after irradiation. Irradiation of the injected bone marrow (500R) abolished its regenerative action. Injection of unirradiated bone marrow into an unirradiated animal boosted regeneration and gave an overshoot at one week (9).

These various experiments demonstrated that the integrity of the bone marrow is essential to normal regenerative growth. (According to our theory, the stem cells of the central growth-control system and their immediate descendents are located in the bone marrow.)

Davies *et al.* (28) studied liver regeneration in the mouse, using radiation, and both allogeneic and syngeneic bone marrow injections. They found that the extent of liver regeneration relates not only to the chimaeric state, with syngeneic chimaeras generally showing better regeneration than allogenic chimaeras, but also to the degree of lymphopenia. Severely lymphopenic animals, whether syngeneic or allogeneic chimaeras, showed poor liver regeneration after two-thirds partial hepatectomy.

Fox and Wahman (29) in experiments designed to test our theory, investigated the effects of radiation and of injected spleen cells on compensatory growth of the kidney in mice. They found that 'sensitized' spleen cells, taken from mice that had undergone unilateral nephrectomy, were the most efficient promoters of compensatory kidney growth (including nucleated cell count) in unilaterally-nephrectomized animals. The increase in whole-body weight of irradiated young animals was, however, better promoted by 'unsensitized', than by 'kidney sensitized' spleen cells. Unirradiated animals showed a sharp increase in the peripheral leucocyte count at 2 days after unilateral nephrectomy. All their findings (29) were consistent with the theory that the lymphoid system, and in particular the spleen, are actively involved in promoting compensatory growth and hyperplasia of the kidney.

CELL SYSTEMS INVOLVED IN GROWTH-CONTROL

Anatomical considerations, together with evidence from (a) transplantation studies and (b) the age-dependence of disease in relation to the anatomical site of the target tissue, indicate that two distinctive systems, at least, are involved in the central control of growth (15,17). When the target tissue - notably endothelium - lies on the blood side of a blood-tissue barrier, the effector cells in growth control, and the primary pathogens in autoaggressive disease, appear to be T-lymphocytes. (These are also the cells that are primarily involved in the rejection of allografts. There are, of course, parallels between certain types of graft rejection and autoaggressive disease. When the target tissue lies behind a blood-tissue barrier - and is therefore normally inaccessible to T-lymphocytes - the effectors (intermediate and/or final) of symmetrical mitosis are necessarily humoral. Circumstantial evidence suggests that these factors are, in some instances at least, a_2- macroglobulins (13,17,19). Recent observations indicate that a_2M is carried on the surface of specific human and mouse lymphocytes and might be synthesized by a special sub-population of bone-marrow-dependent lymphocytes (30,31). Other observations suggest that tissue mast cells are implicated, at least in some instances, in the peripheral part of the effector pathway of the humoral system of growth control (17).

CONCLUSIONS

At the fundamental level our theory is wholly unoriginal in maintaining that growth and the intricacies of morphology - given an adequate environment - are genetically determined. Any theory that denied this well-known phenomenon could, of course, be safely ignored. But in identifying growth-control and morphogenetic genes with those that code for major and minor histocompatibility antigens, classical tissue-specific antigens and subtler recognition polypeptides, our theory has, perhaps, some claim to originality. Moreover it explains, indeed it predicted, the numerous associations discovered recently between autoaggressive disease in man - a breakdown in normal growth-control - and the major HL-A antigens.

Although the fundamental basis of growth control is indisputably genetic the functioning of the control system can be modified in various ways and notably by classical hormonal agents. In principle, an artificial reduction in the concentration of afferent TCFs - brought about, for example, by immunological means - should generally produce an increase in the size of the target tissue that synthesized the affected TCFs. Whether such interference could be easily effected in farm animals, and whether the results would be specific enough to be of economic value, are questions that need to be investigated.

What is more novel about our theory, though still not entirely
new, is the intimate connexion it postulates between growth control
and age-dependent disease. We deduce that diseases with reproducible
age-distributions - *autoaggressive* diseases - are initiated by a
special form of somatic mutation in stem cells of the central
system of growth control. Each autoaggressive disease is confined
to individuals with a specific genetic predisposition: this is
frequently polygenic. Although links between morphology and
disease in man have been recognized since antiquity, I believe our
theory is the first to demonstrate that such associations are
necessary to the biology of complex multi-tissue organisms. The
implications for animal breeding are obvious. Genetic factors
that determine a particular morphological trait may well predispose
to undesired degenerative, neoplastic, or infectious diseases, or
any combination of these. However, these genetic factors are so
complicated they should offer considerable scope to the ingenious
breeder. Because the genetic predisposition to autoaggressive
disease is often polygenic, it should be possible to substitute,
say, one allele for another to preserve the desired morphological
trait while avoiding the particular combination of genes (alleles)
that predispose to the disease. That is to say, in this polygenic
situation, combinations of desirable qualities are not necessarily
excluded.

Experience and theory leave no doubt, therefore, that the
control of normal growth and development has a genetic basis.
Nevertheless, theory and experiment both indicate that, by suitable
manipulation, the central system of growth control can be tricked
into producing over-growth of particular tissues.

REFERENCES

1.   Bullough, W. S. (1962). Biol. Rev. 37, 307.

2.   Bullough, W. S. (1965). Cancer Res. 25, 1683.

3.   Druckery, H. (1959). In: "Mechanisms of Action". Ciba
      Foundation Symposium on Carcinogenesis.
      Eds. G. E. W. Wolstenholme and M. O'Connor. P. 110. (Churchill,
      London).

4.   Rose, S. M. (1958). J. Nat. Cancer Inst. 20, 653.

5.   Tyler, A. (1947). Growth, 10. 7.

6.   Weiss, P. A. (1950). Q. Rev. Biol. 25, 177.

7.   Weiss, P. A. and Kavanau, J. L. (1957). J. Gen. Physiol. 41,
      1.

8. Bucher, N. L. R. (1963). Internat. Rev. Cytol. 15, 245.

9. Czeizel, E.. Vaczo, G. and Kertai, P. (1962). Nature, 196, 240.

10. Sakai, A. (1970). Nature, 228, 1186.

11. Osgood, E. E. (1957). J. Nat. Cancer Inst. 18, 155.

12. Fisher, B., Szuch, P. and Fisher, E. R. (1971). Cancer Res. 31, 322.

13. Burwell, R. G. (1963). Lancet, ii, 69.

14. Weisz, P. B. (1951). Amer. Naturalist, 85, 293.

15. Burch, P. R. J. and Burwell, R. G. (1965). Q. Rev. Biol. 40, 252.

16. Burch, P. R. J. (1963). Lancet, i, 1253.

17. Burch, P. R. J. (1968). "An Inquiry Concerning Growth, Disease and Ageing". (Oliver and Boyd, Edinburgh).

18. Burch, P. R. J. (1970). Nature, 225, 512.

19. Burch, P. R. J. and Milunsky, A. (1969). Lancet, i, 554.

20. Burnet, F. M. (1959). "The Clonal Selection Theory of Acquired Immunity". (Cambridge University Press, London).

21. Burachik, M., Craig, L. C. and Chang, J. (1970). Biochemistry, 9, 3293.

22. Deonier, R. C. and Williams, J. W. (1970). Biochemistry, 9, 4260.

23. Laiken, S., Printz, M.P. and Craig, L. C. (1971). Biochem. Biophys. Res. Comm. 43, 595.

24. Ruttenberg, M. A., King, T. P. and Craig, L. C. (1966). Biochemistry, 5, 2857.

25. Stevenson, G. T. and Strauss, D. (1968). Biochem. J. 108, 375.

26. Davidson, E. H. and Britten, R. J. (1974). Cancer Res. 34, 2034.

27. Moolten, F. L. and Bucher, N. L. R. (1967). Science, 158, 272.

28. Davies, A. J. S., Leuchars, E., Doak, S. M. A. and Cross, A. M.
     (1964). Nature, <u>201</u>, 1097.

29. Fox, M. and Wahman, G. E. (1968). Invest. Urol. <u>5</u>, 521.

30. McCormick, J. H., Nelson, D., Tunstall, A. M. and James, K.
     (1973). Nature New Biol. <u>246</u>, 78.

31. Tunstall, A. M. and James, K. (1974). Clin. Exp. Immunol. <u>17</u>,
     697.

DISCUSSION

Asked for an example of the genetic control of growth, Prof. Burch noted that, whereas the birth weight of monozygotic twins was often different, this difference had usually disappeared by the age of 4 or 5 years. *Prof. Robertson* asked what Prof. Goss meant when he said that something was genetically determined. *Prof. Goss* thought that there were genes determining the existence but not the size of an organ; other genes governed the development of control mechanisms which secondarily regulated the size of organs according to environmental demand. Asked for his own definition, *Prof. Robertson* said that his was an operational one; a characteristic was genetically determined if affected by genetic substitution. *Dr. Dickerson* added that the extent to which such substitution was expressed depended on the environment, and *Prof. Robertson* agreed with this.

Dr. Frisch asked at what stage of the growth cycle fish begin to reproduce. *Prof. Goss* did not know, but noted that fish apparently retain their reproductive capacity throughout life. *Prof. Cahill* thought that the size of viviparous fish at birth was independent of the size of the mother, and wondered how the ovulation rate was regulated in proportion to maternal size. *Prof. Goss* noted that the roe of larger fish was bigger which he thought indicated a higher ovulation rate. *Dr. van Es* noted that the growing carp is sexually mature at $1\frac{1}{2}$-2 years. In its nitrogen and energy metabolism, it is remarkably similar to the growing chicken, although growth apparently continues slowly throughout life. *Dr. Burleigh* asked whether marine mammals might not offer the best possibilities for more rapid meat production, since they are capable of more rapid cell division than land animals.

Dr. Fowler asked how, according to Prof. Burch's theory, a fixed number of comparator cells could allow for both an increase in size and differential development during growth. *Prof. Burch* replied that the kinetics of the growth control circuit varied from one organ to another, and so allowed for differential growth. *Prof. Cahill* asked how, in the mutant hairless mouse which apparently lacks both B and T lymphocytes and therefore immunological surveillance and which develops carcinomas, the proportionality of growth is maintained. *Prof. Burch* replied that, even in the absence of a thymus, there would be a modified form of T lymphocyte system which would allow some growth control to function, though less precisely. He added that B lymphocytes are not involved in normal growth, but are concerned with the growth of forbidden clones and in their absence there is more disease. *Prof. Goss* asked how growth was controlled before immunological competence was developed. There was no theory which yet explained the control of growth in the embryo. *Dr. Frisch* asked

what was the significance of the involution of the thymus at
maturity. *Prof. Burch* noted the parallelism between the output of
lymphocytes from the thymus and the increments of growth; the
normal immunological function of the thymus was complete at
maturity. *Prof. Goss* asked how Prof. Burch's theory explained
the control of erythrocyte number which was known to be determined
by oxygen demand through the mediation of erythropoetin.
Prof. Burch replied that his theory dealt only with the control of
the number of stem or basal cells, and he fully agreed that
functional demand controlled the further hypertrophy or proliferation of these cells. *Prof. Ingram* sought to reconcile the views
of Profs. Goss and Burch. He thought that the important question
asked by Prof. Goss was why growth stopped. This implied the
existence of a comparator which was an essential component of
Prof. Burch's theory.

Environmental Control of Growth

ENVIRONMENTAL CONTROL OF GROWTH:

THE MATERNAL ENVIRONMENT

E. M. Widdowson

Department of Investigative Medicine

University of Cambridge

INTRODUCTION

Nutrition is the dominating environmental influence that determines the rate of growth before birth. The blood of the mother supplies the foetus with all its nutrients, and one might suppose that its composition would be important in determining the rate of foetal growth. This is not so, for the composition of the plasma is much the same throughout the mammalian kingdom. There are a few instances where the composition of the mother's blood can affect the development of the foetus and I shall refer to these later on. Generally speaking, however, it is the quantity of blood reaching the foetus that determines its rate of growth. This is true when we make comparisons between species, and it is also true within any one species. A small species of animal may grow faster than a large species in early foetal life - the foetal rat, for example, grows to be a weight of 5 grams during its 3 weeks gestation, but the human foetus still weighs less than one gram after 8 weeks. On the whole large animals have a longer gestation period than small ones, and their cells go on dividing for longer than those of small ones before they are born. Gain in weight becomes progressively faster as gestation proceeds in all animals, and an animal belonging to a large species must sooner or later have a larger placental circulation and area of placental membrane for maternal and foetal transfer than one belonging to a small one.

REASONS FOR A SMALL BLOOD FLOW TO THE FOETUS

Table 1 sets out the main causes of a small blood flow to the foetus and hence slow growth in utero.

Table 1

Reasons for an inadequate blood supply to the foetus
and intrauterine undernutrition.

A small mother

Placental insufficiency

Large number of feotuses in uterus

Implantation at site where blood supply is poor

Undernutrition of mother

Disease of mother.

Small Size of Mother

Walton and Hammond's (1) classical experiment, in which they
crossed a Shire horse with a Shetland mare, and a Shire mare with
a Shetland pony, was a dramatic demonstration of what the size of
the mother could do for the growth of her foetus. The size of
the foal at birth depended upon the size of the mother, and was
appropriate to it. The Shetland mare, with her smaller circulatory system and uterine blood supply, produced a much smaller foal
than the Shire mare. It weighed 17 kg compared with 53 kg. She
was able to nourish a foetus so that it grew at the same rate as
a pure bred Shetland foetus, but she could do no more. Her foal
was small for nutritional, not for genetic reasons. It was only
after birth that the genetic influence of the large father began
to become apparent. Similar experiments have been made on cattle
(2,3,4), sheep (5), and mice (6). Dogs might be even better
animals for anyone who wishes to extend these experiments.

In the human species it appears that small women tend to have
smaller babies than large ones, whatever the size of the father
has been (7,8). If, however, the father has been tall, then the
small baby of the small mother may well grow into a tall adult.

'Placental Insufficiency'

The placenta is a vital organ so far as nutrition of the
foetus is concerned. Some substances such as oxygen, carbon
dioxide and water diffuse freely across it in both directions;
others such as glucose diffuse less freely, and the concentration
in foetal plasma is lower than in maternal plasma. This has been
shown experimentally to apply over a wide range of maternal levels
in the mare, cow and ewe (9). For other substances there must be

an active transport mechanism, for they are at higher concentrations in the foetal than the maternal plasma. This is true, for example, of amino acids, calcium, phosphorus and potassium. The placenta, moreover, has itself a high rate of metabolism, and it may alter organic substances during their passage through it. Synthesis of proteins, enzymes, nucleic acids, high energy phosphates and hormones all take place in the placenta. Anything, therefore, that interferes with the metabolism of the placenta is likely to hinder the growth of the foetus. Generally speaking, within one species, a small foetus will be found to have a small placenta (10), and of course the placenta, like the foetus, depends on the maternal blood for its nutrients. Pathological changes in the placenta are also likely to lead to foetal undernutrition. Infarcts, haemangiomas, and areas of premature separation or thrombosis of large foetal blood vessels may do this (11).

The placenta is a transient organ. It grows, lives its life and ages like other tissues of the body. One of the signs of its ageing is the calcification of its blood vessels, which must interfere with its circulation. Sometimes the placenta begins to age prematurely, just when the foetal requirements for nutrients are at their highest. Even in normal full term human births there are signs that the placenta is not able to provide the foetus with enough nutrients to maintain its previous rate of growth, which begins to fall off after about the 36th week when the foetus weighs about 3 kg (12,13). This is even more likely to occur if the baby is born post-maturely, and such babies may even lose weight; they are born light for their body length (14).

Large Number of Foetuses in the Uterus

In the early stages of gestation in all species food and accommodation in the uterus are always ample, whatever the number in the litter. In the later stages, however, the number of young sharing the blood supply in the uterus can affect the size of the individual young in some species. This is certainly true of rabbits, mice, rats, guinea pigs, sheep and man (12,15,16,17,18). In man the weight of the foetus is independent of numbers in the uterus till about the 26th week of gestation, but thereafter twins grow more slowly than singletons, triplets still more slowly, and quadruplets more slowly still (12). In some species, for example the rat and pig, the length of gestation is not reduced by litter size so the smaller weight of individuals in large litters must be due to an inadequate blood supply to each foetus and hence to undernutrition and slow growth. In other species, for example human beings and guinea pigs, the length of gestation is reduced if the litter is large, and this further limits the weight of members of multiple births because the more there are the earlier they are usually born (12,19).

Implantation at Site where Blood Supply is Poor

Multiple births do not always produce foetuses of uniform size. Dysmaturity as it occurs spontaneously in animals in the form of runts has been well known to farmers and stockbreeders for centuries. McLaren and Michie (20) were the first to take an interest in runts from the scientific point of view. They worked with the mouse, and showed that runts occurred at specific sites in the uterine horn, particularly at the ovarian end and in the centre of each horn. In both positions the maternal blood supply to the placenta was poorer than elsewhere, the foetus at the ovarian end sharing its supply with the ovary, and the one in the centre being at the junction of the vascular territories of the ovarian and uterine arteries.

Wigglesworth (21) followed this up experimentally by ligating the main uterine blood vessels to one horn on the 16th day of pregnancy in the rat. The animals were killed 4 days later, 24 hours before the litter was due to be born. The foetuses at the vaginal end of the ligated horn were very small, while those at the ovarian end were not affected.

The 'runt pig', with which many of you must be familiar, has clearly got an important place in farming lore. It has a large variety of names from one part of England to another (Fig. 1). Some are terms of contempt and to this category 'runt' belongs and so does 'waster'; others are diminutive, such as didling and winkling while others are terms of endearment – little darling for example.

Perry and Rowell (22) reported that the small foetal pig in a large litter was generally fourth along the uterine horn, again where the blood supply is not as good as elsewhere. In fact the extremes in size of pigs, as of other species, at full term fall at the ends of a normal distribution (23). Within one litter the weights of viable piglets may vary by a factor of four (24). The variation in weight among newborn pigs is greater than that in some other species. Large litters, a long gestation period and relative maturity at birth all contribute to variations in weight of the newborn animal. In rats, for example, which are born after a short gestation period in a very immature state, the weight of the largest in a litter is seldom more than 1.5 times the smallest, but in guinea pigs the weight may vary by 2.5 times.

Undernutrition of Mother

It was shown thirty years ago by Wallace (25) that lambs born to ewes that had been fed on a low plane of nutrition during the last six weeks of pregnancy weighed 40% less than those of well-nourished ewes. The plane of nutrition during the last part of

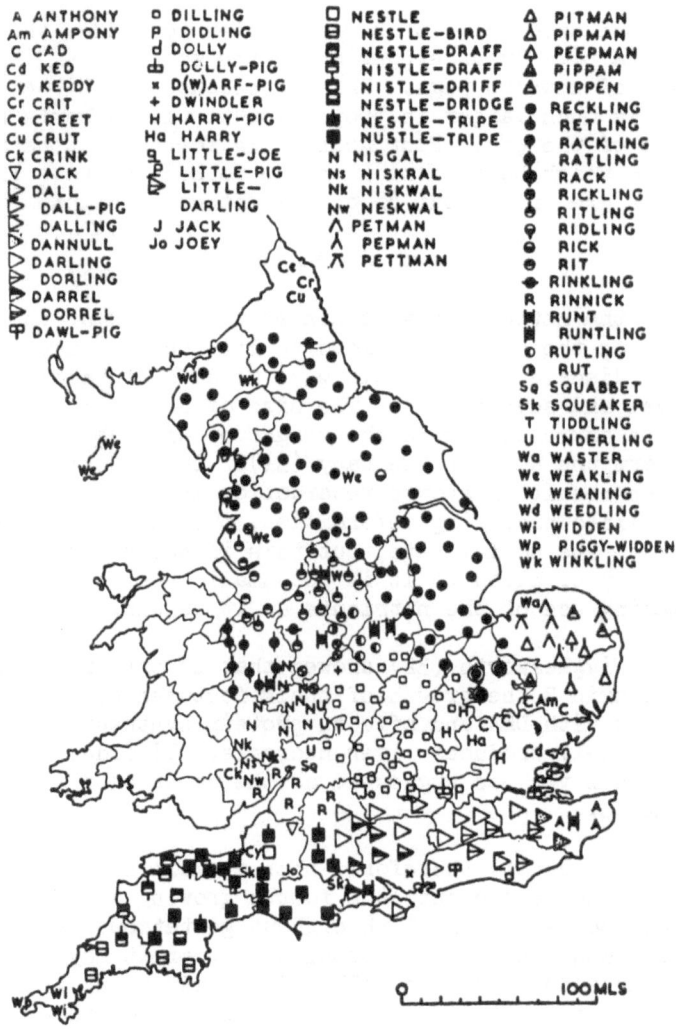

A ANTHONY
Am AMPONY
C CAD
Cd KED
Cy KEDDY
Cr CRIT
Ce CREET
Cu CRUT
Ck CRINK
▽ DACK
▷ DALL
DALL-PIG
DALLING
DANNULL
DARLING
DORLING
DARREL
DORREL
DAWL-PIG

D DILLING
P DIDLING
d DOLLY
db DOLLY-PIG
x D(W)ARF-PIG
+ DWINDLER
H HARRY-PIG
Ha HARRY
LITTLE-JOE
LITTLE-PIG
LITTLE-
  DARLING
J JACK
Jo JOEY

NESTLE
NESTLE-BIRD
NESTLE-DRAFF
NISTLE-DRAFF
NISTLE-DRIFF
NESTLE-DRIDGE
NESTLE-TRIPE
NUSTLE-TRIPE
N NISGAL
Ns NISKRAL
Nk NISKWAL
Nw NESKWAL
Λ PETMAN
Λ PEPMAN
Λ PETTMAN

△ PITMAN
△ PIPMAN
△ PEEPMAN
△ PIPPAM
△ PIPPEN
• RECKLING
RETLING
RACKLING
RATLING
RACK
RICKLING
RITLING
RIDLING
RICK
RIT
RINKLING
R RINNICK
RUNT
RUNTLING
RUTLING
RUT
Sq SQUABBET
Sk SQUEAKER
T TIDDLING
U UNDERLING
Wa WASTER
We WEAKLING
W WEANING
Wd WEEDLING
Wi WIDDEN
Wp PIGGY-WIDDEN
Wk WINKLING

Fig. 1   The weakest pig in the litter (50).

pregnancy, moreover, was much more important than that during the
early part. The rate of growth of the human foetus, too, is
affected by poor nutrition of the mother. Babies born in Holland
and Germany during the periods of severe food shortage during and
after the last war were a little smaller, on a statistical basis,
than those born in the same towns in times of plenty (26,27,28,29).
The extent of the effect of maternal undernutrition on the growth
of the foetus depends not only upon the degree of under-nutrition,

but also upon the species. It is much easier to reduce the size
of newborn guinea pigs by undernourishing the mother than it is to
reduce the size of newborn pigs. This is because the litter of
the guinea pig is much larger in proportion to the mother's size
than the litter of the pig. Four guinea pig foetuses at term -
a usual number - weigh half as much as the mother, whereas 12
new born piglets are only about 8% of the mother's weight.

Disease of the Mother

Disease of the mother can also hinder the growth of the
foetus by reducing the supply of blood to it. Chronic hypertensive
cardiovascular disease acts in this way, as does 'toxaemia', if
it is prolonged and severe (30). The size of a woman's heart has
been found to be correlated with the weight of the baby; the
smaller the heart, the smaller the baby (31). Intrauterine
infections often result in growth retardation in utero. Rubella,
for example, besides producing deformities, prevents the human
foetus growing at its normal rate, but this is not necessarily
because the supply of blood to it is too small; the infection
prevents it making use of the nutrients as it should.

Butler and Wigglesworth (32) showed that if pregnant rats
were given Aflatoxin B the growth of their foetuses was retarded.
Aflatoxin B binds with DNA and affects protein synthesis in the
liver, and ultimately produces liver necrosis. Disordered protein
metabolism in the liver was suggested as the cause of this foetal
growth retardation.

There seems no doubt, therefore, that a small blood flow to
the placenta and foetus prevents the foetus growing as it should.
The supply of blood may be small throughout gestation, may be
small for a time and then become normal again, or it may be normal
until late in gestation and then fall off. However and whenever
it occurs, it is important to consider why blood flow should have
so large an influence on the rate of foetal growth. Many
nutrients reach the foetus in quantities far greater than it
needs, and a large part of what reaches the foetus is returned to
the maternal circulation. Young (33) has investigated the
possibility that amino acids might be the limiting factor. She
showed that less of the non-metabolisable amino acid α amino
isobutyric acid, was transferred to the 'runt' guinea pig in a
litter than to the larger ones. Similarly, the transfer was
less to the small foetuses of undernourished guinea pigs than to
those of well-nourished animals (34). Generally speaking, in
these acute experiments the amount transferred was the same per
unit body weight irrespective of foetal size. This did not solve
the problem, however, for it was impossible to say whether the
rate of transfer of any particular amino acid was limiting the
rate of growth or not.

The rates of transfer of calcium and phosphorus from mother to foetus have been measured in animals by the adminstration of ^{45}Ca and ^{32}P. In small laboratory animals the rate of passage of both elements across the placenta is little more than enough to provide for the needs of the young, particularly if the litter is a large one. Thirteen rabbit foetuses were found to require the whole of the calcium crossing the placenta in order to grow normally, but when there were only 7 foetuses the calcium reaching their placenta was nearly twice that required (35). Wilde, Cowie and Flexner (36) studied phosphorus exchange in guinea pigs and showed that the inorganic phosphorus reaching the foetus from the maternal plasma was approximately equal to the amount required for growth. This was confirmed by Fuchs and Fuchs (37). Similarly the amount of calcium taken up by rat foetuses each hour during the latter part of gestation was shown to be equal to the total amount of calcium in the maternal circulation at any time during the hour (38). The supply of calcium and phosphorus, therefore, may limit the growth of some small species of animals before they are born. Other nutrients, amino acids or glucose, for example, may limit the growth of others. There is probably no one simple explanation in any species, and far less is there a simple explanation that will cover all species. We do not know which nutrient is likely to be the limiting one in man - and it is not likely to be the same one in all cases.

SLOW FOETAL GROWTH CAUSED BY HIGH ALTITUDES, AND SMOKING

Women who live at high altitudes tend to have smaller babies than those who live at sea level (30), and those who smoke a great deal have smaller babies than those who do not (39,40,41). A major function of the placenta is to transfer oxygen to the foetus at the rate required for foetal metabolism and at the PO_2 necessary to provide the requisite diffusion gradient from the foetal peripheral blood to the tissues. If the PO_2 of the foetal mitochondria falls below a critical level then oxygen consumption is reduced and anaerobic metabolism increases. Much work has been done on placental gas exchange, particularly in the sheep. It was pioneered by Barcroft (42) and has been continued by many others since. Reviews have been written on various aspects of the subject, some of which appeared in the Proceedings of the Sir Joseph Barcroft Centenary Symposium (43). There are important species differences, for in the cow and mare, for example, foetal haemoglobin has a smaller affinity for oxygen than maternal; in other species, for example man and the pig, foetal haemoglobin has the greater oxygen affinity (44). Over 100 haemoglobin variants are known in man (45). Some are harmless, others would affect the oxygen supply to the foetus and hence its rate of growth. I do not know how much work has been done on this subject in other animals. Hypoxia probably affects the transfer of other nutrients, for example sodium, though it does not seem to affect blood flow (46). All in all it is not surprising that babies born at high altitudes are a little light in weight. The same presumably applies to the newborn of other species.

Carbon monoxide, produced by tobacco smoke, has a very high affinity for haemoglobin. It diffuses readily across the placenta, and the foetal Hb CO concentration is generally greater than the maternal. Carbon monoxide may, like high altitudes, produce hypoxia in the foetal tissues and so hinder their growth (47). However, the effect of nicotine in constricting the blood vessels is certainly a contributory cause.

PASSAGE OF FATTY ACIDS AND DRUGS ACROSS THE PLACENTA

Fatty acids

It was believed until recently that no fatty acids except the essential polyunsaturated ones were able to cross the placenta, and that the foetus synthesised its own depot fat from glucose. This does not now appear to be the case, at any rate in some species. We have recently found that the whole fatty acid make-up of the body fat of newborn guinea pigs can be altered by feeding the pregnant animal different types of fat; the fatty acid pattern of the body fat of the foetus comes to resemble that of the fat the mother was fed.

Drugs

The teratogenic effect of certain drugs given to the mother at a critical stage of pregnancy was highlighted by the thalidomide disaster. Such drugs cross the placenta and act in a variety of ways, for example by inhibiting mitotic processes, or by interfering with the supply of blood or of oxygen to the foetus (48). In all such instances the effect is the same - to prevent normal foetal development.

LARGE BABIES AT BIRTH

So far I have confined my discussion to growth retardation. There is much less to say about the effect of the maternal environment on the acceleration or prolongation of growth before birth. Infants are sometimes born weighing 4.5 kg or so after a gestation of about 43 weeks (49). They are only able to grow so large because they still have a large and adequate placenta. Mothers who have diabetes also have babies that are heavy and long for their gestational age, though they are often born prematurely. Here the high levels of blood sugar in the mother, and consequently in the foetus, and the insulin the foetus produces to metabolise it, are probably the cause of the rapid rate of growth.

CONSEQUENCES OF A SMALL SIZE AT BIRTH

In the early stages the small animal or baby is at risk because it has a large surface area in relation to its weight, and so it loses heat more rapidly than its larger fellows, but it lacks the stores of glycogen or fat that the larger animal possesses.   The runt pig for example often dies with hypothermia and hypoglycaemia if it is not kept warm and fed. If it is reared it grows well, and becomes almost, but not quite as large as its larger littermate at birth (24). It has less DNA and fewer cells in its organs at birth and it still has fewer after it has grown to maturity. But more important in determining its size are its bones. These do not grow quite so long as those of its littermate that was larger when it was born. While age is the major determinant of the differential development of the skeleton and its parts, the plane of intra-uterine nutrition may have a lasting effect on the growth in length of the bones and hence on the ultimate size of the body.

REFERENCES

1.   Walton, A. and Hammond, J. (1938). Proc. R. Soc. B. 125, 311.

2.   Joubert, D. M. and Hammond, J. (1954). Nature, Lond. 174, 647.

3.   Joubert, D. M. and Hammond, J. (1958). J. Agric. Sci., Camb. 51, 325.

4.   Dickinson, A. G. (1960). J. agric. Sci., Camb. 54, 378.

5.   Hunter, G. L. (1957). J. agric. Sci., Camb. 48, 36.

6.   Brumby, P. J. (1960). Heredity, 14, 1.

7.   Cawley, R. H., McKeown, T. and Record, R. G. (1954). Am. J. hum. Genet. 6, 448.

8.   McKeown, T. and Record, R. G. (1952). Am. J. hum. Genet. 6, 457.

9.   Silver, M., Steven, D. H. and Comline, R. S. (1973). In:"Proc. Sir Joseph Barcroft Centenary Symposium." (Eds. R. S. Comline, K. W. Cross, G. S. Dawes and P. W. Nathanielsz). p.245. (Cambridge University Press, London).

10. Dow, P. and Torpin, R. (1939). Hum. Biol. 11, 248.

11. Gruenwald, P. (1968). In: "Aspects of Prematurity and

Dysmaturity". (Eds. J. H. P. Jonxis, H. K. A. Visser and J. A. Troelstra). p. 37. (H. E. Stenfert Kroese, Leiden).

12.   McKeown, T. and Record, R. G. (1952). J. Endocr. 8, 386.

13.   Lubchenco, L. O., Hansman, C., Dressler, M. and Boyd, E. (1963). Pediatrics, Sprinfield, 32, 793.

14.   Clifford, S. H. (1954). J. Pediat. 44, 1.

15.   Gates, W. H. (1924-25). Anat. Rec. 29, 183.

16.   Eaton, O. H. (1932). U.S. Dept. agric. Tech. Bull. No. 279.

17.   Angulo y Gonzalez, A. W. (1932). Anat. Rec. 52, 117.

18.   Venge, O. (1950). Acta zool. Stockh., 31, 1.

19.   Eckstein, P. and McKeown, T. (1955). J. Endocr. 12, 97.

20.   McLaren, A. and Michie, D. (1960). Nature, Lond. 187, 363.

21.   Wigglesworth, J. S. (1964). J. Path. Bact. 88, 1.

22.   Perry, J. S. and Rowell, J. G. (1969). J. Reprod. Fert. 19, 527.

23.   Widdowson, E. M. (1974). In:"Ciba Foundation Symposium on Size at Birth." In Press.

24.   Widdowson, E. M. (1971). Biol. Neonate, 19, 329.

25.   Wallace, L. R. (1945). J. Physiol., Lond. 104, 34.

26.   Antonov, A. N. (1947). J. Pediat. 30, 250.

27.   Smith, C. A. (1947a). J. Pediat. 30, 229.

28.   Smith, C. A. (1947b). Am. J. Obstet. Gynec. 53, 599.

29.   Dean, R. F. A. (1951). In: "Studies of Undernutrition, Wuppertal 1946-9". p.346. Spec. Rep. Ser. med. Res. Coun. No. 275. (HMSO, London).

30.   Lubchenco, L. O., Hansman, C., and Bäckström, L. (1968). In: "Aspects of Prematurity and Dysmaturity". (Eds. J. H. P. Jonxis, H. K. A. Visser and J. A. Troelstra). p.149. (H. E. Stenfert Kroese, Leiden).

31.   Raiha, C. E. (1964). Guy's Hosp. Rep. 113, 96.

32.   Butler, W. H. and Wigglesworth, J. S. (1960). Br. J. exp.
      Path. 47, 242.

33.   Young, M. (1969). In: "Foetus and Placenta". (Eds. A.
      Klopper and E. Diczfalusy). P. 139. (Blackwell, Oxford).

34.   Young, M. and Widdowson, E. M. (1975). Biol. Neonate.
      In Press.

35.   Wasserman, R. H., Comar, C. L., Nold, M. M. and Lengemann,
      F. W. (1958). Am. J. Physiol. 189, 91.

36.   Wilde, W. S., Cowie, D. B. and Flexner, L. B. (1946).
      Am. J. Physiol. 147, 360.

37.   Fuchs, F. and Fuchs, A. R. (1956-7). Acta physiol. scand.
      38, 379.

38.   Comar, C. L. (1956). Ann. N. Y. Acad. Sci. 64, 281.

39.   Lowe, C. R. (1959). Br. med. J. ii, 673.

40.   Herriot, A., Billewicz, W. Z. and Hytten, F. E. (1962).
      Lancet, i, 771.

41.   Longo, L. D. (1970). Ann. N.Y. Acad. Sci. 174, 1.

42.   Barcroft, J. (1947). Researches on Pre-natal Life.
      (Charles C. Thomas, Springfield, Ill.).

43.   Proc. Sir Joseph Barcroft Centenary Symposium (1973).
      (Eds. R. S. Comline, K. W. Cross, G. S. Dawes and
      P. W. Nathanielsz). (Cambridge University Press, London).

44.   Comline, R. S. and Silver, M. (1974). J. Physiol., Lond.
      242, 805.

45.   Weatherall, D. J. (1974). Br. med. J. ii, 451.

46.   Meschia, G. and Battaglia, G. (1973). In: Proc. Sir
      Joseph Barcroft Centenary Symposium. (Eds. R. S. Comline,
      K. W. Cross, G. S. Dawes and P. W. Nathanielsz). p.272.
      (Cambridge University Press, London).

47.   Forster, R. E. (1973). In:"Proc. Sir Joseph Barcroft
      Centenary Symposium." (Eds. R. S. Comline, K. W. Cross,
      G. S. Dawes and P. W. Nathanielsz). p.223.
      (Cambridge University Press, London).

48.   Robson, J. M. (1970). Br. med. Bull. 26, 212.

49.   Gruenwald, P. (1966). Am. J. Obstet, Gynec. $\underline{94}$, 1112.

50.   Brook, G. L. (1963). In: English Dialects. The Language
       Library. (Ed. E. Partridge). p.37. (Andre Deutsch,
       London).

THE NUTRITIONAL CONTROL OF GROWTH

V. R. Fowler

Rowett Research Institute

Aberdeen, AB2 9SB, Scotland

The fact that nutrition is essential to sustained growth is self-evident since there can be no output from no input. It is evident too that differences in the composition of growth and in its efficiency arise from changes in nutritional status. The contrasts are most stark when expressed in terms of the human diseases of obesity, or, at the other extreme, marasmus and kwashiokor. To these can be added the deforming diseases of specific mineral or vitamin deficiencies such as rickets or cretinism. In animal production, the effects of nutrition tend to be measured in terms of their effect upon efficiency rather than on gross features of the animal but the impact of nutrition on composition is still of considerable importance particularly where animals are subjected to grading for fatness or leanness.

The immensity of the subject means that only selected aspects can be dealt with in a paper of this length. Its objective, therefore, is to consider only some of the principles involved in attempts to modify the composition of the animal by nutritional means. The species forming the majority of the examples is the pig since it has been the target of dietary manipulation more than any other farm animal.

GENERAL CONCEPTS

Simplifications of complex phenomena are often misleading or, worse still, can retard the development of thought for several years if accepted as rules. With this reservation in mind it is essential to attempt to find integrating concepts in an area so complex as the interaction of growth and nutrition. A useful

starting point is to consider the adaptive changes which the animal
might be expected to show to nutritional situations as a result of
mechanisms established during natural selection prior to domestication.   Changes in nutritional status represent a challenge to the
ability of the animal to maintain physiological homeostasis.   The
concept of homeostasis, first formulated by Claude Bernard, is of
considerable value throughout biology and especially so in nutrition.

The wild herbivore or, in the case of the pig, omnivore is
subjected to considerable variations in the availability of its
food supply and, as Dr. Fuller showed in his paper at this meeting,
this is very considerably affected by climate.   Against this background of varying nutrition the animal has developed the ability
both to store nutrients and also to synchronise periods of maximum
nutrient demand such as lactation with the peak in food supply.
Inevitably, however, a species has had to survive crisis years in
which fire, drought, flood, or disease have damaged or virtually
destroyed the food supply.   The adaptations that the animal can
make must be consistent with the preservation of itself as a sound
individual.   This concept can be formally expressed in the
proposition

'The animal tends to adjust to environmental (nutritional)
changes in such a way that the vital functional relationships between essential body components are preserved, or
modified to a form which gives the animal its best chance
of survival and successful reproduction.' (1)

Although this statement must be true generally, the concept
may not be applicable to specific agricultural contexts.   New
factors such as artificial selection, elimination of the natural
cycle of food supply, intensification of production methods by the
provision of high quality diets and efforts to improve reproductive
performance sometimes by a factor of two bring new stresses to the
animal.   In response to these it may not react in a way that can
be interpreted teleologically.   Examples of such gross changes
are the feeding of ruminants with concentrates, the early-weaning
of piglets onto semi-synthetic diets soon after birth, and the
acceleration of the breeding cycle in sheep by the use of hormones
or by the manipulation of day length patterns.   It is important
therefore to achieve a proper balance between generalisations that
are based on fundamental biological principles honed by millennia
of natural selection and empirical findings which describe a
particular situation at a given stage in the adaptation of a
species to the needs of man.

NUTRITION AND BODY PROPORTIONS

The early workers on growth and development at Cambridge
observed that during undernutrition the cranial region of the
animal became proportionately larger than the rest (2).   Since
the head was regarded as one of the earlier developing parts of
the body, they formed the opinion that as a general rule early
maturing parts of the body were more resistant to undernutrition
than were the later maturing parts.   In addition to this it was
quickly noted that a late-maturing tissue, subcutaneous fat, was
readily affected by controlling daily feed intake in the later
stages of the growing period.   The further generalisation was,
therefore, made that food deprivation tended to affect those parts
which were growing most actively during that period.   Although
both observations were correct, the hypotheses which were developed
were too sweeping since they postulated that the same principles
applied to most other tissues and components of the body.   It was
later shown using the same data that very few of the components of
the body were disproportionately affected by different nutritional
patterns.   Those that were tended to be the head, the fat depots
and the organs associated with the digestion of food and with
sustaining the metabolism of the animal (2,3).

An example of the effect of nutrition on proportionality is
provided by data obtained at the Rowett Institute (4).   This
experiment was designed to provide serial killing data on entire
males, barrows and gilts which were fed on one of three feeding
regimes, a high scale of a high protein diet H,HP, a high level
of a low protein diet H,LP and a low level of the high protein
diet L,HP.   The high rate of feeding was 120 g per kg live-
weight$^{0.75}$ per day and the low rate 60 g/kg$^{0.75}$d.   The high protein diet contained 22% crude protein and the low protein one only
8%.   These regimes were intended to span the extremes which would
be likely to be encountered in practice.   Animals were slaughtered
at regular intervals between 20 and 120 kg liveweight;   carcasses
were dissected and chemically analysed.

The treatments resulted in profound differences between the
groups in the amount of fatty tissue.   The most affected component
was the subcutaneous fat and results for the different groups
adjusted to a standard carcass weight are given in the first row of
Table 1.

Once a difference has been established in such a major component as fat, it is important to avoid transmitting its effect
automatically to the comparison of other components of the body.
The relative weights of all organs or tissues other than subcutaneous fat could appear as the mirror image of the degree of
fatness if comparisons were made on the same basis.   Several
approaches to overcome this problem are possible, but all have in

Table 1

Weights of subcutaneous fat adjusted to standard carcass weight and weights of different components of the body adjusted to a standard weight of the 'basic animal' (means of male, castrate and female pigs)

Dependent variate	Independent variate with selected value (kg)	Treatment			Overall significance of treatments
		High Plane High Protein n = 20	High Plane Low Protein n = 21	Low Plane High Protein n = 18	
Subcutaneous fat plus skin (kg)	80 c.w.	20.8	27.2	15.17	p < 0.001
Head, less superficial soft tissues (kg)	50 b.a.	2.08	2.04	2.41	p < 0.001
Forelimbs (kg)	50 b.a.	10.87	9.93	10.83	p < 0.05
Hindlimbs (kg)	50 b.a.	14.0	13.3	14.27	N.S.
Ribs plus intercostal tissue (kg)	50 b.a.	3.75	3.90	3.79	N.S.
Abdominal wall (kg)	50 b.a.	2.35	2.22	1.91	p < 0.01
Liver (kg)	50 b.a.	1.92	1.99	1.7	N.S.
Lungs and trachea (kg)	50 b.a.	1.18	1.32	1.25	N.S.
Heart (g)	50 b.a.	317	338	318	N.S.

common the need to exclude from the independent or adjusting
variable the effect of changes in fatness per se (3,4). The
method selected in this case was to exclude blood, the internal
organs, perinephric and subcutaneous fat, from the adjusting
variable. The sum of all other components was then described as
the 'basic animal' and all dependent variates were then adjusted
to a constant basic animal weight of 50 kg. Some results calculated on this basis are given in Table 1.

The results show that in these terms the limbs, ribs and major
internal organs were not affected disproportionately by the treatments but there were significant effects on the cranial parts and
on the weight of the abdominal wall.

The treatments in this experiment were relatively severe and
it is one example of many which show that within the range of
husbandry practices it is very unlikely that nutritional manipulation will greatly affect the proportions of organs and parts
which co-ordinate closely with one another in their function.
The exceptions tend to be those organs associated with energy
storage, i.e. the adipose depots, the head, which has relatively
discrete functions, and the abdominal wall, which tends to respond
to the level of feeding and the associated differences in gut
volume and size.

Another experiment in which large differences in nutrition
were imposed was that of Fowler and Ross (5). The objective was
to examine the changes in the various fat depots associated with
feeding pigs either on a diet designed to promote fat deposition
or on one intended to reduce fatness without affecting the total
amount of lean tissue. The pigs were selected for the experiment
when they weighed 80 kg prior to which they had been fed on a
conventional diet and feeding scale. The fattened group were
given a diet containing only 4% crude protein at a daily rate
which was close to four times their maintenance requirement for
energy. The slimmed group received a very high protein diet (30%)
at a rate which was estimated to provide only enough energy for
about half their maintenance requirement. After four and eight
weeks a pair of pigs were slaughtered from each treatment and some
results are given in Table 2.

The effect of the fattening treatments was very great and
there was about a three-fold increase over eight weeks in the
amounts of subcutaneous, perinephric and mesenteric fat. Inter-
and intra-muscular fat responded also but to a lesser extent.
More interesting, from a practical viewpoint, were the results for
the 'slimming' treatment. There was little impact after four
weeks slimming on any of the fat depots but a substantial loss in
fat-free body. After eight weeks slimming the loss in fat-free
body for the final four weeks appeared rather less than that for

Table 2

Amounts of chemical fat in different adipose depots
expressed as proportions of the fat-free body

	Control	4 weeks fattened	8 weeks fattened	4 weeks slimmed	8 weeks slimmed
Weight of fat-free body (kg)	59.7	59.3	61.3	55.3	52.6
Chemical fat (g/kg FFB) Total body fat	310	593.5	777.5	313.5	274
Subcutaneous fat	157	340.5	484.5	161	134.5
Perirenal fat	15	35.5	50	17	9
Mesenteric fat	7.5	19.5	24.5	14.5	9
Intermuscular fat (limbs only)	14.5	19.5	20	13	10.5
Intramuscular fat (limbs only)	18	21.5	21	16	15.5

the first four weeks and there were signs that depot fat was being
mobilized from the subcutaneous and perinephric regions.   The
loss of fat-free body during the slimming period, despite the use
of a high protein diet, suggests that some fraction of this component is highly labile and is quickly drawn upon during starvation
or severe food restriction.   Similar effects have been noted in
human undernutrition (6) and in other experiments with pigs (7).

GROWTH CHECKS AND COMPENSATORY GROWTH

In animal production in the United Kingdom prolonged periods
of undernutrition are unlikely to occur except perhaps for animals
grazed on hills during the winter.   In other parts of the world,
however, such as the extensive grazing areas of Australia and
Africa, a store period is relatively common and the way in which
animals recover is an important practical topic.   The classical
experiments of McCance and his collaborators on the severely
retarded pig (8) provide a valuable model of the resilience of
animals to quite profound disturbance of the food supply.   The
pigs which were kept at a constant weight of only 5 to 6 kg for

12 months were able after rehabilitation to grow almost to the
mature size of control litter-mates and reproduce successfully.
This example illustrates the important principle that the genetic
make-up of the pig ensures, even under very adverse conditions,
that the competence of the animal to grow and reproduce is retained
and that physiological homeostasis is maintained.

A number of comprehensive reviews on compensatory growth in
farm animals have been made (9,10).   It is an aspect of animal
growth, however, which has not always given rise to precise experimentation, and there is still some controversy about the formula-
tion of hypotheses which should be tested.   In general terms, the
difficulties of interpretation arise from the fact that during
rehabilitation the appetite is often enhanced so that a comparison
is made between animals with dissimilar food intake.   Two other
causes of confusion are that animals are not always compared over
the same range of liveweight and increases in liveweight caused
solely by differences in gutfill are often interpreted as compensatory growth.

In a recent experiment on the phenomenon of catch-up growth
in the pig conducted at the Rowett Institute, entire male and
female pigs were checked at 60 kg liveweight and fed a maintenance
diet for either 28 or 112 days.   At the end of the check period
they were rationed according to a scale based on liveweight and
adjustments were made twice weekly depending on the gains.   The
scale used was a high one, 120 g/kg $W^{0.75}$, and though it allowed
very rapid rates of gain, it did not allow any expression of
differences in appetite potential between the pigs.   Some preliminary results giving the rates of net accretion of protein, water,
fat and ash are given in Table 3 and were calculated from regression
equations for animals slaughtered serially.

The interest of these results lies in the fact that they
demonstrate an apparent compensation in the rate at which protein
and water were deposited.   The performance of the animals,
including the controls, was outstanding and the compensating groups
gained at well over 1 kg per day.   Another feature of the data is
the apparently improved energetic efficiency of the group which was
checked for 112 days demonstrated in the increased rate of fat
deposition.   The effect could plausibly be explained by a reduction
in metabolic rate during maintenance feeding, an adaptation which
was then carried over into the compensatory period.   Although the
results suggest that an improved utilization of feed for growth is
possible during recovery, the overall efficiency is lower than
that of the controls because of the feed cost of maintenance for
the much longer overall period of growth from 20-100 kg.

Table 3

Rates of deposition (g/day) of chemical components of
body for 6 week period following growth checks at 60 kg
liveweight of 28 and 112 days (n = 10)

	Control	Checked 28 days	Checked 112 days	s.e. of means
Protein	130	153	132	13
Water	428	512	517	41
Fat	329	395	480	35
Ash	22.6	21.1	23.3	4.4

PRACTICAL IMPLICATIONS OF MANIPULATING
GROWTH BY NUTRITION

Level of Feeding

The main practical use of dietary manipulations is to alter
the proportion of fat in the carcass of the animal.   As explained
in the paper of Dr. Rhodes, excess fat in carcasses is wasteful
since it is energetically expensive to produce and is regarded as
undesirable or even as a health hazard by the consumer.   Its
presence in the carcass is in some respects a carry over of the
life-support system of the wild ancestors of our stock.   At
first sight, it would appear that since fat is a means of storing
energy surplus to the immediately requirements of the animal, all
that is necessary to reduce it is to restrict the animal to the
amount of feed which just meets its requirements for the growth of
lean tissue.   This is the basic idea behind the practice of
restricting the feed intake of pigs destined for bacon production.
In reality, however, many pigs which have been subjected to
restricted feeding are still unacceptably fat when slaughtered.
The reason for this is that the genotype of many pigs still dictates
that provided food is available at more than subsistence level a
substantial proportion of growth should be in the form of an energy
insurance or in other words adipose tissue.   This concept is
illustrated by data from some experiments conducted at the Rowett
Institute by my colleagues Dr. Fuller and Dr Houseman.   In Fig. 1
data are shown from an experiment (11) in which pigs were fed the
same diet at rates ranging from 130 g/kg $W^{0.75}$ per day down to
65 g/kg $W^{0.75}$ in eight equal steps.

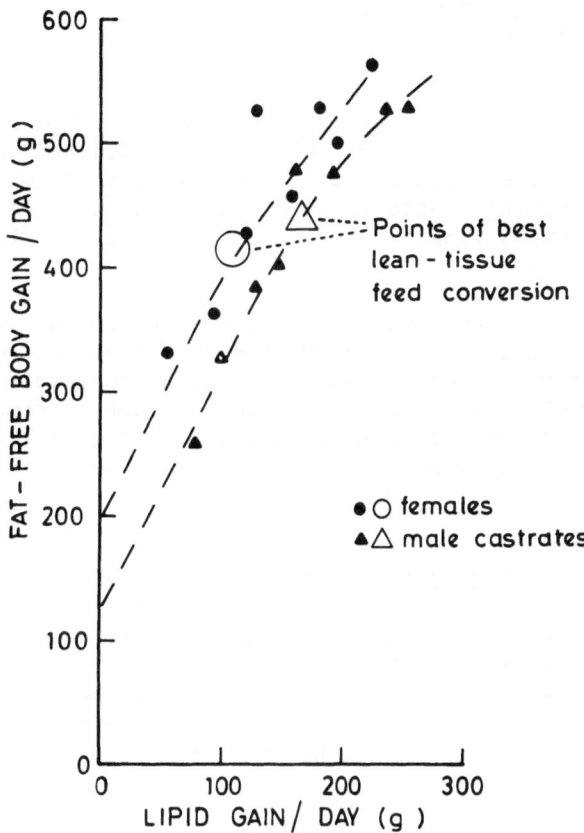

Fig. 1.   Relationship between rates of growth (g day) of fat-free body and lipid for female and castrated male pigs on different levels of feeding.   Data from Fuller and Livingstone (11).

The graph shows the relationship between chemical lipid and fat-free body calculated from measurements of specific gravity on the carcasses.   Extrapolation of the curves suggests that at low levels of intake, approaching maintenance, fat deposition ceased but the fat-free tissue continued to grow at a very slow rate. As growth rates increased, however, the fat deposition occurs and continues in a relatively constant ratio over a wide range, the value being about 0.65 kg fat to each kg of fat-free tissue.   In this example the differences between male and female appear to be more related to differences of intercept rather than to differences of slope.

In the second example, data were derived from two experiments in which the weights of fat-free body and of lipid were deter-mined directly.   The objective of the nutritional treatments was to produce pigs having a wide range of composition of the body and protein concentration ranged from 10 to 22% (12,13).   The results are shown in Fig. 2.

The regression line drawn on the graph is for the two highest concentrations of protein 18 and 22% and represents approximately the limit to the reduction in fat which can be achieved with this genotype.   The slope of this line is rather steeper than that of the previous example and indicates that about 400 g of lipid is inevitably associated with each kg of fat-free body gain at normal rates of growth.

PROTEIN CONCENTRATION IN THE DIET

For non-ruminants and for young ruminants there is an extensive literature showing that as the concentration of protein in the diet is increased there is an increase in the proportion of lean tissue in each unit of gain until a certain level is reached (14,15). After this point, further increments may cause a decline in per-formance (16).

The biological implications of this are interesting since the animal is prepared to allow a considerable disturbance of its fat to lean ratio without attempting to compensate.   This was particu-larly true of the pig experiment mentioned in the previous section where a 4% protein diet caused a three-fold increase in fatness. Dr. Webster at this symposium has suggested that rats may eat to maintain a constant rate of growth of the lean tissue and this could be true to a certain extent for farm animals attempting to maintain a growth timetable to fit the seasonal requirements.   The fact remains, however, that whatever protein level is given to pigs they still choose to lay down a high proportion of fat.   As discussed earlier, this insurance factor is possibly genetically determined and is greater in the female than in the entire male.

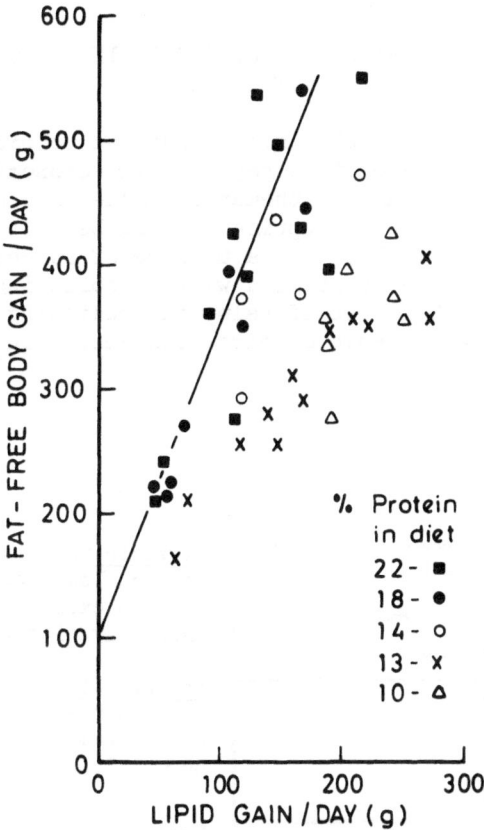

Fig. 2. Relationships between rates of growth (g/day) of fat-free body and lipid for pigs given a range of protein concentrations in the diet. Data from Houseman (12,13).

The teleological reasons for this are obvious and a similar case
has been made out for data on human fatness by Dr. Frisch at this
symposium.   In conceptual terms, it is convenient to describe
this fat deposition inevitably associated with normal rates of lean
deposition as the 'target' fat.   It is deposited whatever protein
level is given and can be eroded only when the feeding level
approaches maintenance.   The concept of 'target' fat will be
referred to in the following integrating section.

BIOLOGICAL PRINCIPLES OF GROWTH IN RELATION
TO NUTRITION

The adaptations of generalized principles to economic situations is not the purpose of this paper.   The economic exercise is
primarily one of producing a mathematical model and attempting by
its use to arrive at economic optima.   A number of attempts have
been made to do this (17,18) with a measure of success.   The
present argument has been concerned with establishing those biological concepts which may be of value in predicting the response
of the animal to a wide range of circumstances.   In Fig. 3 is set

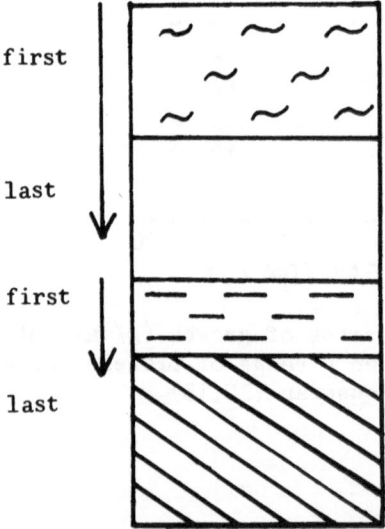

Fig. 3.   A classification of different components of the animal
body in terms of their responsiveness to changes in nutritional
status.

out a basis for categorizing the different parts of the animal in relation to their responsiveness to changes in nutrition. The divisions between the compartments are not absolute but the main blocks can form the basis for discussions about the nature of differences between animals of different genotype of sex.

The most stable component is the essential tissues which comprise bone, muscle, nervous tissue, and essential lipid. A relatively labile part of the lean tissue which is utilized during undernutrition must also be considered and possibly consists of parts of the liver and of non-crucial skeletal muscle. Above this lies the 'target' fat which is the 'insurance' energy which the animal prefers to deposit when it is accreting muscle at a normal rate. It will include fat in the subcutaneous depot which also has an insulative function. Finally one can consider the remaining fat as the variable fat which is deposited during periods of growth when protein or critical amino acids are in short supply and energy is in excess of the requirement for growth of muscle and target fat, or in periods where the level of nutrition is so high that there is a surplus of both energy and protein. Changes within the essential tissues are difficult to achieve by nutritional means but are possible by long term genetic programmes. For example, the Pietrain pig which is genetically very different from most of our breeds appears to have a relatively low proportion of bone to each unit of muscle (19). Changes in the variable lean tissue would only occur on sub-maintenance diets and changes in target fat only when the level of feed intake fell well below conventional production levels.

It is suggested, therefore, that from the point of view of maximising biological efficiency the only practicable means of modifying body composition by nutrition is to concentrate on reducing variable fat to the point where lean tissue feed conversion is maximised. This can be achieved by controlling feed intake and by supplying adequate, but not excessive, concentrations of protein or essential amino acids in the diet.

In attempting to improve the efficiency of production of our stock we must not lose sight of the fact that the animal has a heritage of functional resilience, and cannot be expected to conform to generalized formulae unless conditions of genotype, sex, environment and disease are also taken into account. Given the nutritional opportunity, young animals will quickly store surplus energy in their 'bank account' (20) although theory would suggest that fat is a late developing tissue. On the other hand, after nutritional deprivation animals appear able to break through genetic ceilings for lean tissue deposition in order to catch up with their 'biological clock'. It is this resilience and individuality that should warn us to remember that our livestock are not merely an aggregation of predictable chemical reactions

but highly developed survival kits with a whole range of integrated and sophisticated systems for resisting environmental
insults.   Our attempts to modify body composition nutritionally
in a favourable way can only succeed to the extent that we appre-
ciate and understand the limits imposed by the biological constraints
developed by the animal for the benefit of its own species.

ACKNOWLEDGEMENTS

I should like to thank Dr. M. F. Fuller and Dr. R. A. Houseman
for permission to use the data given in Figs. 1 and 2.

REFERENCES

1.   Fowler, V. R. (1968). In: "Growth and Development of Mammals"
     (Eds. G. A. Lodge and G. E. Lamming) (Butterworth, London)

2.   Pålsson, H. (1955). In: "Progress in the Physiology of Farm
     Animals", Vol. 2, P. 130 (Ed. J. Hammond) (Butterworth, London)

3.   Elsley, F. W. H., McDonald, I. and Fowler, V. R. (1964).
     Anim. Prod. 6, 141

4.   Fowler, V. R. and Livingstone, R. M. (1972). In: "Pig
     Production" (Ed. D. J. A. Cole) (Butterworth, London)

5.   Fowler, V. R. and Ross, W. (1974). Proc. Nutr. Soc. 33, 94A

6.   Pomeroy, R. W. (1941). J. agric. Sci. Camb. 31, 50

7.   Keys, A., Brozek, J., Henschel, A., Mickelsen, O. and
     Taylor, H. L. (1950). In: "The Biology of Human Starvation",
     Vol. 1, p. 398 (University of Minnesota, Minneapolis)

8.   Lister, D. and McCance, R. A. (1967). Br. J. Nutr. 21, 787

9.   Wilson, P. N. and Osbourne, D. F. (1960). Biol. Rev. 35, 324

10. Allden, W. G. (1970). Nutr. Abstr. Rev. 40, 1167

11. Fuller, M. F. and Livingstone, R. M. (1975). Proc. Br. Soc.
     Anim. Prod. 4 (in press)

12. Houseman, R. A. and McDonald, I. (1973). Anim. Prod. 17, 295

13. Fuller, M. F., Houseman, R. A. and Cadenhead, A. (1971). Br.
     J. Nutr. 26, 203

14. Black, J. L. (1974). Proc. Aust. Soc. Anim. Prod. 10, 211

15. Cooke, R., Lodge, G. A. and Lewis, D. (1972). Anim. Prod.
     14, 35

16. Lodge, G. A., Hardy, B. and Lewis, D. (1972). Anim. Prod.
     14, 229

17. Dent, J. B., Blair, R., English, P. R. and Raeburn, J. R.
     (1970). J. agric. Sci. Camb. 75, 189

18. Whittemore, C. T. and Fawcett, R. H. (1974). Anim. Prod.
     19, 221

19. Davies, A. S. (1974). Anim. Prod. 19, 367

20. Elsley, F. W. H. (1964). J. agric. Sci. Camb. 61, 233

CLIMATE AND SEASON

M. F. Fuller

Rowett Research Institute

Bucksburn, Aberdeen AB2 9SB, Scotland

Although man depends ultimately for his food on the photo-
synthetic activity of green plants, animals are useful inter-
mediates in his food chain for several purposes. These include
the conversion to high quality foods of those plant products which
are unpalatable or of low nutritive value to him, and those in
seasonal excess which it is impracticable or uneconomic to store
for his own future use.

Man's food requirement is more or less constant the year
round, whereas plant growth varies with the seasons. The production of food crops by plants is particularly intermittent, with
only short harvest periods. Man's need has, therefore, always
been, and will continue to be, to even out a spasmodic food supply
to meet an unvarying requirement. The intermediate storage of
nutrients in the tissues of meat animals represents a valuable
means of doing so.

Animals also represent in many circumstances the only
satisfactory means of utilising the products of photosynthesis
over the large areas of the earth's surface where the cultivation
of crop plants for direct human consumption is limited by their
inaccessibility, by their poor soil or by seasonal insufficiencies
of water, sunlight or temperature.

These uses of animals, it may be argued, are supplementary to
primary food production by crop plants in that they increase the
sum total of food available to man. They are to be contrasted
with an expanding industry in which animals are produced to
satisfy a consumer demand for meat, in which the upgrading of
food quality is a subsidiary aspect, and in which they compete

directly for agricultural resources with primary food crops. Arguments about the use of animals for food production should preserve this distinction.

CLIMATIC VARIATION

The changing seasons are but one of a number of patterns of climatic change to which animals are subject: the seasonal pattern is important in the present context because it is on a time scale of the same order as the major processes of plant and animal growth. Animals are subject also to climatic changes on time scales measured in minutes and in millennia. In the very short term animals respond to changes of climate or weather with rapid alterations of their behaviour, insulation and metabolism. These homeostatic adjustments are the subject matter of much of environmental physiology, which it is not the purpose of this paper to discuss in detail. On a longer time scale minute-to-minute changes in environment are seen to be superimposed on a diurnal rhythm which is in turn superimposed on a seasonal pattern of climatic change. The word 'climate' usually refers to the long-term average pattern measured over a number of years, in which seasonal and shorter-term fluctuations are supposed to be evened out. However, the variability of a climate is as important in characterising that climate as are average meteorological data.

The main purpose of this paper is to consider in broad terms the limitations imposed by climate on the growth and productive efficiency of meat producing animals. The most important of these limitations concerns energy. Fig. 1 shows in a simplified scheme the main ways in which climate may affect the energy budget of animals. The direct effects of climate on the animal's energy exchanges are shown on the left, the indirect effects, by way of its food supply, on the right. There is no doubt that the predominant effect of climate on animal production is through the growth and maturation of plants. In the humid tropics plant growth may, with optimum husbandry, approach rates achieved under artificial conditions, and be virtually uninfluenced by season. But according to FAO such areas comprise only 9% of the world's agricultural land; on the remainder, plant growth is primarily limited by temperature (36%), water (31%) or both (24%). The direct effects of climate, though of lesser importance in the total animal economy, are nevertheless a major determinant of animal productivity in many practical situations.

To this must be added the fact that the season of poorest plant growth usually coincides with the greatest direct climatic stress. In temperate and subarctic regions, winter combines the direct effects of cold with the virtual cessation of plant growth. In the wet/dry tropics, lack of rain in the dry season not only

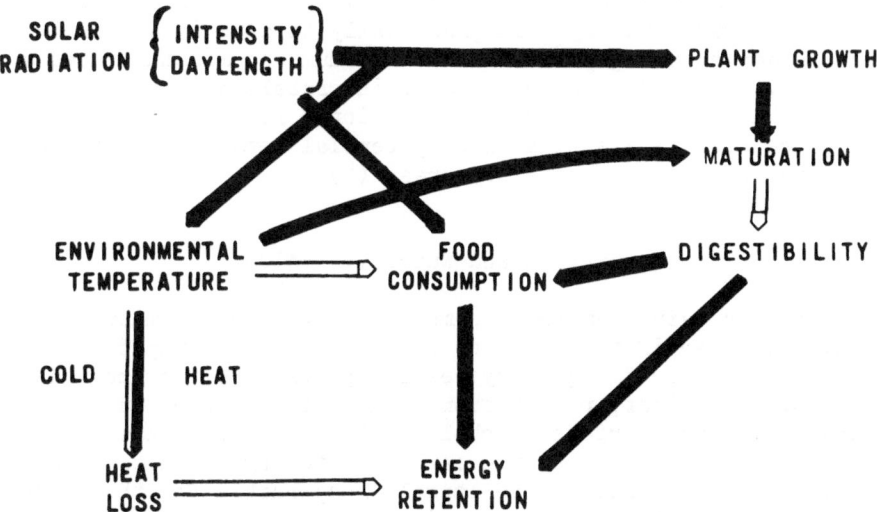

Fig. 1. Outline of major pathways by which seasonal change affects the growth of domestic animals. Black arrows indicate positive relationships, white negative.

halts plant growth but limits the amount of water available to animals for evaporative heat dissipation at a time when their heat load is usually at its greatest.

INDIRECT EFFECTS: PLANT GROWTH

De Wit (1) calculated the potential net assimilation of a closed plant canopy as a function of season in the Netherlands. A peak rate of carbyhydrate production in June of 290 kg/ha.d was 5-6 times the potential rate in December and January. Sibma (2), comparing various crop plants during their closed canopy phase, concluded that rates of total dry matter production in the different species were very similar, about 250 kg/ha.d and that differences in crop yield were attributable mainly to differences in the percentage of total production harvested and to the length of the effective growing season. In these temperate climates, because of incomplete crop cover, low temperature and poor nutrient supply, the growth of herbage plants is in practice much lower for much of the year, and may fall practically to zero in winter, further widening the gap between the maximum and minimum rates of production.   In addition, crop cover in grassland is often incomplete as a result of drought and winter kill (3).   Fig. 2 compares de Wit's (1) estimates of potential net assimilation with the growth curve of perennial ryegrass (4).

Maturation and Nutritive Value

With increasing temperature many herbage species mature more rapidly (5) (even if they grow no faster).   A more rapid deterioration of forage quality results partly from an increased ratio of stem to leaf, partly from changes in the chemical composition of the various morphological parts, principally the increased replacement of simple carbohydrates by cellulose and lignin.   These effects may be largely responsible for the poorer quality of tropical, as compared with temperate forages (6).

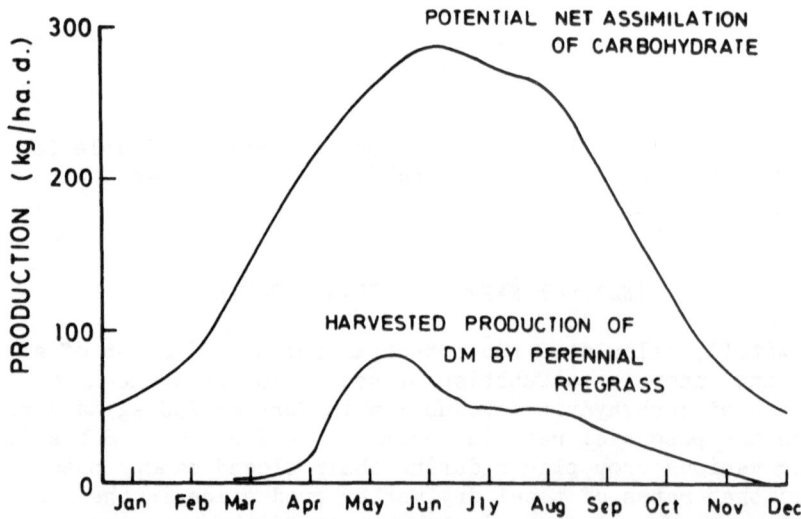

Fig. 2.   Seasonal variation at latitude 52°N in the potential net assimilation of a closed plant canopy (1) and the aerial growth of perennial ryegrass (4).

As a result of the changing rates of growth and maturation, the digestibility of grazed herbage varies considerably through the year, reflecting not only seasonal effects on plant composition, but also varying degrees of selection as plant growth exceeds or falls behind consumption. Fig. 3 shows the seasonal changes in digestibility of three types of grassland. The intensive management of grassland, whether by grazing or cutting, attempts to equate plant growth and removal to maintain a constant herbage quality, as seen in line a.

A major feature of herbage utilization by ruminants is the cascade effect whereby a reduced digestibility of the food results in a disproportionately large fall in productivity. As the season progresses and herbage digestibility declines, voluntary intake falls so that the digestible energy intake is rapidly reduced. Furthermore, the efficiency with which the digestible energy is used, both for maintenance and to an even greater extent for productive functions, is also reduced. To illustrate the magnitude of these summated effects for an animal under natural

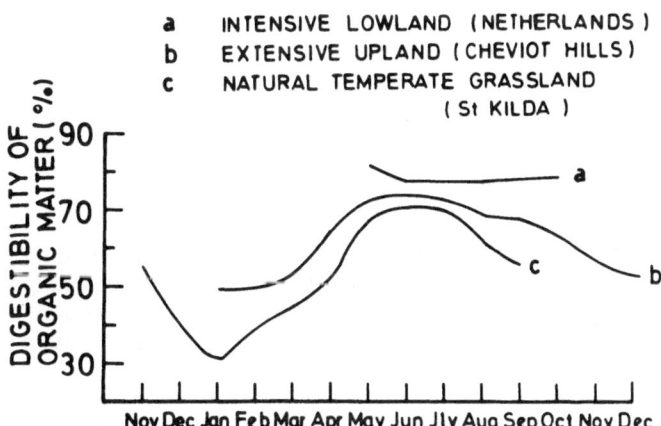

Fig. 3.   Seasonal changes in the digestibility of herbage on three types of grassland. a. Intensive lowland (Netherlands; 7), b. Extensive upland (Cheviot hills; 8), c. Natural temperate grassland (St. Kilda; 9).

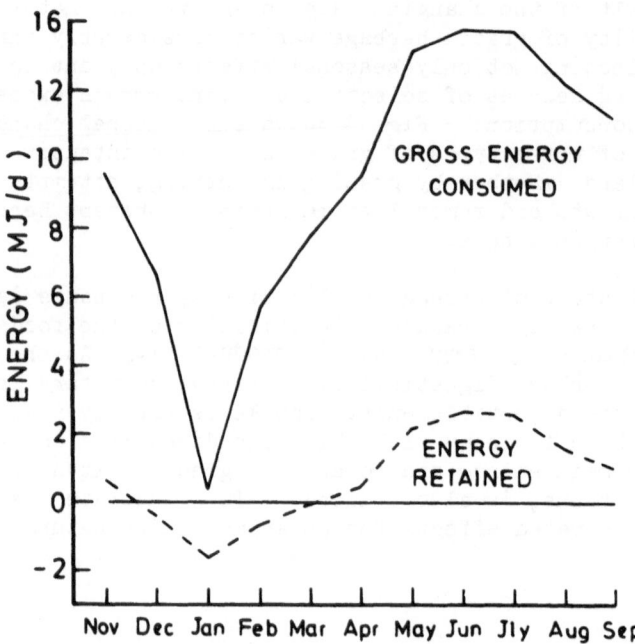

Fig. 4. Estimates of the seasonal changes in the intake and retention of energy by feral Soay sheep grazing the natural grassland of St. Kilda. Calculated from the estimates of organic matter digestibility and intake of Milner and Gwynne (9) and the equations of the ARC (10). ME was assumed to be 81% of digestible energy; digestible organic matter was assumed to have 18.5 MJ/kg.

conditions, Fig. 4 shows estimates of the seasonal changes in the energy intake and retention of a feral Soay sheep grazing its native pasture, which was included in Fig. 3. It would seem that for four months of the year such an animal is unable even to meet its maintenance requirement of energy.

In all but the most primitive systems of animal husbandry these fluctuations are reduced by various techniques of grassland management, crop conservation and the use of specialised crops for winter feeding so as to assure the animals a more uniform year-round intake of nutrients.

DIRECT EFFECTS: ANIMAL METABOLISM

The major routes by which climate directly affects an
animal's growth are by way of its heat loss, its voluntary food
intake, and its protein metabolism which determines the partition
of its retained energy between protein and fat.

Heat Loss

The rate at which an animal loses heat to its environment
depends on its size and shape and on the thermal insulation of its
tissues and coat. It also depends on its climatic environment
which comprises air temperature, radiation, both from the sun and
the environment at large, wind speed, humidity and precipitation.
The effects of these factors on the heat losses of domestic
animals have been reviewed recently (11). In indoor environments
air temperature is the predominant variable, but for animals kept
outdoors wind, rain and snow, which tend to destroy the insulation
of their coats, solar radiation and the radiant heat sink of the
night sky assume major importance. In tropical environments high
rates of direct solar radiation are often combined with ambient
temperatures within a few degrees of body temperature. In these
circumstances, when evaporation becomes the most important route
of heat loss, humidity becomes a critical factor in the maintenenance
of thermal balance.

Food Intake

The food intake of an animal may be affected by climate and
season in any of three ways. First, as already mentioned,
seasonal changes in the digestibility of their food may be the
overriding factor in determining the food intake of herbivores.
Second, there appears to be a general response to environmental
temperature, illustrated in Fig. 5 by results obtained with growing
chicks, pigs and cattle. The increase in food intake in the cold
appears to be of a similar magnitude in these three species. In
all three there is a markedly steeper decline above thermal
neutrality. Sheep also exhibit this general response to cold (16),
but show in addition a fall in food intake in autumn which is
independent of food quality (17). Red deer, especially the stags,
also show this effect (18); work with simulated lighting regimes
has shown that the stimulus is provided by shortening day length
(19).

Since heat production varies with food intake, a further
consequence of reduced voluntary intake is to reduce heat
production and thereby raise the critical temperature. Herbi-
vorous animals outwintered without supplementary feed are therefore

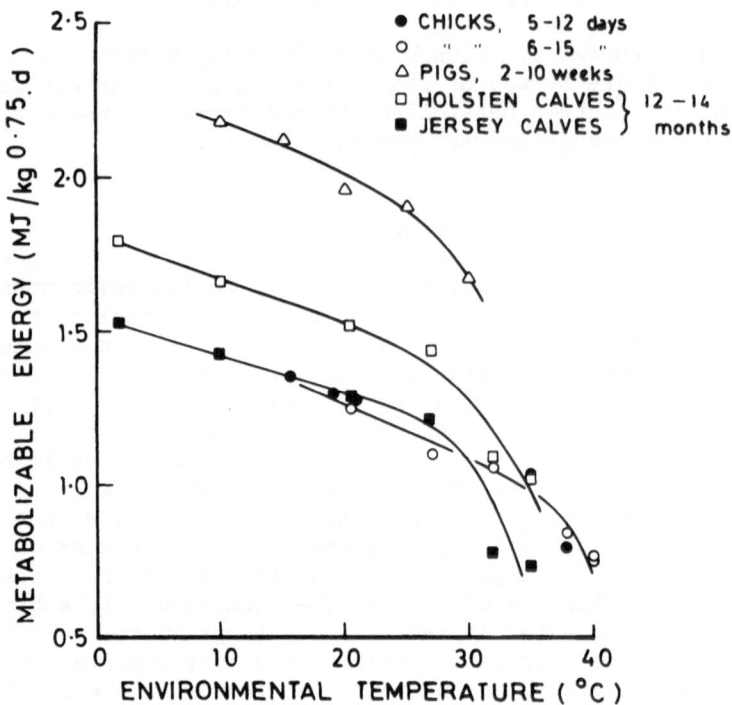

Fig. 5.   Environmental temperature and the voluntary food intake
of growing animals, expressed as metabolizable energy: chicks
5-12 d (12), chicks 6-15 d (13), pigs 2-10 wks (14), calves
12-14 mo (15).

particularly vulnerable to this combination of the direct and
indirect stresses of winter.   In contrast, sheep and cattle given
sufficient food for rapid growth are rarely subject to climatic
stress even in very cold climates and with little shelter (20).

Energy Retention

The rate at which an animal retains energy represents the
balance of its energy budget.   Since the various components of
climate affect both the inputs and outputs of energy, the rate of
energy retention may be regarded as an integration of the diverse
effects of climatic variation.   In hot environments, above
thermal neutrality, energy retention is reduced both through a
decline in food intake and to a lesser extent an increase in heat
production.   Whether or not it is reduced below thermal neutrality
depends on the extent to which the cold-induced increase in food

intake can compensate for the increased heat loss.   Evidence with growing chicks (13) suggests that moderate cold may elicit an overcompensation of food intake so that energy retention may be greatest at a temperature below that at which heat production is minimal (Table 1).   Unfortunately, comparable evidence with larger species is lacking.

Protein Metabolism and Body Composition

Since protein may be used either for the growth of new tissue or as an energy source, the extent of its contribution to thermo-regulatory heat production largely determines what effect adverse environments have on growth and body composition.

Cold thermogenesis in adult sheep (21,22) and cattle (23) was shown not to involve an increase in protein katabolism, but was met by an increased oxidation of fat.   Nitrogen excretion was, however, increased at high temperatures; Graham's results (22) suggest that this may apply only to the non-pregnant animal.   In growing animals on the other hand, with high rates of protein accretion accounting for a substantial part of their energy reten-tion, it might be expected that protein metabolism would be more sensitive to changes in energy expenditure, and experiments have shown that this is so.   Pigs, for example, on a fixed food intake, retain less nitrogen in the cold (24,25,26).   When growing animals are fed _ad libitum_, however, the possibility arises that they may

Table 1

Effects of environmental temperature on chicks
6-15 days of age.   Data of Kleiber & Dougherty (13)

Temperature ($^\circ$C)	21	27	32	38	40
Food intake (g/d)	15.0	13.3	11.7	8.7	7.9
Energy retention (kcal/d)	6.9	10.4	11.8	8.7	6.7
Protein deposition (g/d)	1.10	1.08	0.97	0.79	0.68
Fat deposition (g/d)	0.06	0.44	0.67	0.44	0.30
Weight gain (g/d)	4.88	4.64	4.39	2.97	2.91

compensate for the increased energy expenditure by eating more, as
already mentioned, and that part at least of the extra protein
they consume may be used for growth.   This has been shown to occur
with chicks (Table 1) and in rats, especially on a low protein
diet (27).   Indeed, diets very low in protein, which do not
support the life of animals at thermal neutrality, may do so when
given to animals in the cold which consume more of them and thereby
raise their protein intake to the minimum necessary for survival
(28).   Similarly, diets grossly imbalanced in amino acids on
which rats cannot survive at room temperature, may maintain animals
kept in the cold, which eat more (29).   The increased katabolism
of amino acids in the cold is in this case necessary to survival.
Young pigs, given a normal diet _ad libitum_ in the cold ate more,
at the same body weight, than those in the warm, and utilised the
dietary protein with the same efficiency (14,30).

The body composition of a growing animal, in terms of its fat
and protein contents, is determined by the relative rates of fat
and protein accretion, virtually all the energy retained being in

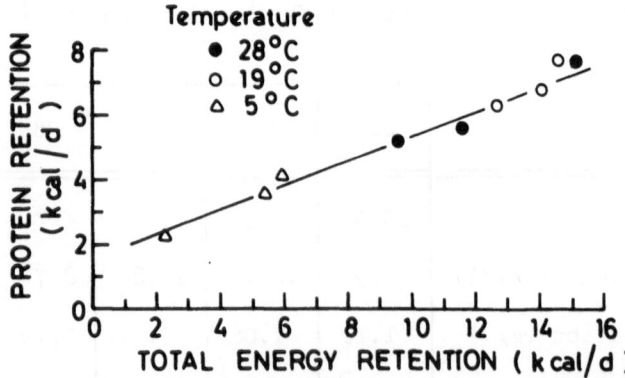

Fig. 6.   Relation between the retentions of protein and energy by
rats growing at different environmental temperatures (31).

these forms.   If, as suggested above, cold affects fat deposition
proportionately more than protein, animals raised in the cold
would have less fat than those in the warm.   If food is available
ad libitum, a reduced rate of fat deposition at high temperatures
is primarily a consequence of a reduced food intake: in the cold,
of the failure of voluntary food intake to increase sufficiently
to match the increased energy expenditure.

For any growing animal on a given diet there is, irrespective
of temperature, a characteristic relationship between protein
accretion and total energy retention;   at a very low rate of
energy retention protein accounts for all the retained energy and
this proportion falls progressively as total energy retention
increases.   A lower body fat content may therefore arise simply
from a lower rate of energy retention with a corresponding increase
in the proportion of the energy retained as protein.   In a typical
experiment with growing rats (31) shown in Fig. 6, though food
intake was increased in the cold, energy retention was reduced,
and nitrogen retention with it, though the relation between the

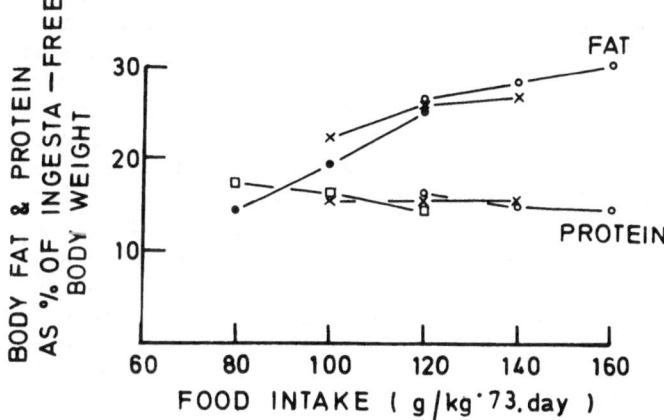

Fig. 7.   Fat and protein contents of the whole bodies of 90 kg
pigs kept during their growth from 20 kg at different environmental
temperatures and given graded quantities of food (26).

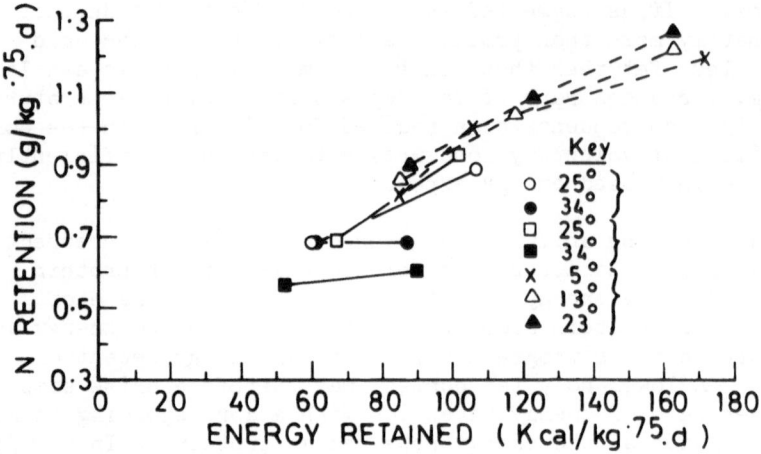

Fig. 8.   Relationships between the retentions of nitrogen and energy in pigs growing at 5°, 13° and 23° (26,34) and at 25° and 34° (32,33).

two appeared to be undisturbed by cold _per se_.   Similarly, in experiments with growing pigs (26), body composition was altered by the animals' food intake, but not by the temperature at which they had been kept during their growth (Fig. 7).   There is some evidence that high temperatures may produce a different result; Holmes' results (32,33) suggest that whereas in a moderately warm environment an increased food intake resulted in the expected increase in both nitrogen and energy retention, at a high temperature there was a smaller increase in energy retention, as a result of a higher rate of heat production, and no increase in nitrogen retention.   These results are presented with our own (26,34) in Fig. 8 and suggest that whereas cold produces no change in the partition of retained energy between protein and fat, high temperatures result in a distinctly lower relative rate of nitrogen retention.

Composition of body fat.   When animals are raised in the cold their body fat is softer, containing more unsaturated long chain fatty acids than when they are kept in the warm (35,36,37).   It has been suggested (35) that the gradient of unsaturation through the subcutaneous fat is a result of the gradient of temperature, which is steeper in animals acclimatized to cold.   In spite of this parallel, the finding (36) that the composition of peri-nephric fat was affected by external cold to no less an extent than subcutaneous fat suggests that the effect must be, at least in part, systemically mediated, though the details of the mechanism are as yet unknown.

ADAPTATIONS TO CLIMATIC AND SEASON CHANGE

As a result of the seasonal changes in plant growth and senesence the energy available to animals may be many times greater in summer than in winter - or in the rainy season than in the dry.   The energy requirements, on the other hand, of an animal population with a random seasonal distribution of repro-duction, tend to be greatest in winter or the dry season.

The adaptations of animals to seasonal climatic change are therefore concerned with means to even out the fluctuations in either the supply or the demand for food, so as to match require-ments more closely to availability.   In intensive systems of animal husbandry these ends are achieved by protecting animals from direct climatic stress, and by various systems of conserva-tion or the use of specialized crops for winter feeding.

Under natural conditions, some degree of natural conservation occurs when animals do not eat plant material as soon as it is produced.   Under extensive grazing conditions, the quantity of standing herbage shows less seasonal variation than does the rate of plant growth.   Although of poorer quality than fresh, dead herbage may be the only direct source of food for grazing animals in winter.   But the most important buffer of variations in food supply is the animal's ability to store in body tissues nutrients in excess of its immediate requirements in time of plenty and to mobilize them in times of scarcity.   A 70 kg sheep may, in normal growth, have 30 kg of body fat, representing 150 times its daily resting energy expenditure.

The growth of animals under natural conditions is therefore characterized by the cyclic depletion and replenishment of body reserves.   The example in Fig. 9 (38) is again provided by the Soay sheep of St. Kilda, maintained for the last four decades without human interference.   The figure charts the growth of male and female sheep over their first six years.   The character-istic saw-tooth growth pattern, which McDowell (39) also showed in cattle on unimproved tropical grazing, summates the effects of plant growth and climatic stress on the energy reserves of the animal.   The consequence of the greater energy requirement of the ewes in pregnancy and particularly in lactation is clearly seen.

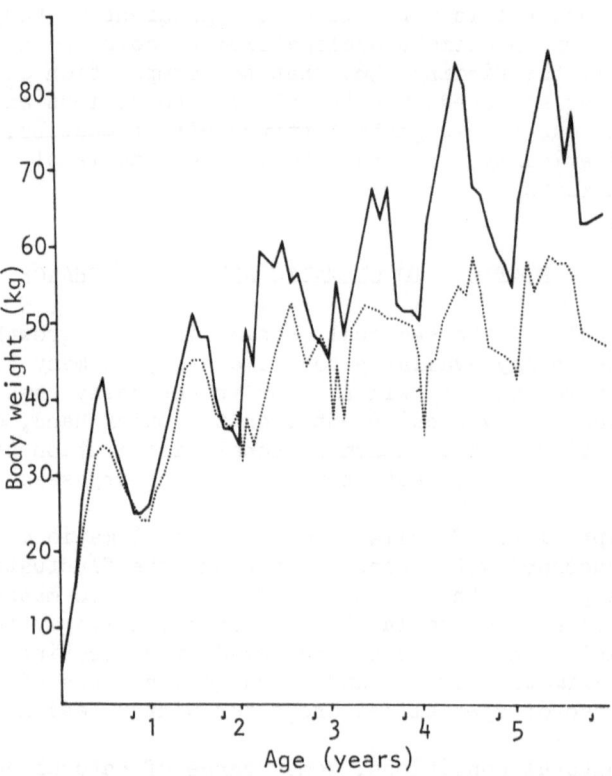

Fig. 9. Growth of feral Soay sheep on St. Kilda over the first
five years of life (38).   Solid line: rams, dotted line: ewes.
J indicates January of each year.

      Although body fat is overwhelmingly the animal's most
important energy reserve, much of the weight loss of sheep, in the
early stages of undernutrition, consists of protein and water (40,
41), with the result that animals which have lost weight may
contain more fat than they did when at the same weight during their original growth. An early and rapid loss of unessen-
tial lean body mass may be a significant adaptation to reduce
energy requirements in undernutrition.   In this respect an energy
deficit resulting from undernutrition apparently differs from that
resulting from cold.

                        Timing of Reproduction

      Apart from cold stress, the major increases in energy demand
are occasioned by pregnancy, and more especially by lactation.
The synchronization of reproduction with season in species subject
to large seasonal variations in either food supply or climatic

stress, therefore represents an important adaptation to the seasonal influence.   The proximate stimuli, i.e. the signals which initiate reproduction, vary widely according to the chosen optimum season, the length of gestation and other factors (42).   Brody remarked (43) that 'domestication tends to free animals from the seasonal influence'.   Apart from the question of taming, this might be a good definition of domestication, for the provision of food and shelter, by relieving animals of the necessity to provide for their own survival in times of energy deficit, allows man to select for characteristics, such as continuous breeding, which would not favour survival in the wild.   Thus, the European wild pig is reported to have a sharply defined breeding season in mid winter (44), but the domestic pig of temperate climates appears to have lost even a residual tendency to seasonal breeding.

Flexibility of the Growth Schedule

It is so obvious as almost to be forgotten that animals have no fixed food requirements for growth.   If growth were a process rigidly programmed in time, animal populations would require very favourable environments to assure their survival.   When given unlimited access to food, young animals may eat 3 to 4 times as much food as they require for their maintenance, but, as the experiments of McCance and his colleagues (45) clearly showed, young animals can survive for years on very low levels of food intake, remaining at infantile weights while their littermates have grown up and produced young themselves, yet retain considerable capacity to grow when food is at last made available.   Thus animals born outside the optimum season to which the population has adapted, have possibilities of adapting as individuals.

Adaptations to Reduce Climatic Stress

Physiological adaptations to reduce the impact of climatic stress include changes in behaviour, in thermal insulation, in metabolism and in the morphology of growth, as well as the changes in food intake which have already been considered.

Behaviour.   Behavioural thermoregulation may in some cases be regarded as modifying the animal's local environment, in others as altering its external insulation.   Huddling, sheltering, nest building in the cold, the search for shade or water in which to wallow in the heat, are in many cases vital to survival.   To some extent they are instinctive, though animals learn to regulate their environment by conscious action (46,47,48).

Insulation.   In most of the animals which habitually live in cold climates the hair coat represents the largest component of thermal insulation.   In sheep control of the seasonal growth and shedding of the fleece is exercised solely by the daylength, not by temperature.   There is a direct effect of temperature on the hair coat of cattle (20) and even of pigs (14), a species in which the hair is of little value for insulation.   Low temperatures are associated with a reduced rate of hair shedding (20,26).

It is commonly assumed that subcutaneous fat necessarily contributes to thermal insulation, but in fact the insulation of subcutaneous fat may be all but abolished by increasing peripheral blood flow (49).   Only by restricting the cutaneous circulation so as to allow skin temperature to fall can more of the temperature gradient between the deep body and the environment be taken up within the tissues.   Enhanced tissue insulation implies peripheral cooling and the development of an increased toleration of low skin temperature.   This is particularly important in relatively hairless species such as pigs (50), but acclimatization of sheep also leads to a greater degree of peripheral cooling (51).   From this it is clear that although a large amount of tissue insulation cannot be expected without subcutaneous fat, the mere presence of fat does not imply that it necessarily performs an insulative role. It is therefore not surprising that pigs kept in the cold do not deposit more of their body fat subcutaneously (14).

Metabolism.   Seasonal changes in basal metabolism are well known in small mammals in which they assist in enlarging the limits of survival.   Large animals are generally said to rely on insulative changes, but measurements made many years ago suggest that seasonal changes in metabolism may also occur.   Fig. 10 shows the seasonal changes in the thermoneutral metabolism of an adult wether sheep (52) and an adult sow (53).   The measurements with the sow were made after an adequate fast, but those with the sheep may reflect to some extent concomitant seasonal changes in food intake.

Morphology of growth.   Phylogenetically, one may point to extreme differences in the conformation of animals which suit them to their native climates.   Within a domestic species one may similarly contrast breeds native to temperate and tropical environments.   Even within the lifetime of the individual there exist considerable possibilities for morphological adaptations which change the animal's ratio of surface area to mass.   These include the enlargement of the ears, legs and tail (14,54,55).   However, since, according to McDowell (56), amputation of the characteristic appendages of Zebu cattle did not reduce their heat tolerance, the physiological significance of these adaptations may be in doubt.

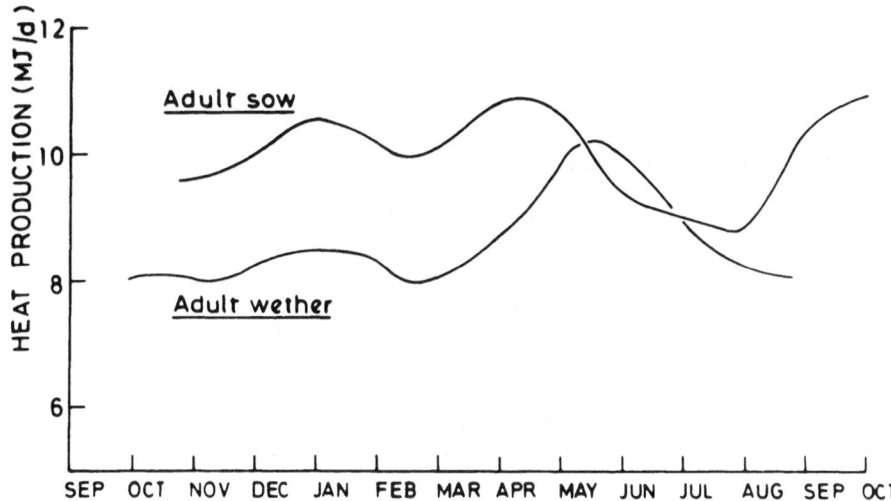

Fig. 10.   Seasonal variation in the metabolism of an adult wether
sheep (52) and of an adult sow (53), measured under fasting conditions in a thermoneutral environment.

THE USE OF LIVESTOCK IN RELATION TO PAST
AND FUTURE CLIMATES

        For most of their domesticated history domestic animals have
been kept primarily for purposes other than the production of lean
meat: sheep, largely for their wool, cattle for draft or milk.
Insofar that their meat was valued, it was as an alternative source
of energy to the starches and oils of plants.   Human aversion to
animal fat is a recent phenomenon of a few advanced Western cultures.   Man's need, like that of his animals, has been to buffer
variations in his food supply resulting from the seasonal nature
of plant growth and fruition.   This is still so in many parts of
the world today.   The propensity of animals to respond to a
seasonal glut of food by depositing large amounts of body fat was
favoured not only for their own survival but for man's also.   In
Britain, it was only with the development, in the 18th century, of
a more effective technology of winter keep, especially the introduction of the turnip crop, that selection for larger and faster
growing animals was made possible.   Lord Ernle (57):

'Without the aid of turnips the mere support of livestock had
been in winter and spring a difficult problem; to fatten sheep
and cattle for the market was in many districts a practical
impossibility ...... The introduction of turnip and clover husbandry doubled the number and weight of the stock which the land
would carry, and the early maturity of the improved breeds enabled
farmers to fatten them more expeditiously.'

Small size, slow growth and large fat reserves favoured survival under these conditions. Only by assuming responsibility for
the winter survival of his animals could man begin genetic selection for the opposite characteristics: a further stage in domestication. It seems unlikely that genetic selection for increased
growth rate and leanness in, for example, hill sheep will be
successful while natural selection still favours those animals
which, by virtue of their small size and extensive fat deposits,
are best fitted for winter survival.

The future. Agriculture is a marginal business, for man,
like other animals, tends to push out to his ecological limits.
Every year, somewhere in the world, there is a savage reminder that
very small changes in climate, especially as they affect the distribution of rainfall, can have far-reaching repercussions: in
recent years, the inundation of Bangladesh, the drought in the
Sahel, the large scale crop failures of Asia. Whatever the
future pattern of world climate, it is clear that the first half
of this century, when so much of modern agricultural technology
was developed, was a period of unusually good climates for much of
the world (58), and that we must expect such large scale irregular-
ities to recur, perhaps with increasing frequency over the next
50 years (59,60).

Such occurrences bring enormous human tragedy; to avoid
being overtaken by events requires the intelligent allocation of
agricultural resources in space and time, with adequate global
reserves of food, as well as of seed, fertilizer and other inputs
necessary for rapid recovery of agricultural production.

Animal production may well face periodic scarcities and
surpluses of feed grains resulting from the vagaries of climate
in the principal feed grain growing areas. If such fluctuations
are expected, there is no reason for them to have the disruptive
effects that we have seen in the past three years. But to cope
effectively requires the development of an increased flexibility
in animal production, to use livestock in the way most appropriate
to the circumstances.

REFERENCES

1. de Wit, C. T. (1959). Neth. J. agric. Sci. 7, 141.

2. Sibma, L. (1968). Neth. J. agric. Sci. 16, 211

3. Munro, J. M. M. and Davis, D. A. (1973). J. Br. Grassld Soc.
    28, 161.

4. Anslow, R. C. and Green, J. O. (1967). J. agric. Sci., Camb.
    68, 109.

5. Deinum, B. (1966). Meded. Landb. 66, 11.

6. Deinum, B. and Dirvan, J. G. P. (1972). Neth. J. agric. Sci.
    20, 125.

7. Van Es, A. J. H., Van der Honing, Y., Vogt, J. E., Bangma,
    G. A., Terlun, R. and Nijkamp, H. J. (1974). Z. Tierphysiol.
    Tierernahr. Futtermittelk. 33, 193.

8. Eadie, J. (1967). 4th Rep. H.F.R.O.

9. Milner, C. and Gwynne, D. (1974). In: "Island Survivors"
    (Eds. P. A. Jewell, C. Milner and J. M. Boyd), p. 273 (The
    Athlone Press, London).

10. Agricultural Research Council (1965). "The Nutrient Require-
    ments of Farm Livestock." No. 2. Ruminants (London, HMSO).

11. Monteith, J. L. and Mount, L. E. (Eds.) (1975). "Heat Loss
    from Animals and Man" (Butterworths, London).

12. Winchester, C. F. and Kleiber, M. (1938). J. agric. Res.
    57, 529.

13. Kleiber, M. and Dougherty, J. E. (1934). J. gen. Physiol.
    17, 701.

14. Fuller, M. F. (1965). Br. J. Nutr. 19, 531.

15. Johnson, H. D. and Yeck, R. G. (1964). Res. Bull. Mo. agric.
    exp. Stn. 865.

16. Weston, R. H. (1970). Aust. J. exp. Agric. Anim. Husb. 10,
    679.

17. Gordon, J. G. (1964). Nature, Lond. 204, 798.

18. Blaxter, K. L., Sharman, G. A. M., Kay, R. N. B., Cunningham,

J. M. M. and Hamilton, W. J. (1974). "Farming the Red Deer" (London, HMSO).

19. Kay, R. N. B. (1974). Unpublished results.

20. Webster, A. J. F., Chlumecky, J. and Young, B. A. (1970). Can. J. Anim. Sci. 50, 89.

21. Graham, N. McC., Wainman, F. W., Blaxter, K. L. and Armstrong, D. G. (1959). J. agric. Sci., Camb. 52, 13.

22. Graham, N. M.C. (1964). Aust. J. agric. Res. 15, 982.

23. Blaxter, K. L. and Wainman, F. W. (1961). J. agric. Sci., Camb. 56, 81.

24. Piatkowski, B. (1958). Arch. Tierernähr. 8, 161.

25. Moustgaard, J., Nielsen, P. B. and Sørensen, P. H. (1959). Årsberetn. Inst. Sterilitetsforsk., p. 173 (K. Veterinaer-og Landbohøjskole, Copenhagen).

26. Fuller, M. F. and Boyne, A. W. (1971). Br. J. Nutr. 25, 259.

27. Jacob, M. and Payne, P. R. (1964). Proc. Nutr. Soc. 23, v.

28. Andik, I., Donhoffer, S. Z., Farkas, M. and Schmidt, P. (1963). Br. J. Nutr. 17, 257.

29. Anderson, H. L., Benevenga, N. J. and Harper, A. E. (1969). J. Nutr. 99, 184.

30. Fuller, M. F. (1966). Proc. Nutr. Soc. 25, vi.

31. Fuller, M. F. and Smart, R. (1966). Unpublished results.

32. Holmes, C. W. (1971). Anim. Prod. 13, 521.

33. Holmes, C. W. (1973). Anim. Prod. 16, 117.

34. Fuller, M. F. and Boyne, A. W. (1972). Br. J. Nutr. 28, 373.

35. McGrath, W. S., Jr., Vander Noot, G. W., Gilbreath, R. L. and Fisher, H. (1968). J. Nutr. 96, 461.

36. Fuller, M. F., Duncan, W. R. H. and Boyne, A. W. (1974). J. Sci. Fd Agric. 25, 205.

37. Marchello, J. A., Cramer, D. A. and Miller, L. G. (1967). J. Anim. Sci. 26, 294.

38. Doney, J. M., Ryder, M. L., Gunn, R. G. and Grubb, P. (1974). In: "Island Survivors" (Eds. P. A. Jewell, C. Milner and J. M. Boyd) (The Athlone Press, London).

39. McDowell, R. E. (1968). Nature, Lond. 218, 641.

40. Russell, A. J. F., Gunn, R. G. and Doney, J. M. (1968). Anim. Prod. 10, 43.

41. Burton, J. H., Anderson, M. and Reid, J. T. (1974). Br. J. Nutr. 32, 515.

42. Sadleir, R. M. F. S. (1969). "The Ecology of Reproduction in Wild and Domestic Mammals" (Methuen, London).

43. Brody, S. (1945). "Bioenergetics and Growth," p. 207 (Reinhold, New York).

44. Bertin, L., rev. Burton, M. (1967). In: "Larousse Encyclopaidia of Animal Life", p. 589 (Paul Hamlyn, London).

45. McCance, R. A. and Widdowson, E. M. (1974). Proc. R. Soc. Lond. B. 185, 1.

46. Mount, L. E. (1963). Nature, Lond. 199, 122.

47. Baldwin, B. A. and Ingram, D. L. (1967). Physiol. Behav. 2, 15.

48. Baldwin, B. A. and Ingram, D. L. (1968). Physiol. Behav. 3, 395.

49. Miller, A. I., Jr. and Blyth, C. S. (1958). J. appl. Physiol. 12, 17.

50. Irving, L., Peyton, L. J. and Monson, M. (1956). J. appl. Physiol. 9, 421.

51. Sykes, A. R. and Slee, J. (1968). Anim. Prod. 10, 17.

52. Brody, S. (1945). In: "Bioenergetics and Growth", p. 227 (Reinhold, New York).

53. Ritzman, E. G. and Colovos, N. F. (1941). Tech. Bull. New Hamps. agric. Exp. Stn. 75.

54. Weaver, M. E. and Ingram, D. L. (1969). Ecology 50, 710.

55. Johnson, H. D., Ragsdale, A. C., Sikes, J. D., Kennedy, J. I., O'Bannon, E. B., Jr. and Hartman, D. (1961). Res. Bull. Mo.

agric. exp. Stn. 770.

56. McDowell, R. E. (1972). "Improvement of Livestock Production in Warm Climates"(Freeman, San Francisco).

57. Ernle, Lord (1922). In: "English Farming, Past and Present", p. 176, 3rd ed. (Longmans, Green & Co., London).

58. Lamb, H. H. (1965). In: "The Biological Significance of Climatic Changes in Britain"(Ed. C. G. Johnson) (Academic Press, London).

59. World Meteorological Organization (1962). "Climatic Fluctuations and the Problems of Foresight." Unpublished report.

60. Hay, R. F. M. (1974). Span 17, 104.

DISCUSSION

   Dr. Pomeroy wondered to what extent the slower postnatal
growth of animals born small as a result of maternal undernutrition
was due to the reduced milk yield of the dam and to the poorer
sucking ability of the young. In reply to this and to a question
by Dr. Rerat, *Dr. Widdowson* confirmed that pigs and guinea pigs
which were born smaller than average, remained so throughout life
even when fostered on to adequately-nourished mothers. *Prof.
Ingram* stressed the importance of maternal undernutrition, the
timing and severity of which could affect both birth size and
adult size. *Prof. Lucas* noted that the growth of the placenta
could be restricted in undernourished pregnant animals. Dr.
Widdowson said that her results with runt pigs suggested that the
growth setback occurred early in pregnancy for the runts had
fewer body cells at birth than normal foetuses of the same body
weight. Referring to Dr. Widdowson's results on the limitation
of foetal growth by calcium and phosphorus, *Dr. Wilson* pointed out
that, in the dairy cow, the flux of calcium and phosphorus during
lactation was considerably greater than the requirement of the
foetus at full term. He thought that in cases such as the Shire x
Shetland experiment the physical limitation of the body cavity
must be of importance. *Dr. Widdowson's* answer was that, if
nutrition were not important, one would expect the foetus of the
Shetland mare to develop rapidly and be born prematurely, whereas
in fact it grew more slowly than the Shire's foetus and was born
after the normal gestation period. In reply to Dr. Moody, *Dr.
Widdowson* said that she had no information on the cellularity of
adipose tissue of the runt pig, either at birth or subsequently.
It was her impression that both runts and pigs undernourished
postnatally had, when subsequently well nourished, more fat than
normally-reared animals.

   Dr. Rhodes asked what was the physiological necessity for a
target fat mass. *Dr. Fowler* thought that there were examples in
certain species where survival was favoured by a particular
pattern of fat deposition during early growth and, in general,
animals tended to maintain a proportionality between protein and
fat deposition. In supporting Dr. Fowler's concept of a critical
fat mass, *Prof. Elsley* said that, in normally-growing pigs, the
ratio of fat to protein deposited rarely fell below 1. *Dr. Rhodes*
asked if this implied a limit to genetic selection for leanness.
Prof. Elsley said that perhaps such selection might lead to a
disturbance of the physiological integrity of the animal, such as
that seen in PSE meat. *Prof. Ingram* wondered whether the pig's
tendency to lay down large amounts of body fat indicated that it
evolved in temperate regions subject to a large seasonal variation
in food supply. *Dr. Fuller* replied that pigs were found wild in
both temperate and tropical regions. He thought it likely that
the characteristic tendency to deposit fat had been intensified by

genetic selection during domestication. The pig had been kept
primarily for its meat for longer than other species and its
ability to convert surplus perishable foods to fat had probably
been considered a valuable character for most of its domesticated
history. *Prof. Goss* thought that study of the control of the
seasonal deposition of fat in hibernating animals might yield
results of relevance to animal production.

Physiological Significance
of Differences in Body Composition

THE PHYSIOLOGICAL BASIS OF REPRODUCTIVE EFFICIENCY

Rose E. Frisch

Harvard University

Center for Population Studies

It is commonly observed that human beings, like other mammals, first must "grow up" before becoming capable of reproduction. But on what scale is the "growing up" best measured: age, height, weight, skeletal age, or some other measure of the body? And how is the synchronization of the "grown upness" and reproductive ability brought about? (1).

Recent findings that the onset and maintenance of regular menstrual function in the human female are each dependent on the maintenance of a minimum weight for height, apparently representing a critical fat storage (2), imply that a particular body composition of fat/lean, or fat/body weight may be an important determinant for female reproductive ability (2,3,4). Undernutrition and energy-requiring activities would then be expected to affect reproductive ability as has been observed in both animals (5) and human beings (3,6). Genetic traits such as those controlling the tempo of growth, which affects the rate of deposition of fat (4,7) or those causing obesity (8,9) or extreme lack of adiposity (10) also would be expected to affect reproductive ability, as has been observed (8-10).

The importance of a particular level of body fatness for female reproductive ability calls attention to physiologic indices, such as metabolic rate, blood pressure, and body temperature in relation to the onset and maintenance of reproductive ability, instead of the usually emphasized skeletal age and secondary sex characters.

CRITICAL WEIGHTS

The idea that relative fatness is important for female reproductive ability follows from earlier findings that the events of the adolescent growth spurt in boys and girls, and particularly menarche in girls, are each closely related to a critical body weight (11,12). This finding was unexpected for human beings, although it was well known for other mammals that sexual maturity (defined by vaginal opening, or more precisely by first estrus) is weight dependent: e.g., rats (1,13), mice (14,15), pigs (16) and cattle (17,18).Dickerson et.al. state that female pigs did not ovulate until they were approaching the body weight at which ovulation normally occurs (16). Kennedy (19) observed: "Puberty is determined by weight rather than age."

There is also an example from primates: female rhesus monkeys treated with androgens gained weight rapidly and had menarche at age one year instead of the normal age of two years, but at the weight and length characteristic of the two-year-old animals (20).

The weight findings in relation to sexual maturation in girls and boys were first indicated by an analysis of cross-sectional body weight data of Asian and Latin American peoples in relation to calorie supplies. When the age of fastest growth in weight of the adolescent growth spurt (hereafter termed peak weight velocity) was studied in relation to calorie intake, it was found that undernutrition delayed the age of peak weight velocity, and high levels of nutrition advanced the age of this event (21) as had already been observed for the adolescent spurt in general (22). Unexpectedly, however, peak weight velocity, which normally precedes menarche, seemed to take place at the same mean weight for a particular racial group, regardless of whether the age of the event was advanced or delayed in accordance with the calorie supplies (21).

To pursue this interesting finding, each event of the adolescent growth spurt was analyzed using longitudinal growth data of three completed, comparable United States studies. A velocity curve of height and weight growth from birth to age 18 years (Figs. 1 and 2) was plotted for each of the 201 girls and 209 boys of the three studies.

We found that the mean weight of girls at the time of initiation of the adolescent growth spurt (30 kg) (23), at the time of maximum rate of weight gain (39 kg) (24) and at menarche (47 kg) (11,12) did not differ for early and late maturing girls, whereas their mean height at each of these events increased significantly with age of the event. (Fig. 3).

These results accounted for the many observations in the

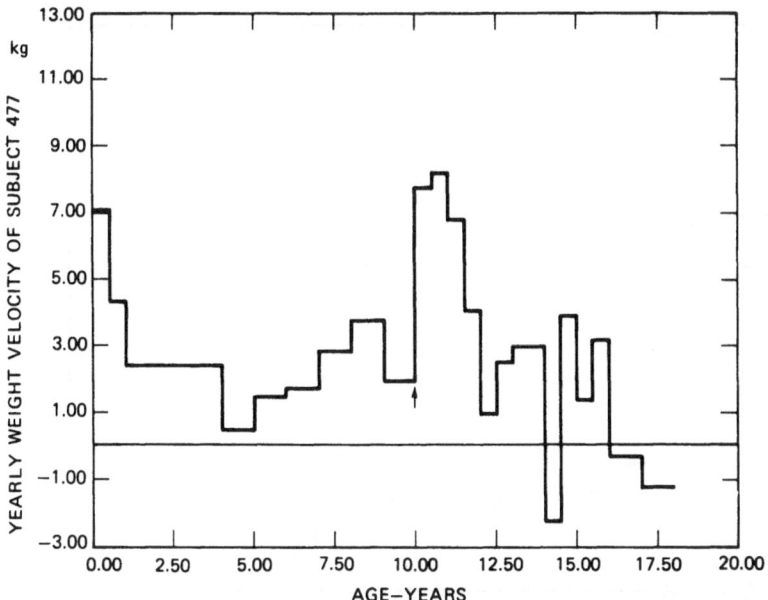

Fig. 1.   Computer-plotted yearly weight velocity versus age of
a CRC girl showing initiation of weight spurt at age 10. Menarche
is at 12.3 years. CRC, and BGS and HSPH (see later Figs.) refer to
3 studies of ref. (23). (Reprinted with permission from Human Biology.)

literature (22) that early maturers have more weight for height
than late maturers at spurt initiation and throughout the adolescent spurt, including menarche.

     Our basic finding was also true for boys, but at different
ages, weights and heights from the girls (Tables 1 and 2), with
the exception that the latest maturing boys were slightly but significantly heavier at both spurt initiation (23) and at peak
weight velocity (24). Boys attained each event about two years
later than the girls, at a mean weight about 6 kg heavier and a
mean height about 11 cm taller than that of the girls at the
corresponding event (Table 2). This suggests that "genarche" in
boys, the ability to reproduce, comparable to menarche in girls,
may be attained at a mean age of 14.9 years, at a mean weight of
about 55 kg (121 lb), and a mean height of about 169 cm (66.5
inches).

     Based on these findings of an invariant mean weight in girls,
we proposed that there is a direct relation between critical body
weight and menarche. The mechanism proposed, adapted from that of
Kennedy and Mitra (1) assumes that the attainment of the average
critical weight, which we now know represents a critical level of
body fatness (2,4,25), attained at varying heights and weights by

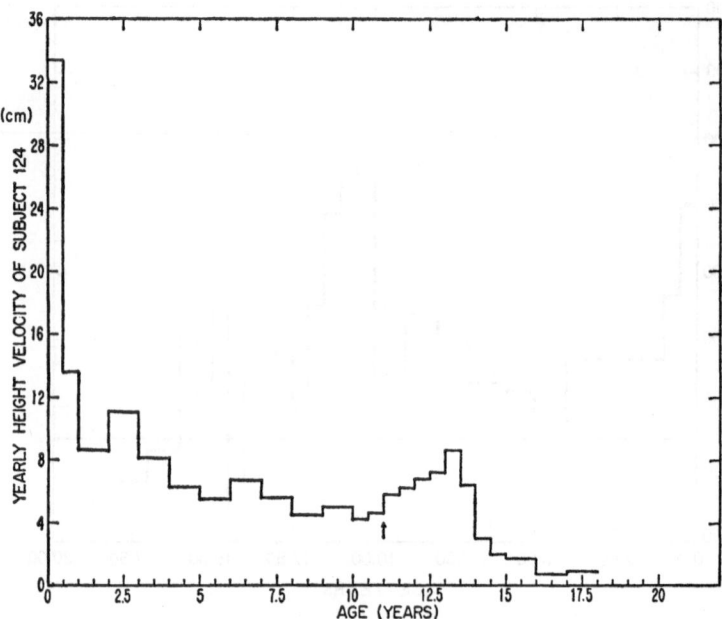

Fig. 2.   Computer-plotted yearly height velocity versus age of
CRC girl no. 124, a late maturer. Age H$_i$ is 11.0 years. Age of
menarche is 14.6 years. (Reprinted with permission from Human
Biology.)

girls within a population, causes a change in metabolic rate per
unit mass, which in turn affects the hypothalamus-ovarian feedback
by decreasing the sensitivity of the hypothalamus to estrogen.
The feedback is then reset at a level high enough to induce the
maturation resulting in menarche (12,24). There is evidence for
such a change of sensitivity of the hypothalamic "gonadostat" in
girls and boys (26).

       Whatever the mechanism, the assumption that a critical weight
is a signal for menarche explained simply many unexplained observations associated with early or late age of menarche. Observations
of earlier menarche are associated with attaining the critical
weight more quickly. The most important example is the secular
trend to an earlier menarche of about 3 or 4 months per decade in
Europe in the last one hundred years (27). Our explanation is
that children now are bigger sooner, and therefore girls, on the
average, reach 46-47 kg, the mean critical weight of United States
and English populations, more quickly. According to our hypothesis
also, the secular trend should end when the weight of children of
successive cohorts remains the same because of the attainment of
maximum nutrition and child care, (11) which now may have happened
(28).

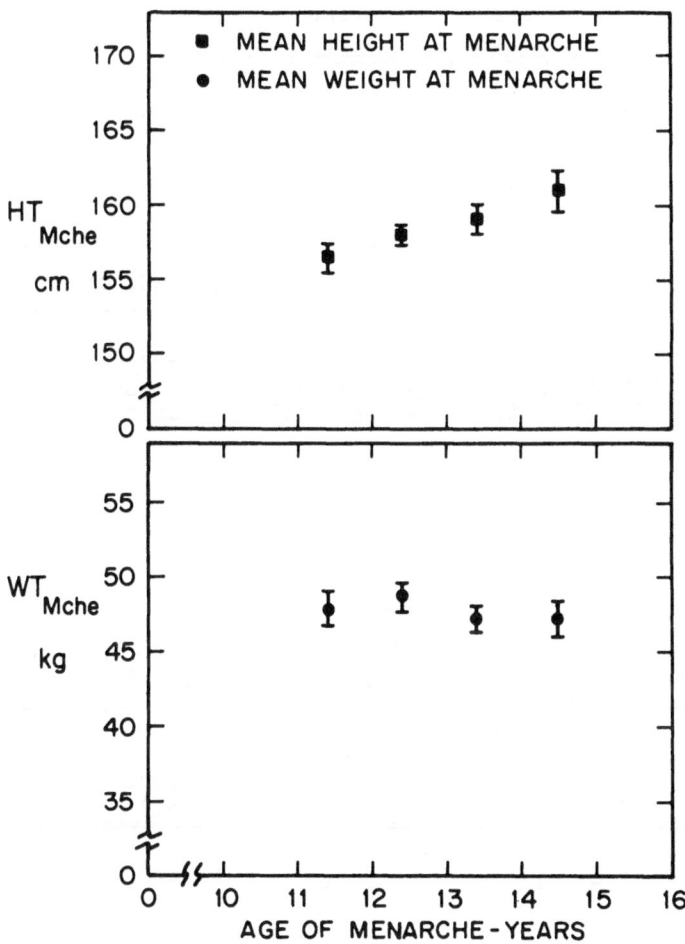

Fig. 3. Mean height (± standard error) at menarche versus mean age of menarche; and mean weight (± standard error) versus mean age of menarche, of CRC, BGS and HSPH girls grouped by age of menarche (12).

Table 1

Mean age, height, and weight of girls at initiation of the
adolescent spurt, peak velocity, menarche, and age 18 (60)

Adolescent event	No.	Age of height event (yr)	Age of weight event (yr)	Height[a] (cm)	Weight[b] (kg)
Initiation of spurt	184	9.6±0.1	9.5±0.1	136.5±0.84	30.6±0.30
Peak velocity	170	11.8±0.1	12.1±0.1	146.5±0.50	39.3±0.45
Menarche	181	12.9±0.1		158.5±0.50	47.8±0.51
Age 18	181	-		165.6±0.48[b]	57.1±0.57[c]

[a] Increases significantly (p < 0.01) with increasing age of event.
[b] Does not change significantly with increasing age of event.
[c] Decreases significantly (p < 0.02) with increasing age of event.
± = SE.

Table 2

Mean age (± SE), height, and weight of boys at initiation of
the adolescent spurt, peak velocity, and age 18 (60)

Adolescent event	No.	Age of height event (yr)	Age of weight event (yr)	Height (cm)	Weight (kg)
Initiation of spurt	179	11.7±0.1	11.6±0.1	147.3±0.49[a]	36.9±0.36[b]
Peak velocity	189	14.0±0.1	14.1±0.1	158.3±0.48[b]	47.3±0.52[b]
Age 18	179			178.1±0.46[c]	68.2±0.69[d]

[a] Increases significantly (p < 0.01) with increasing age of event.
[b] Latest maturers slightly but significantly (p < 0.01) heavier
than earlier age groups
[c] Does not change with increasing age of event
[d] Decreases significantly with increasing age of event

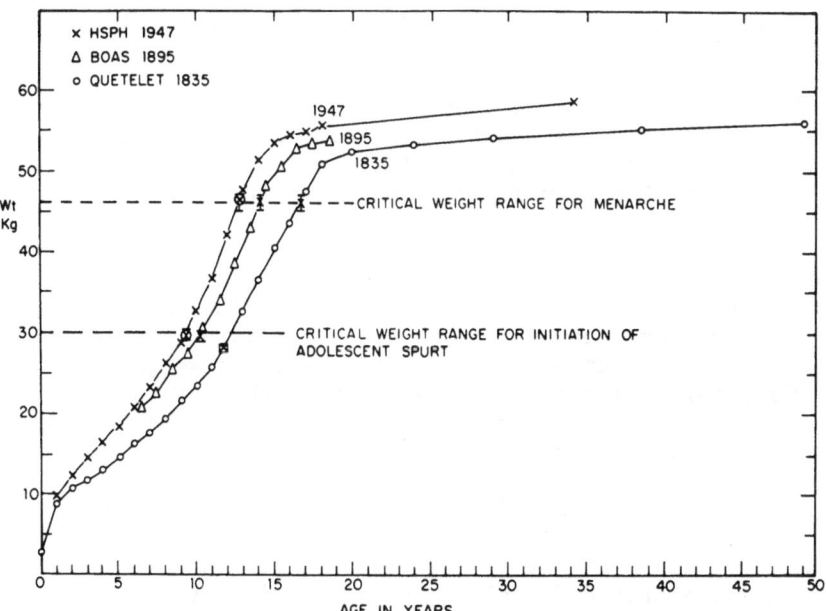

Fig. 4.   Weight growth with age of Belgian girls in 1835, and
United States girls in 1895, and 1930 to 1950 (average year of
menarche, 1947), showing age of attainment of critical weight
range,XX at initiation of the adolescent spurt, and at menarche:
X computed mean $\bar{-}$ S.E., Frisch and Revelle (12) X "avant la
puberté," Quetelet, 1869. Reprinted with permission from
<u>Pediatrics</u> (6) from which additional details may be obtained.

There is evidence to support this explanation of the secular
trend (Fig. 4). The mean weight at menarche for contemporary
United States, English, Dutch and Finnish girls is the same as the
mean weight for girls of three decades ago (29). Historically,
Boas' weight data for California girls in 1895 show that 47 kg
would be attained at about 14 years, which is consistent with
menarcheal ages recorded for about 1900 in the United States. And
Quetelet's growth data of 1835 show that Belgian girls of average
social class attained a weight of 46 kg at about age 16.5 years,
which is consistent with the existing data on age of menarche of
a century and a half ago (6).

Another example of more rapid attainment of the critical
weight is the earlier menarche of most obese girls (30,31).

Conversely, a late menarche is associated with body weight
growth that is slower prenatally, postnatally, or both, so that
the critical weight is reached at a later age: malnutrition delays

menarche, 6,21) and twins have later menarche than singletons of the same population (12).

In the case of undernourishment, the mean weight at menarche of control girls and undernourished Alabama girls did not differ, although the mean age of menarche of the underfed girls was two years later, and at a significantly taller height than that of controls (6).

For the effects of altitude, the well-nourished upper middle class girls of the CRC Denver (altitude 5,280 feet) study attained menarche at the same mean weight as the comparable California (BGS) sea level subjects, but at a later age; the birth weights of the Denver girls were significantly lighter than that of the California girls, and the Denver girls grew more slowly up to the time of initiation of the adolescent spurt (12,23).

COMPONENTS OF THE CRITICAL WEIGHT

The variability of the critical weight at menarche, 47.8 kg, was large; the standard deviation is 6.9 kg, (coefficient of variability 14.0%) (Fig 5). In order to make the notion of a critical weight meaningful for an individual girl we looked at components of the weight.

Weight is considered a reasonably good measure of metabolism in the normal child, but total body **water** (TW) and lean body weight (LBW, TW/0.72) (32), are more closely correlated with metabolic rate than is body weight, since they represent the metabolic mass, as a first approximation (33,34). These components and fat (body weight minus LBW) were calculated for each girl at menarche and at the initiation of the spurt. Study of the body composition of girls, all of whom are at the same stage of the adolescent growth spurt, or at menarche, gives more precise data on the changes that take place from spurt initiation to menarche. Also, since the data were longitudinal, body composition changes could be followed in the same girls (4).

Total body water was calculated for each girl, using the previously determined height and weight of each girl at menarche and at spurt initiation in a regression equation of Mellits and Cheek (35) from deuterium oxide measurements.

Equation (1) $TW = -10.313 + 0.252 (Wt_{kg})$ and $0.154 (Ht_{cm})$, when height > 110.8 cm

and also, for comparison, at menarche, by the equation of Moore et al.(36) from deuterium oxide measurements:

Equation (2) $TW = 11.63 + 0.318 (Wt_{kg})$

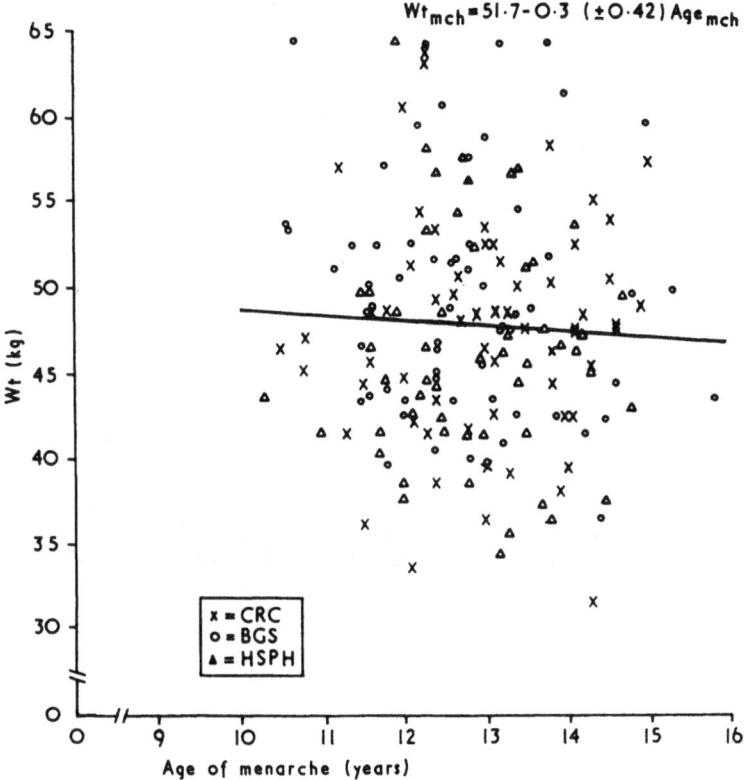

Fig. 5.   Weight at menarche (Wt_{mch}) vs. age of menarche (Age_{mch}) for CRC, BGS, and HSPH girls. Slope of regression line of Wt_{mch} does not differ significantly from zero (P>0.50).
Reprinted with permission from Archives of Disease in Childhood (12).

The total water at menarche for all subjects calculated by equations (1) and (2) were comparable, 26.2 $\pm$ 0.18 (S.D. 2.4) liters, and 26.8 $\pm$ 0.16 (S.D. 2.2) liters respectively.

        Total water calculated by either equation does not change significantly with increasing age of menarche, and the variability is 36% less than that of weight at menarche. (Fig. 6). Since lean body weight is calculated by TBW/0.72 lean body weight also is invariant with increasing age of menarche (4).

        All further results given here were calculated by equation (1), which was preferred because the range of ages of the subjects covered all of the adolescent spurt, and because the use of height

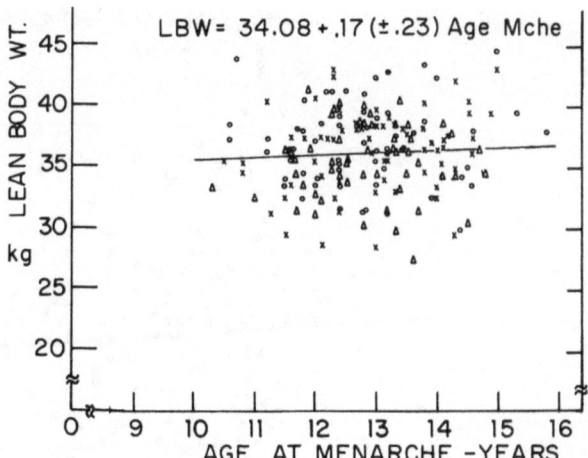

Fig. 6.   Lean body weight (LBW) vs. age of menarche for CRC, BGS, and HSPH girls. Slope of regression line of LBW on Age$_{mche}$ does not differ significantly from zero.   X - CRC; O - BGS; Δ - HSPH.

and weight, rather than weight alone, usually gives the lowest variance (37).

The mean lean body weight at menarche, 36.3 ± 0.3 kg, and mean fat at menarche, 11.5 ± 0.3 kg, are similar to those obtained at ages 12.5 - 13 years from ^{40}K counting (38).

The greatest change in body composition of both early and late maturing girls during the adolescent growth spurt is a very large increase in fat, from about 5 kg to 11 kg, a 120% increase compared to a 44% increase in LBW. There is thus a change in ratio of LBW to fat from 5:1 at initiation of the spurt to 3:1 at menarche (4). A fall in metabolic rate/kg body weight would be expected from this change in body composition, particularly the large increase in fat, because the internal organs, which con- tribute the most heat to the basal metabolism (33,39,40) become a

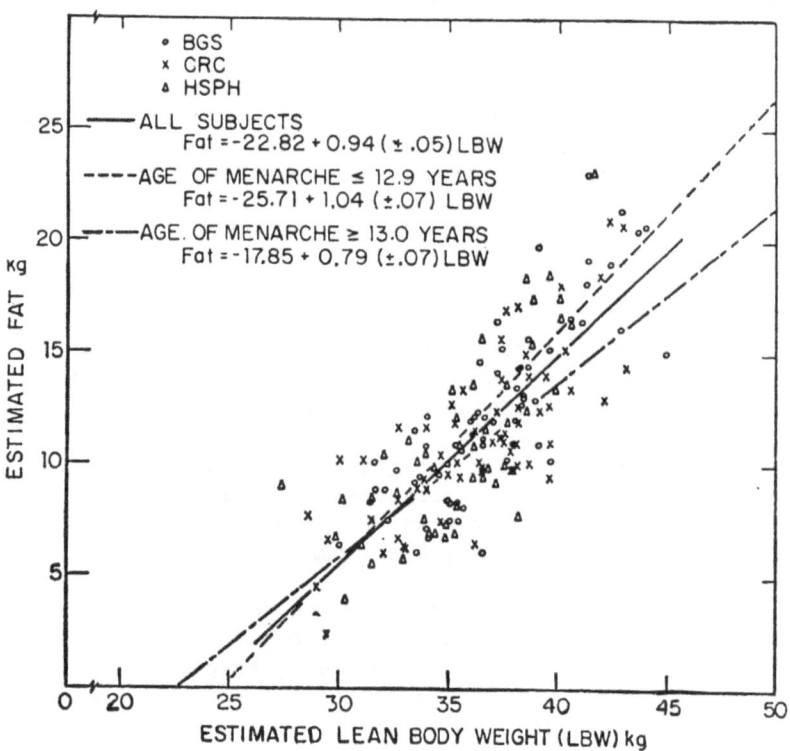

Fig. 7.   Fat versus lean body weight (LBW) at menarche for CRC,
BGS and HSPH girls. The slope of the regression line of fat on
LBW for early maturers is significantly greater (p<.01) than that
of the late maturers. Reprinted with permission from Human
Biology (4).

smaller proportion of the body weight (33,39). In fact, the
BMR/kg by Talbot's (41) standards is 35 kcal/kg per day at the
mean weight (30 kg) of initiation of the adolescent spurt in girls
and the BMR/kg is 28 kcal/kg per day at the mean weight (47 kg) of
menarche. This decrease of BMR/kg as body weight and fat content
increases "has the biological advantage of diminishing heat
production as the surface to volume ratio decreases" (33).

    An alternative explanation for the fall in metabolic rate/kg
is the probability that adipose tissue is heat-producing (42,43).

    Another finding about body fat during the adolescent spurt
is of special interest. Fat increases linearly with increasing

lean body weight for all subjects at menarche and at spurt initiation, but at both events fat increases at a slower rate with increasing lean body weight in late maturers than in early maturers (Fig. 7). This explains why late maturers have less fat on the average at each event than do early maturers, although they do not differ in lean body weight. Widdowson and McCance (7) observed this difference in fat gain between fast and slow growing rats, and it is also found in early and late maturing domestic sheep, pigs and cattle (47).

TOTAL BODY WATER AS PERCENT OF BODY WEIGHT, AN INDEX OF FATNESS

Total water as percent of body weight (TW/BWt%) is an even more important index than the absolute amount of total water because it is an index of fatness (37) (Table 3)

When girls are grouped by height at menarche rather than by age at menarche, the shortest, lightest girls and the tallest, heaviest girls differ in height by 20 cm and in weight by 12 kg; they certainly do not have weight in common. (Table 4).

But these two extreme groups have the same relative fatness, as shown by their similar percentages of total water/body weight, 56.3 ± 0.5 percent, and 55.3 ± 0.5 percent, respectively. Both these values are similar to the mean for all subjects, 55.1 ± 0.3 percent (Fig. 8 and Table 5).

Further, although the shortest, lightest girls at menarche

Table 3

Total Water/Body Weight Percent as an Index of Fatness

	Female	Male
Weight (kg)	65	65
Total Water (liters)	33	40
LBW (kg) (TW/0.72)	46	56
Fat (kg)	19	9
Fat/BWt %	29	14
TW/BWt%	51	62

$$\text{Fat/Body Wt \%} = 100 - \frac{\text{TW/BWt\%}}{0.72}$$

Table 4

Weight (mean ± S.E.) at various heights (mean ± S.E.) with increasing age of menarche, and at all ages of menarche. Categories of height are by rounded standard deviation (6 cm) from the rounded mean, 158 cm.

Height category (cm)	Menarche ≤12.9 years			Menarche ≥13.0 years			All ages of menarche		
	No	Height (cm)	Weight (cm)	No.	Height (cm)	Weight (kg)	No	Height (cm)	Weight (kg)
<152.0	17	148.6+0.56	40.9+0.84	8	146.0+1.8	38.8+1.1	25	147.8+0.72	40.2+0.69 *
152.1-158.0	32	155.2+0.32	48.5+1.2	26	155.1+0.37	44.8+1.2†	58	155.1+0.24	46.9+0.87
158.1-164.0	37	161.2+0.24	50.6+0.96	27	160.7+0.30	48.0+1.2	64	161.0+0.19	49.5+0.75
>164.1	9	165.9+0.64	53.4+2.2	25	167.9+0.74	51.4+1.1	34	167.4+0.59	51.9+0.98 *
All subjects	95	157.4+0.57	48.4+0.71	86	159.7+0.78	47.2+0.72	181	158.5+0.48	47.8+0.51

* Differs from mean for all subjects at P<.05.

From Frisch, Revelle and Cook (61)

Table 5

Height, weight, total water/body weight (TW/BW) percent, lean body weight (LBW), fat, fat/body weight percent, and ratio of LBW to fat of girls grouped by height at menarche

Height category (cm)	No.	Height (cm)	Weight (kg)	TW/BWt %	LBW (kg)	Fat (kg)	Fat/body weight %	Ratio LBW to fat
		All	a g e s	o f	m e n a r c h e			
<152.0	25	147.8	40.2	56.3	31.4	8.9	21.8	3.5:1
		SD 3.6	3.5	2.4	1.7	2.0	3.3	
152.1–158.0	58	155.1	46.9	54.7	35.2	11.6	24.6	3.0:1
		SD 1.8	6.6	4.2	2.4	4.3	5.8	
158.1–164.0	64	161.0	49.5	54.8	37.5	12.1	23.8	3.1:1
		SD 1.5	6.0	3.7	2.2	3.8	4.6	
>164.1	34	167.4	51.9	55.3	39.7	12.3	23.2	3.2:1
		SD 3.4	5.7	3.2	2.3	3.7	4.4	
All subjects	181	158.5	47.8	55.1	36.3	11.5	23.5	3.2:1
		SD 6.5	6.9	3.6	3.4	3.9	4.8	

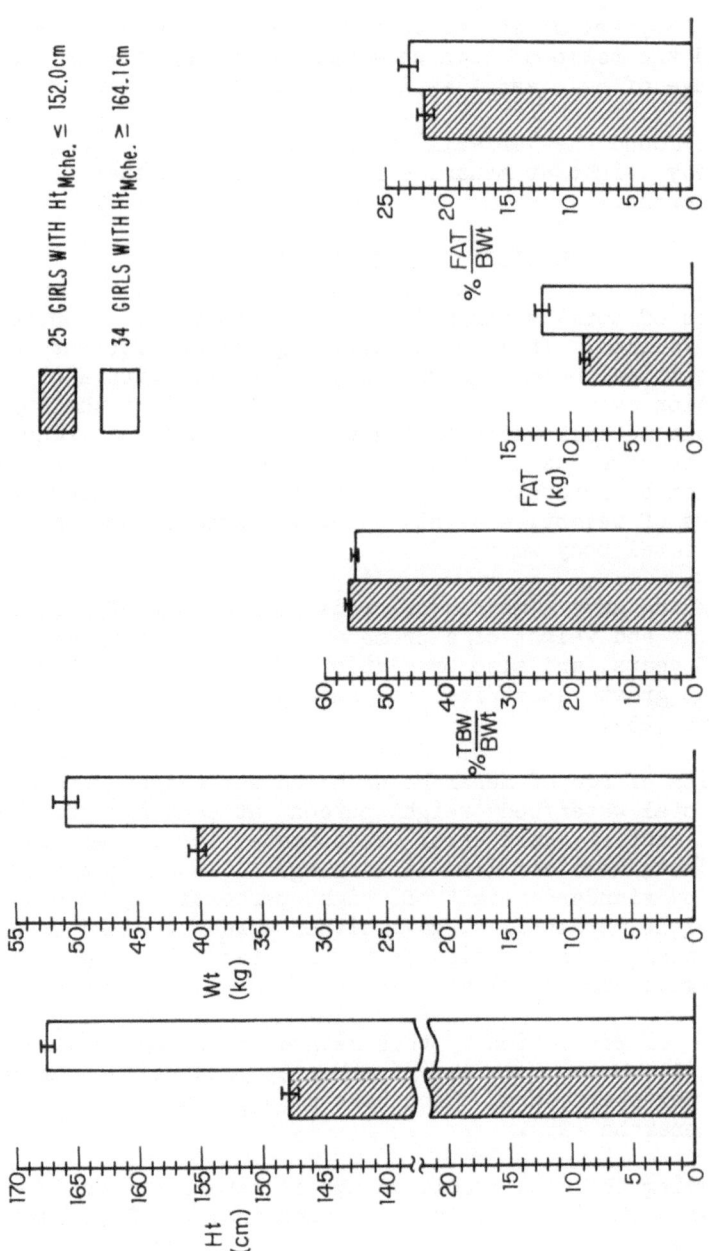

Fig. 8. Height, weight, total water/body weight percent (TBW/BWt%), fat, fat/body weight percent (fat/BWt%) of girls at the extremes of the height range at menarche (61).

have a smaller absolute amount of fat, 8.9 ± 0.4 kg, compared to that of the tallest, heaviest girls, 12.3 ± 0.6 kg, (the mean for all subjects is 11.5 ± 0.3 kg), both extreme groups have about 22% of their body weight as fat at menarche, as do all subjects (Table 5), and the ratio of lean body weight to fat of both groups is in the range of 3:1, as it is in all subjects (4) (Table 5).

Thus, we found the variability of total body water as percent of body weight at menarche is 55% less than that of weight at menarche.

PREDICTION OF MENARCHE

Quartiles of total water/body weight percent are essentially quartiles of fatness. It is especially significant for our hypothesis of a critical metabolic rate/kg, associated with a critical body composition as a signal for menarche, that 82% of the 169 girls who could be followed from spurt initiation to menarche remained in the same quartiles of total water/body weight percent from initiation to menarche, compared to only 47% remaining in the same quartiles of weight, and only 39% remaining in the same quartiles of total body water (25).

This finding gave a method of prediction of age of menarche from the height and weight of a premenarcheal girl at ages 9, 10, 11, 12 and 13 years, and also prediction of age of initiation of the adolescent growth spurt from the height and weight of a girl at age 8 years (25).

Regression of age of menarche on height or weight within a quartile of total water/body weight percent at each age gave the lowest significant standard error of estimate, lower even than when all subjects were combined at each age. Classification of the subjects by standard height of weight percentiles, or by weight for height percentiles, gave either worse standard error of estimates than classification by total water/body weight percent, or insignificant results.

The error of prediction by this method at all ages is less, in some quartiles by as much as 65%, than by prediction from stage of secondary sex characters, which have a quite variable association with menarche (25).

Girls having the same predicted age of menarche should be more homogeneous physiologically, and endocrinologically, than girls classified by chronological age, as Shock (44) actually observed for the physiological criteria of oxygen consumption, pulse rate and blood pressure. It would be useful, therefore, to classify premenarcheal girls by predicted age of menarche in studies of the endocrinological and growth changes of adolescence.

EXCEPTIONALLY EARLY MATURING GIRLS

At each age from 9 to 12 years, prediction of age of menarche is better for the heavier girls, who are found in the lowest total water/body weight quartiles, (inversely to the weight quartiles) than for the lighter weight girls. This is because of an exceptional group of early maturing girls (18% of early maturers, 9.5% of all subjects), who are very short and light weight at menarche (mean height 147.8 ± 0.7 cm; mean weight 40.9 ± 0.8 kg) (25). Normally at menarche, the short girls have more weight for their height than do tall girls, (11,12). These short, light, early maturing girls have only about 9 kg of fat at menarche compared to the average of 11.5 kg found for all subjects, but their relative fatness is the same, about 22% fat/body weight percent.

The small number of short, light, late maturing girls (9.3% of late maturers, 4.4% of all subjects), are also exceptional since late maturers at menarche are usually taller than early maturers (11,12). These short, late maturers have only about 8 kg fat at menarche, but their relative fatness is about 21% of body weight at menarche (25).

These two exceptional groups of girls may represent different metabolic or endocrinological patterns (45) since they attain the same fat percentage of body weight usually found at menarche, but at lower weights for height, or shorter heights for weight than are usual for the population at menarche. The data of Osler and Crawford (46) on weights at menarche of ambulatory and bed ridden patients support this explanation (25).

Comparison of the growth and body composition of the short, light, late girls with wild type breeds of animals suggest that these females may represent the "wild type" primeval female. These slow-growing, smaller individuals would have greater survival ability in times of fluctuating food supply (47).

The short, light early girls might be the equivalent of the meat type now being sought by Australian sheep breeders: fast growing, early maturers with a small leg joint and not too heavily marbled with fat!

FATNESS AS A DETERMINANT OF MINIMAL WEIGHTS FOR MENSTRUAL CYCLES.

The total water/body weight percent data of each of the same 181 girls followed from menarche to the completion of growth at ages 16-18 years provided a method of determining a minimal weight for height necessary for the onset of menstrual cycles (menarche) in primary amenorrhea and for the restoration of menstrual cycles in cases of secondary amenorrhea, when the amenorrhea is due to undernourishment (2).

Percentiles of total water/body weight percent, which are percentiles of fatness, were made at menarche and at age 18 years, the age at which body composition was stabilized. Each set of percentiles was then drawn on a height-weight grid and the weights at the cessation and restoration of regular menstrual cycles (two or more) of 9 patients with amenorrhea due to weight loss, other possible causes having been excluded, and 8 cases cited in the literature, were studied in relation to the weights indicated by the diagonal percentile lines in Fig. 9 and Fig. 10.

We found that 56.1 percent of total water/body weight %, the 10th percentile at age 18 years, which is equivalent to about 22% fat of body weight, indicates a minimal weight for height necessary for the restoration and maintenance of menstrual cycles. For example, a 20-year-old woman whose height is 160 cm should weigh at least 46.3 kg before menstrual cycles would be expected to resume.

The weights at which menstrual cycles ceased or resumed in post-menarcheal patients ages 16 and older (Fig. 10) are about 10 percent heavier than the minimal weights for the same height observed at menarche (Fig. 9).

In accord with this finding, the data on body composition show that both early and late maturing girls gain an average of 4.5 kg of fat from menarche to age 18 years. Almost all of this gain is achieved by age 16 years, when mean fat is 15.7 $\pm$ 0.3 kg, 27 percent of body weight. At age 18 years mean fat is 16.0 $\pm$ 0.3 kg, 28 percent of the mean body weight of 57.1 $\pm$ 0.6 kg. Reflecting this increase in fatness, the total water/body weight percent decreases from 55.1 $\pm$ 0.2 percent at menarche (12.9 $\pm$ 0.1 years) to 52.1 $\pm$ 0.2 percent (S.D. 3.0) at age 18 years.

Because girls are less fat at menarche than when they achieve stable reproductive ability, the minimal weight for height for the onset of menstrual cycles in cases of primary amenorrhea due to undernutrition is indicated by the 10th percentile of fractional body water at menarche, 59.8 percent, which is equivalent to about 17 % of body weight as fat   (Fig. 9). The standards of Fig. 9 would be used also for girls who become amenorrheic as a result of weight loss shortly after menarche, as is often found in cases of anorexia nervosa in adolescent girls (2,48).

The absolute and relative increase in fatness from menarche to age 16 to 18 years is of special interest because this interval coincides with the period of adolescent sterility. During this time there is rapid growth of the uterus and the ovaries (2).

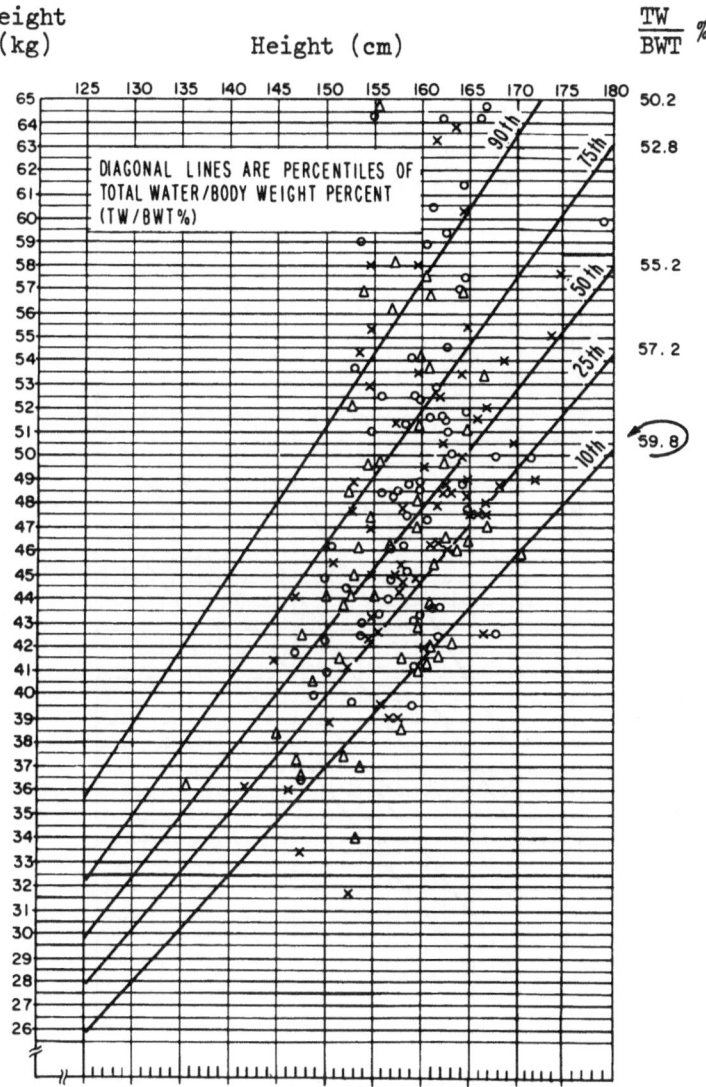

Fig. 9. The minimal weight necessary for a particular height for onset of mentrual cycles is indicated on the weight scale by the 10th percentile diagonal line of total water/body weight percent, 59.8 percent, as it crosses the vertical height lines. Height growth of girls must be completed, or approaching completion. For example, a 15-year-old girl whose completed height is 160 cm (63 inches) should weight at least 41.4 kg (91 lb) before menstrual cycles can be expected to start. Symbols are the height and weight at menarch of each of the 181 girls of the Berkeley Guidance Study, O; Child Research Council Study, X; and Harvard School of Public Health Study, Δ. Reprinted with permission from Science (2).

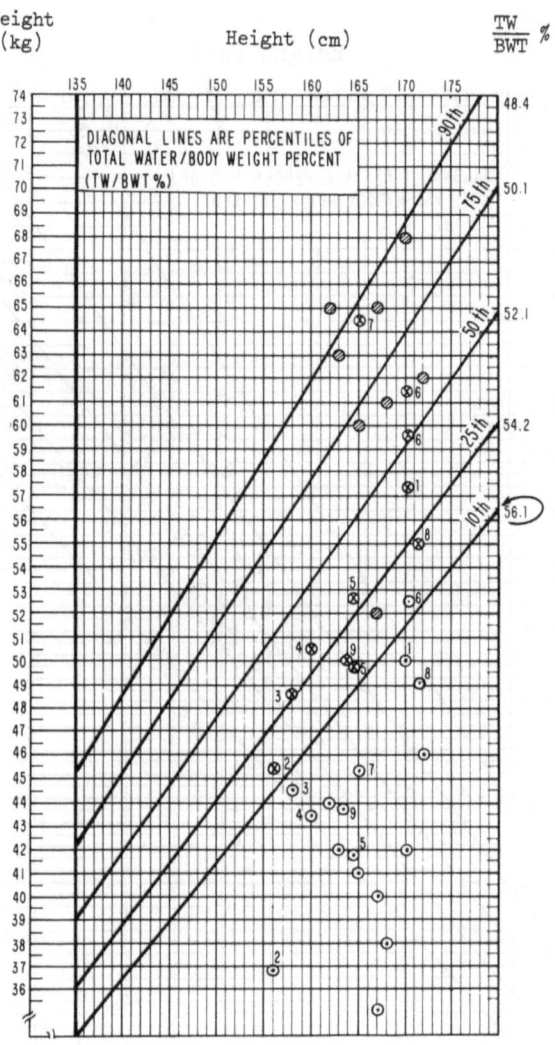

Fig. 10. The minimal weight necessary for a particular height for
restoration of menstrual cycles is indicated on the weight scale
by the Loth percentile diagonal line of total water/body weight
percent, 56.1 percent, as it crosses the vertical height line. For
example, a 20-year-old woman whose height is 160 cm should weigh at
least 46.3 kg (102 lb) before menstrual cycles would be expected to
resume.   O[1-9], Weights while amenorrheic of patients of one of us
(J.W.M.); ⊠[1-9], their weights at resumption of regular cycles.
When two weights are given for a patient, the lower weight is at
first resumed cycle. O, The weights before occurrence of amenorrhea of subjects cited by Lundberg et al., and ⊙ are their weights
while amenorrheic. Reprinted with permission from Science (2).

REPRODUCTIVE EFFICIENCY

The weight changes associated with the cessation and restoration of menstrual cycles are in the range of 10 to 15 percent of body weight. Weight loss or weight gain of this magnitude is mainly loss or gain of fat (49). This suggests that a minimum level of stored, easily mobilized energy is necessary for ovulation and menstrual cycles in the human female (2).

If a minimum of stored fat is necessary for normal menstrual function, one would expect that women who live on marginal diets would have irregular cycles, and be less fertile, as had been observed, and that poorly nourished lactating women would not resume menstrual cycles as early after parturition as well-nourished women, as also has been observed (2,3). The main function of the 16 kg of fat stored on average by early and late maturing girls by age 18 years may be to provide easily mobilized energy for a pregnancy and for lactation; the 144,000 calories would be sufficient for a pregnancy and 3 months' lactation (2). The human brain, it should be noted, grows most rapidly during the last trimester of pregnancy and the first months after birth (50).

Irrespective of any causal relationship, the weight dependency of menarche in human beings, as in animals (15,19) operates as a "compensatory mechanism" for both environmental and genetic variation. The result is a reduction in the variability of body size at sexual maturity, and therefore, a reduction in the variability of adult body size. As in many other animals, human body size at sexual maturity is close to adult size; weight and height at menarche are 85% and 96% of adult weight and height respectively (12). The regulation of female body size has obvious selective advantages for the species since birth weight is correlated with the prepregnancy weight of the mother, and infant survival is correlated with birth weight (3,6).

An example of compensation for poor nutrition is the significantly later age of menarche of undernourished girls compared to controls (6). An example of compensation for genetic variation is the longer time interval of the adolescent growth spurt of normal, slow growing late maturers (Fig. 2), compared to the more rapid advancement from initiation to menarche of the early maturers (Fig. 1). Boas noted that the "tempo of growth" was inherited (51). Thus the well-established genetic component of variation in age of menarche (52) may be in part through the genetic control of growth rate. Tanner notes that age of menarche is a very convenient measure of "tempo of growth", (52), which is to be expected from the association of menarche with a critical weight representing a particular body composition of relative fatness (29).

Different racial groups have different critical weights (21).

The inheritance of the absolute value of a critical menarcheal weight, independently of tempo of growth, could also be a component of the genetic control of age of menarche. We do not know as yet whether the different critical weights of different races also represent different body compositions (29).

ENVIRONMENTAL EFFECTS ON REPRODUCTIVE EFFICIENCY

The findings on minimal weights for heights necessary for the onset and maintenance of reproductive ability indicate that ordinary environmental factors which affect physical growth, such as nutrition and disease, can affect the time of attainment and level of function of each reproductive event in the female, thus affecting the length of the reproductive span and reproductive efficiency. For example, undernourished girls have later menarche: a longer period of adolescent sterility, a higher incidence of irregular and anovulatory cycles than normal, amenorrhea when weight loss is in the range of 10-15% of body weight, higher pregnancy wastage; longer lactational amenorrhea, and therefore longer birth intervals, and a shorter time to menopause (3) (Fig. 11).

These findings for the human female are consistent with the effects of nutrition on other domestic and wild mammals: subnormal nutrition disrupts menstrual cycles in monkeys (53), and suppresses oestrus in non-primates (54,55). In these animals, as in the human female, a gain in body weight is followed after varying periods of time by the resumption of cycles and ovulation (2,53, 54,55).

It must be added that, of course, other factors, such as emotional stress, affect the maintenance or onset of menstrual cycles in human beings. Therefore, menstrual cycles may cease without weight loss, and may not resume in some subjects even though the minimum required weight is attained (2).

SIGNIFICANCE OF FLUSHING

The importance of fatness for the development and maintenance of reproductive ability, a late maturing character, in the human female, is consistent with Hammond's "growth gradients" in the development of domestic animals (56). Some stimulus of fatness may also be involved in a positive effect of nutrition on ovulation: flushing in sheep (57) and pigs (58). The increase in the rate of twinning with increased caloric intake before mating, even if the sheep is already in good condition (57), is a very interesting and suggestive finding to human biologists: it has been shown that the rate of human dizygotic twinning, but not monozygotic twinning, fell during war-time restrictions in nutrition and the rate returned to normal after the return of normal food supply (59). Even more suggestive of a direct relation between control

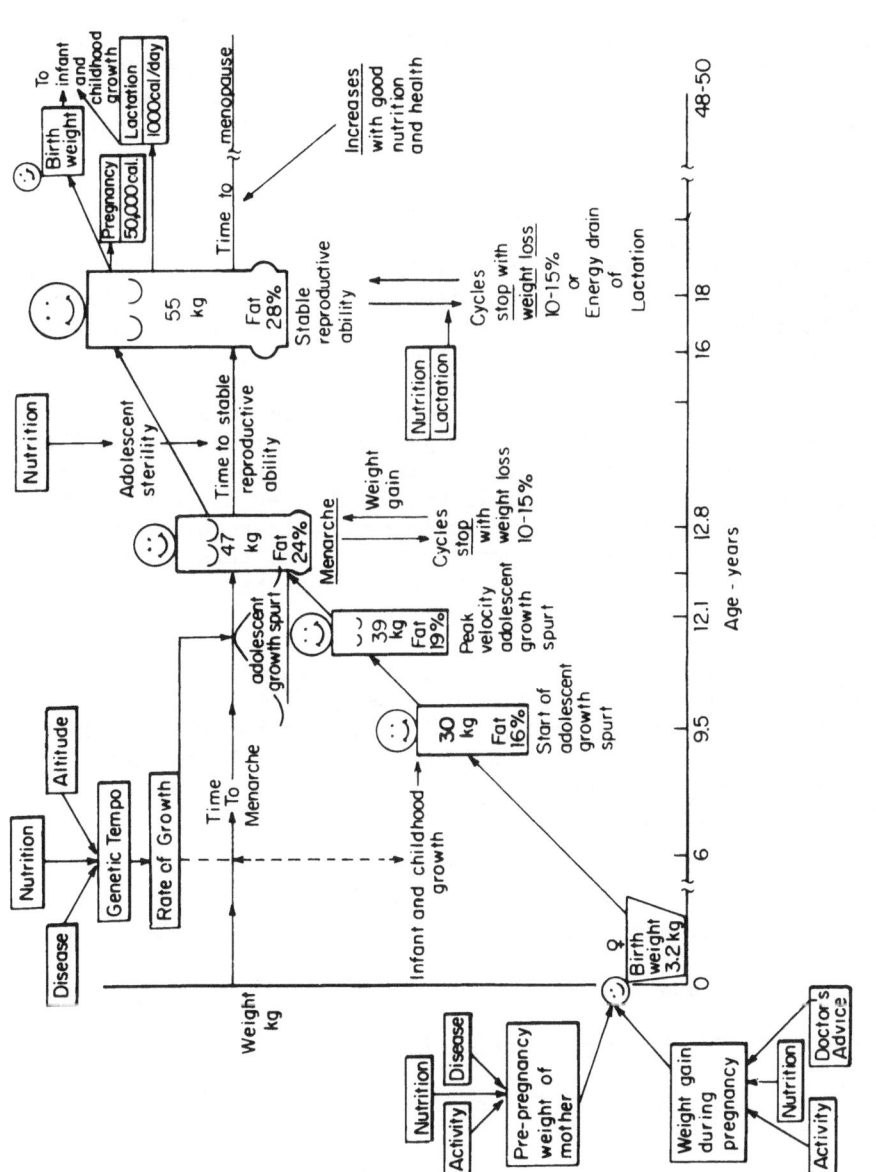

Fig. 11. Environmental effects on reproductive efficiency.

of food intake, caloric control, and ovulation is the finding
that when gilts were fed lard at approximately 150% of the caloric
intake of those fed glucose, the ovulation rate exceeded that of
the control gilts by 4.1 ova, and the glucose and low fat gilts
by 3.3 and 2.2 ova respectively (58).

One wonders if the adolescent growth spurt, which is unique
to primates, may make the sexually mature primate female less
responsive to such direct ovarian stimulation from high caloric
intake, as well as protecting her for some time from the harmful
effects of low caloric intake.

RELATIVE FATNESS AS A DETERMINANT OF EARLY OR LATE MATURATION

The difference in fat deposition found between early and late
maturing girls could be one of the determinants of early or late
sexual maturation. One possibility is that the storage of estrogen
in fat depots affects blood levels of estrogen, and/or other
steroids, or their secretion rates. Brown and Strong (62) found
differences in estrogen production and metabolism as a function of
body weight and hence, fat; they suggest that estrogen metabolism
may be influenced by factors involved in fat metabolism, including
thyroid hormones. A third possibility, which does not exclude the
others, is that fat concentration is important in regulating
energy balance by the hypothalamus (63,64,65) and thus determines
the setting of the gonadostat, which determines the output of
gonadotrophins, and therefore, estrogen.

If the number of cells of the adipose tissue in human beings
is determined by early nutritional experiences (66) as it is in
the rat, (67,68,69), and there is an interaction between adipose
tissue and gonadal hormones, or metabolic rate, or both, early or
late maturation might be determined by differences in fat depots
established very early (60).

REFERENCES

1.   Kennedy, G. C. and Mitra, J. (1963). J. Physiol. <u>166</u>, 408.

2.   Frisch, R.E. and McArthur, J. W. (1974). Science <u>185</u>, 949.

3.   Frisch, R. E. In press, Social Biol.

4.   Frisch, R. E., Revelle, R. and Cook, S. (1973) Human Biol.
      <u>45</u>, 469.

5.   Hammond, J. Ed. (1955). Progress in the Physiology of Farm
      Animals. Vol. 2 (Butterworths, London).

6.   Frisch, R. E. (1972).   Pediat. <u>50</u>, 445.

7.   Widdowson, E. M. and McCance, R.A. (1960). Proc. Roy. Soc.
      (B) 152, 188.

8.   Runner, M. N. and Gates, A. (1954). J. Heredity 45, 51.

9.   Lane, P. W. and Dickie, M. M. (1954). J. Heredity 45, 56.

10.   Hammond, J. Jr., Mason, I. L., and Robinson, T. J. Eds.
      (1971). "Hammond's Farm Animals," 4th Edition (Arnold,
      London).

11.   Frisch, R. E. and Revelle, R. (1970). Science 169, 397.

12.   Frisch, R. E. and Revelle, R. (1971). Archiv. Dis. Childh.
      46, 695.

13.   Widdowson, E. M., Mavor, W. O. and McCance, R. A. (1964).
      J. Endocrinol. 29, 119.

14.   Barnett, S. A. and Coleman, E. M. (1959). J. Endocrinol.
      19, 232.

15.   Monteiro, L. S. and Falconer, D. S. (1966). Animal Product.
      8, 179.

16.   Dickerson, J. W. T., Gresham, G. A. and McCance, R. A. (1964).
      J. Endocrin. 29, 111.

17.   Crichton, J. A., Aitken, J. N. and Boyne, A. W. (1959).
      Anim. Prod. 1, 145.

18.   Joubert, D. M. (1963). Animal Breed. 31, 295.

19.   Kennedy, G. C. (1969). Ann. N.Y. Acad. Sci. 157, 1049.

20.   van Wagenen, G. (1949). Endocrinol. 45, 544.

21.   Frisch, R. E. and Revelle, R. (1969). Human Biol. 41, 185.

22.   Tanner, J. M. (1962). "Growth at Adolescence." 2nd ed.
      (Blackwell, Oxford).

23.   Frisch, R. E. and Revelle, R. (1971). Human Biol. 43, 140.

24.   Frisch, R. E. and Revelle, R. (1969). Human Biol. 41, 536.

25.   Frisch, R. E. (1974). Pediatrics 53, 584.

26.   Kulin, H. E., Grumbach, M. M. and Kaplan, S. L. (1972).
      Pediat. Res. 6, 162.

27. Tanner, J. M. (1966). Tijd. Geneesk. 44, 524.

28. Damon, A. (1974). Social Biol. 21, 8.

29. Frisch, R. E. (In press). In: "Biosocial Interrelations in
     Population Adaptation." (Eds. E. Watts, F. Johnston, and
     G. Lasker) (Mouton, The Hague).

30. Donovan, B. T. and van der Werff ten Bosch, J. J. (1965).
     "Physiology of Puberty." (Arnold, London).

31. Zacharias, L., Wurtman, R. and Schatzoff, M. (1970).
     Am. J. Ob. Gyn. 108, 833.

32. Pace, N. and Rathbun, E. N. (1945). J. Biol. Chem. 158, 695.

33. Holliday, M. A., Potter, D., Jarrah, A. and Bearg, S. (1967).
     Pediat. Res. 1, 185.

34. Keys, A. and Brozek, J. (1953). Physiol. Rev. 33, 245.

35. Mellits, E. D. and Cheek, D. B. (1970). Monogr. Soc. Res.
     Child. Develop. 35, 12.

36. Moore, F. K., Olesen, H., McMurrey, J. D., Parker, V., Ball,
     M. R., Boyden, C. M. (1963). "The Body Cell Mass and its
     Supporting Environment." (Saunders, Philadelphia).

37. Friis-Hansen, B. J. (1956). Acta Paediat. Suppl. 110, 1.

38. Forbes, G. B. (1972). Growth 36, 325.

39. Brozek, J. and Grande, F. (1955). Hum. Biol. 27, 22.

40. Holliday, M. A. (1971). Pediat. 47, 169.

41. Talbot, B. F. (1938). Am. J. Dis. Childh. 55, 455.

42. Cahill, G. Jr. (1962). In: "Fat as a Tissue" (Eds. K. Rodahl
     and B. Issekutz) p. 169 (McGraw-Hill, New York).

43. Benedict, F. G. (1938). "Vital Energetics." (Carnegie,
     Washington, D. C.).

44. Shock, N. W. (1943). Am. J. Physiol. 139, 288.

45. Reichlin, S., Martin, J. B., Mitnick, M., Boshans, R. L.,
     Grim, Y., Bellinger, J., Gordon, J. and Malacara, J. (1972).
     Recent Progr. Hormone Res. 28, 229.

46. Osler, D. and Crawford, J. (1973). Pediat. 51, 675.

47. Palsson, H. (1955). In:"Progress in the Physiology of Farm Animals," Vol. 2. (Ed. J. Hammond) p. 430 (Butterworths, London).

48. Frisch, R. E. Unpublished data.

49. Kleiber, M. (1961). "The Fire of Life." (Wiley, New York).

50. Frisch, R. E. (1972). In: "Annual Progress in Child Psychiatry and Child Development." (Eds. S. Chess and A. Thomas). p. 361 (Brunner/Mazel, New York).

51. Boas, F. (1935). Proc. Nat. Acad. Sci. 21, 413.

52. Tanner, J. M. (196). In:"Human Growth" Vol. III, Sympos. Soc. Study Human Biol. (Ed. J. M. Tanner) (Pergamon, London).

53. Gupta, S. R. and Armand, B. K. (1971). Endocrin, 89, 652.

54. Marrian, G. F. and Parkes, A. S. (1930). Proc. Roy. Soc. Series B. 105, 248.

55. Sadlier, R. M. I. S. (1969). "The Ecology of Reproduction in Wild and Domestic Animals".(Methuen, London).

56. Hammond, J. (1952). "Farm Animals" 2nd Edition (Butler and Tanner, London).

57. Coop, I. E. (1966). J. Agr. Sci. Camb. 67, 305.

58. Zimmerman, D. R., Spies, H. G., Self, H. L. and Casida, L. E. (1960). J. Animal Sci. 19, 295.

59. Pearl, R. (1939). "A Natural History of Population".(Oxford New York).

60. Frisch, R. E. (1974). In:"Control of Onset of Puberty"(Eds. M. M. Grumbach, G. D. Grave, F. E. Mayer) p. 403 (Wiley, New York).

61. Frisch, R. E., Revelle, R. and Cook, S. (1971). Science 174, 1148.

62. Brown, J. B. and Strong, J. A. (1965). J. Endocrin, 32, 107.

63. Hervey, G. R. (1969). Nature 222, 629.

64. Hervey, G. R. (1971). Nutr. Soc. Proc. 30, 109.

65. Kennedy, G. C. (1953). Proc. Roy. Soc. London (B) 140, 578.

66. Hirsch, J. and Knittle, J. L. (1970). Fed. Proc. 29, 1516.

67. Knittle, J. L. and Hirsch, J. (1968). J. Clin. Invest. 47, 2091.

68. Knittle, J. L. (1971). Bull. N.Y. Acad. Med. 47, 579.

69. Knittle, J. L. (1972). J. Nutr. 102, 427.

HORMONAL INFLUENCES ON THE GROWTH, METABOLISM AND BODY COMPOSITION OF PIGS

David Lister

Meat Research Institute

Langford, Bristol BS18 7DY, England

INTRODUCTION

Pig producers have been aware for some time that as the growth rates and leanness of their stock increase the quality of the lean, in terms of its paleness and wetness, appears to decline.   Over the same period of time the scale of operation and rate of through-put of packing plants have also increased and the view is often taken that the decline in meat quality is, in part, attributable to this.   It is striking, however, that meat quality is not necessarily maintained even where technological expertise and humanitarian principles are combined optimally in the slaughter procedures, and animal scientists have been quick to explore other possibilities.

Ludvigsen (1) was among the first to examine the notion that the physiological means whereby the growth of animals was altered might also influence the quality of the meat they produced.   This view is now widely held, but the arguments in its support have often been teleological and it is only lately that a coherent picture has started to emerge.   Its development provides a useful model for the role of hormones in a wider context.

THE CONTROL OF MEAT QUALITY

Rigor mortis develops within 4-8 hours after slaughter in pigs, 8-12 in sheep, and 12-24 in cattle.   At this time the muscle is fully acidified to about pH 5.5-5.8 and stiff.   The rate and extent of acidification are largely responsible for determining the quality of meat and are themselves altered by the severity and

duration of stress of any kind which an animal might suffer prior
to death (2).   If an animal is subjected to prolonged stress prior
to slaughter the pH of muscle in rigor may be 6.5 or higher and
the meat will retain a dark appearance, the so-called dark cutting
or dark, firm and dry (DFD) condition.   Some pig carcasses
will develop full acidity within a few minutes of slaughter and,
because of the extensive denaturation of muscle proteins which
results from the interaction of acidity and temperature, the meat
appears pale, soft in texture and exudes fluid from the cut
surfaces.   This is seen typically as the pale, soft and exudative
(PSE) condition,   and is frequently found when animals suffer
severe stress prior to slaughter.

      The quality of the meat of pigs slaughtered under the same
conditions is apt to vary with the breed of pig.   Thus the
Pietrain breed consistently produces meat of inferior quality
whereas the Large White produces, on the whole, meat of acceptable
quality.   These differences are reflected in the rate of muscle
acidification, measured in terms of its pH at 30 or 45 min. after
slaughter (Fig. 1).   But it is also evident from Fig. 1 that pigs
which have been tested in schemes designed to identify stock with
superior performance tend to have faster rates of muscle acidification than their 'commercial' counterparts.

                    The Effects of Muscle Stimulation

      The importance to meat quality of an animal's being quiet
and rested at slaughter has long been appreciated by butchers and
the specific role of neuromuscular stimulation in the control of
glycolysis and acidification of muscle post mortem has been
examined empirically (3,4).   However, those breeds of pig which
consistently produce meat of poorer quality do not necessarily
produce better meat even when they are anaesthetised and paralysed
by the same amounts of neuromuscular blocking drugs which prove
beneficial in, say, Large Whites (5,6).   This is to suggest that
some pigs will not respond in terms of meat quality to the most
careful and quite impracticable handling at slaughter.

      There are at least two explanations for this.   Firstly, if
the same degree of neuromuscular blockade is achieved in all
animals by a given dose of, for instance, curare, then any observed
differences in muscle metabolism are likely to result from
'spontaneous' activity beyond the motor end plate i.e. in the
muscle fibres themselves.   Such activity is found in various
clinical disorders in which the flux of sodium, potassium or calcium across membranes is disturbed (7).   Secondly, it is possible
that in pigs which are sensitive to stimulation in the way that
Pietrains, and some Poland Chinas and Landraces are, there are
genotypic differences in the characteristics of neuromuscular

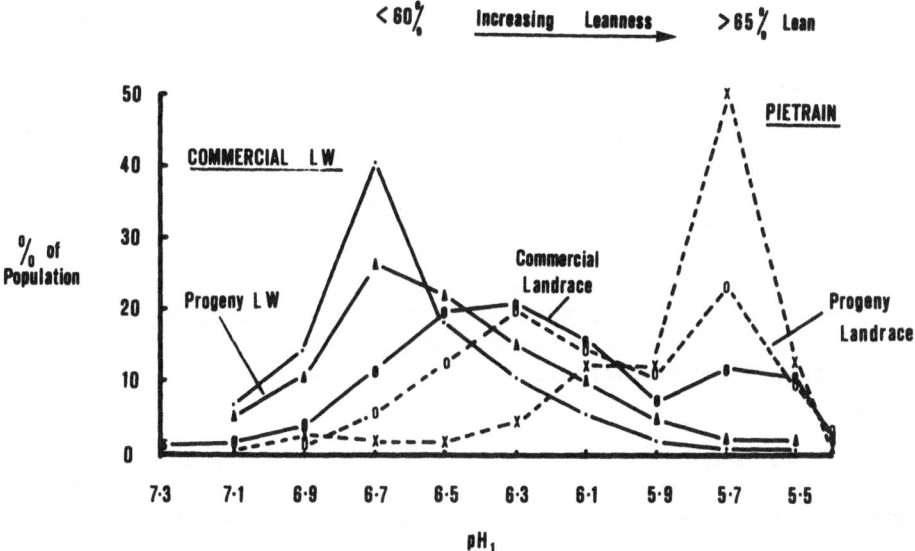

Fig. 1.   The distribution of pH₁ values (i.e. pH at 45 min. after
slaughter) in Large White, Landrace and Pietrain pigs.

transmission.   It is thought, for instance, that the motor end-
plate in the muscles of such pigs is more extensive (8) which
suggests that more transmitter substance is released per stimulus
and more receptor sites are involved than can perhaps be successfully blocked with conventional doses of curare.

The correction of certain disorders of muscle metabolism in
human patients suffering from electrolyte imbalances has frequently
been accomplished by treatment of the electrolyte disturbance (9).
Such findings have prompted investigations, especially of the metabolism of sodium and potassium, in pigs of a stress sensitive type.
Sodium loading or drug-induced potassium retention lead to some
alteration of muscle metabolism post mortem and meat quality (10,
11,12) but it seems unlikely that such effects contribute substantially to the control of meat quality in untreated animals.

The ability of the sarcoplasmic reticulum or of the mitochondria in muscle to handle calcium have also been considered as
important post end-plate phenomena which may affect muscle metabolism or the provision of energy (2).   Reductions in the calcium
accumulating ability of these organelles have been observed in PSE

muscle, but when care is taken to prevent the extensive denaturation of proteins which would normally occur in potentially PSE
muscle through the acidity/temperature interactions, the apparent
loss of function can be explained (13,14,15).

Investigation of the physiology of the neuromuscular junction
has received more attention largely because of the widespread
interest in the condition described as Malignant Hyperthermia (MH)
which occurs commonly in stress sensitive pigs and uncommonly in
human patients when they are treated with some anaesthetic drugs
(16).   The condition, which shows many of the metabolic features
of a normal response to extreme muscle or physical stimulation
(17), such as might be encountered during slaughter, has provided a
useful model for the investigation of the PSE condition of pig
muscle.

The two drugs commonly implicated as triggering agents for
the condition are halothane and suxamethonium which in pigs of the
Pietrain breed promptly induce the characteristic metabolic
sequelae, fever and death (18).   There is a fall in the Free
Thyroxine Index (FTI) (19,20) in the serum (21) which occurs simultaneously with and may be the result of a substantial (50-100 fold)
increase in plasma catecholamines of which noradrenaline predominates (22).   Infusion of the pigs with the α-adrenergic
blocking drug phentolamine will prevent a response; β-blockade
on the other hand does not prevent the usual fatal outcome.   By
using the same degree of α- and β-blockade of Pietrain pigs prior
to slaughter, we have now been able to show that the same mechanism
can be invoked to explain this breed's capacity for producing PSE
meat (23,24).

It has been our experience that adequate α adrenergic blockade
is achieved in Pietrains only by the administration of large
amounts of phentolamine, and that the block is labile and easily
reversed.   The same, we now know, applies to the neuromuscular
blockade produced by curare, which may well explain the limited
effects which this drug had on post mortem change in muscle and
meat quality of Pietrain and Poland China pigs (5,6).   McLoughlin
(personal communication) has recently observed a significant
retardation of muscle glycolysis post mortem in Pietrain pigs
given large doses of curare prior to death.   We (25) have recently
confirmed this and have shown that pancuronium may be the preferable agent to establish neuromuscular blockade in Pietrain-like
pigs for its hypotensive effects are not so marked as those of
curare and in consequence large doses may be given with less risk.

All these results demonstrate the important differences in
the characteristics of neuromuscular transmission which are to be
found from one pig to another and, characteristically, from one
breed to another.   A primary feature of the latter is the

involvement of catecholamines in the reactions of stress sensitive pigs and especially the α effects which are known to contribute to the augmentation of transmitter release, the potentiation of the action of suxamethonium and anti-curare effects in muscle (26).

GROWTH AND BODY COMPOSITION OF STRESS SENSITIVE PIGS

Extreme mesomorphism and leanness of the carcass are the characteristics most associated with sensitivity to stress or the production of PSE meat in pigs (27). Some authors (28) consider that the association is not simply the result of a reduction in body fat, but more particularly with the alteration of the ratio of muscle to bone. Most of the observations have been in cross-sectional studies of pig populations where differences in apparent leanness may emerge as a result of feeding practices, particularly feed restriction, through genotypic differences in ability to partition dietary constituents into body protein or fat, or reduced capacity to deposit fat. Simple assessment of the carcass composition will not identify the route whereby a particular composition has been achieved. Associations between body composition and meat quality derived from surveys of carcasses in packing plants must be viewed with caution for, apart from the wide variation in the treatment of individual animals at slaughter, it is not possible to discriminate between the contributions of genetics and environment to body composition.

We (29,30) have suggested that genotypic differences in ability to deposit and mobilise fat may be particularly important in determining compositional differences between animals which show consistent differences in meat quality. Our conclusions were based on experiments in which Large White pigs were pair fed to the lower voluntary food intake of Pietrains of the same weight. During the same period of time when all the animals had consumed the same amount of food, the Pietrains doubled their weight but the Large Whites gained even more. Both breeds deposited the same amount of muscle during the experiment and the greater gain in liveweight of the Large Whites was attributable to the extra fat which they had deposited. At 90 kg liveweight, the carcasses of Pietrains are leaner than those of Large Whites because of the smaller amount of fat which they have retained, not because of a superior ability to partition dietary energy and protein into body tissues in an energetically efficient way. They are also older because of their slower overall growth and in consequence they have accumulated more lean, but only that which Large Whites would have in a similar period of time. Lean et al. (31) have reported similar findings in comparisons between Pietrain and Landrace pigs.

Thus so far as Pietrains and Large Whites are concerned one can conclude that the percentage of lean in a carcass is deter-

mined primarily by the rate at which fat is deposited and not by
the rate of lean deposition.   Thus since lean deposition is not
taking place at such a prodigious or different rate, one might
look more profitably into the reasons for the differences in fat
metabolism for possible links between the composition of the
carcass and the quality of the meat.   It might also be added that
the production of the lean Pietrain carcass is energetically
inefficient for on the same amount of food Large Whites retain at
least as much protein and more energy as stored fat.

HORMONAL FUNCTION IN STRESS SENSITIVE PIGS

The identification of a key role for fat in determining the
different production characteristics of Pietrain and Large White
pigs also identifies a potentially useful basis for the interpretation of the information on hormone function which has been
collected over many years.   In the past, investigators have
looked with limited success for associations between muscle metabolism, meat quality and hormones and have only hinted at associations with body composition.

The hormones of the hypothalamic-pituitary-adrenal (H-P-A)
axis and of the thyroid have long been implicated in the syndrome
of stress sensitivity.   So far as the H-P-A axis is concerned,
there is evidence (32) that its responsiveness to the usual stimulation procedures is similar in both Pietrain and Large White
pigs.   There are reports, however, that the turnover of some of
the adrenal corticosteroids is increased (33).   This might also
be the feature of thyroid hormone metabolism which explains the
conflicting evidence on the role of this hormone in stress sensitive animals.

Many reports on the associations between thyroid status and
meat quality rely on the measurement of Protein Bound Iodine (PBI)
in serum collected when pigs are slaughtered.   Most investigators
have found lowered values for PBI (34) and have tended to conclude
that pigs which show low PBI in slaughter blood are relatively
hypothyroid.   It is, however, a matter of common clinical observation that hypothyroid individuals tend to be fat, not lean as the
pigs are, and have low heat production.   Both of these characteristics are in line with the known metabolic properties of the
hormone.   It is established also that the thyroid secretion rates
of the most sensitive breeds tend to be high (35,36,37).

If the usual resting levels of hormone in blood are to be
maintained, then a high rate of secretion will need to be matched
by a high rate of utilisation of the hormone.   The rate of peripheral utilisation of thyroid hormone is increased by a variety of
stimuli including cold, exercise, excitement and fatigue (38).

Our results from studies of MH showed that the extreme muscle
stimulation induced by the triggering agent caused a fall in the
circulating levels of thyroid hormone (as measured by the FTI
of the serum) which again suggests the increased peripheral
utilisation of the hormone.   Moreover α-blockade prevented the
muscle stimulation and the fall in FTI despite the massive rise
in catecholamines with which it was simultaneously associated.
All these characteristics can be observed as part of the response
of animals to the stress of slaughter and presumably explain the
low levels of PBI and FTI in blood taken at slaughter and the
suggestions of hypothyroidism.

The raised turnover of thyroid hormones in Pietrain pigs can
be seen also when the thyroid status of resting unrestrained
animals is assessed by the isotopic labelling technique of Nicoloff
(39).   Iodine (^{125}I) labelled thyroxine and iodine (^{131}I) are
injected into the subject to label the body's reserves of thyroxine
and the hormone synthesised by the thyroid gland respectively.
The daily urinary output of the two isotopes is measured thereafter
and their pattern of excretion is thought to reflect thyroid status.
A typical excretion pattern for Large Whites and Pietrains can be
seen in Fig. 2.   There appears to be little difference in overall

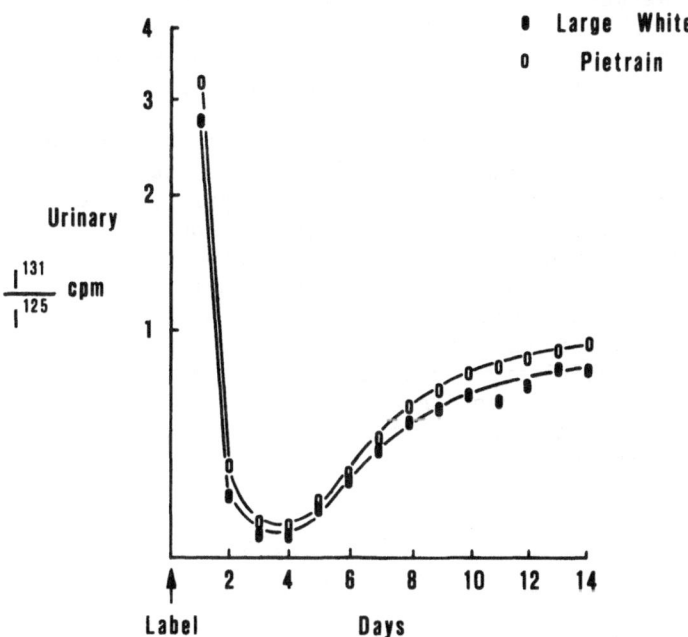

Fig. 2.   The excretion patterns of ^{125}I and ^{131}I in the urine of
Pietrain and Large White pigs.

thyroid status of the two breeds of pig but the Pietrains excreted
approximately 40% more isotope which roughly represents the
observed difference in thyroid secretion rate between the two
breeds.   Moss (40), however, postulates that the iodine-trapping
mechanism of the thyroids of Pietrain pigs is impaired and this
may additionally influence conventional assessments of thyroid
status.   But it seems clear that, like those of the adrenal
cortex (33), the hormones of the thyroid show a greater turnover
in stress sensitive pigs.

Several of the hormones of the pituitary-thyroid-adrenal axis
are lipolytic (41) and, in animals in which the turnover of hormones is increased, it would not be unreasonable to suppose that
lipolysis and lipogenesis could be altered.   Empirical observations (42) bear this out.   The rates of utilisation of fatty acids
both during feeding and fasting are higher in Pietrains than in
Large Whites and fasting plasma insulin levels are lower in
Pietrains.

The lipolytic response of pigs to infused noradrenaline (42)
provides additional evidence of genotypic differences in fat metabolism.   Infusion of conscious untrained Pietrains and Large
Whites reveals only small differences between the breeds in the
extent to which plasma free fatty acids (FFA) or glucose increase.
However, anaesthetised Large Whites mobilise FFA to only a limited
extent but carbohydrate extensively.   The anaesthetised Pietrain
on the other hand continues to mobilise FFA to approximately the
same extent as the conscious animal (Fig. 3).

Analogy of these results with those from studies on muscle
metabolism and meat quality is striking.   It seems that in
Pietrains both muscle metabolism and lipolysis are easily if not
constantly stimulated and are not retarded appreciably by the same
doses of anaesthetic which readily slow them in Large Whites.   In
short, the threshold of sensitivity to stimulation appears to be
much lower in Pietrain than in Large White pigs and low resting
values are achieved only with the greatest difficulty.   This
differential sensitivity might also present serious complications
in the measurement of resting heat production if the animals are
not carefully trained and acclimatised to the conditions under
which measurements are made.

SOME IMPLICATIONS FOR THE ANIMAL BREEDER AND
MEAT PRODUCTION

It is clear from the preceding account that important
physiological differences can be identified between two breeds of
pig to which their peculiar developmental and metabolic characteristics can be attributed.   But it is pertinent to ask whether

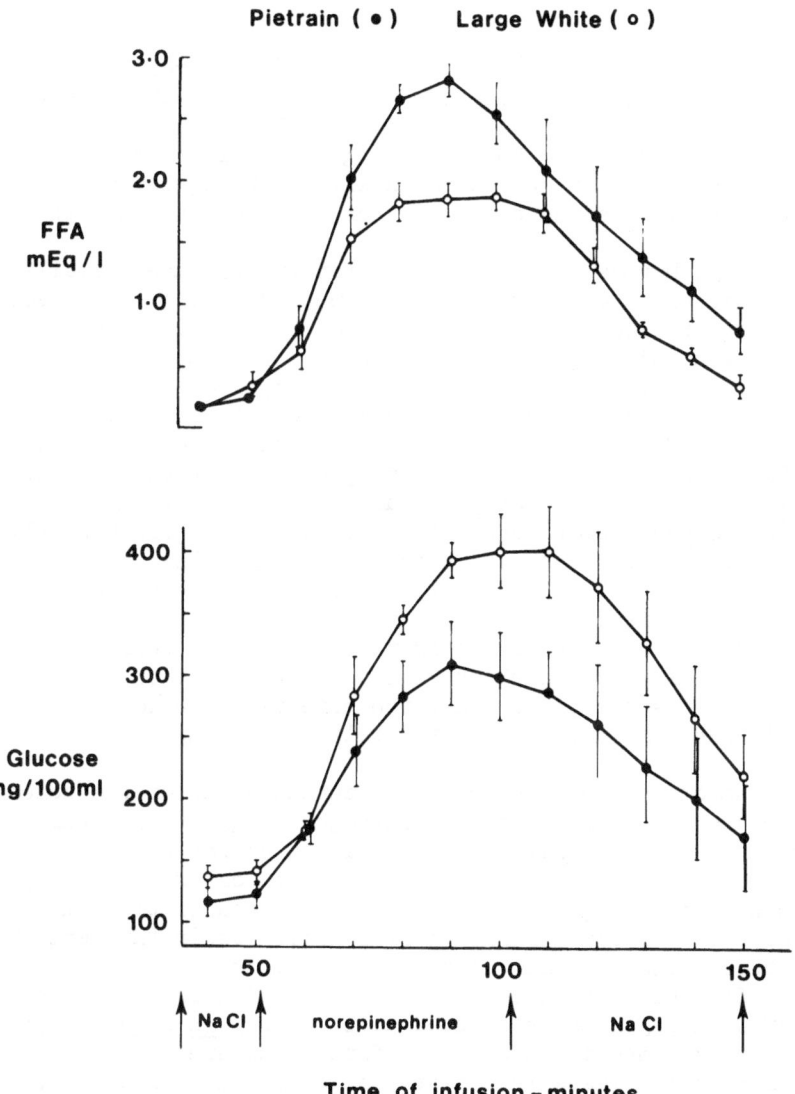

Fig. 3. Plasma concentrations of glucose and FFA during infusions of norepinephrine in anaesthetised pigs.

such a model is relevant to pig production generally.   Is it
likely, for instance, that extreme leanness can only be attained
within a breed of pigs in effect by changing their metabolism
from 'Large White' to 'Pietrain'?   If it is, what can we infer in
relation to the efficiency of meat production?

There are two series of experiments which support the use of
the Large White-Pietrain comparison as a model for the effects of
selection for leanness.   Hetzer and his colleagues (43) in the
U.S.A. selected Duroc and Yorkshire pigs, over 10 or more generations, for increased or decreased backfat at 80 kg liveweight.
The results they obtained differed from one breed to the other.

At 80 kg, Durocs are younger and fatter and have a smaller
skeletal size than Yorkshires of the same body weight.   Selection
for decreased backfat thickness in Durocs results in an increased
rate of growth and larger skeletal size.   On the other hand
selection of the same criteria in the already leaner Yorkshire
induces a slower overall rate of growth and only small increase in
skeletal size (Table 1).   This is to say that selection in the
fatter Duroc led to faster growth and greater leanness whereas the
leaner Yorkshire appeared only to curtail further its deposition
of fat in a manner reminiscent of the Pietrain.   Unfortunately
the experiments were concluded before it became clear whether or
not Durocs develop the growth pattern of the Yorkshires once their
body composition reaches that with which the Yorkshires started
out.   In practice, however, it is those pigs with performances
similar to the Yorkshires which receive most attention from
breeders and it seems likely that the effects of selection against
backfat in these animals will ultimately prove similar.

The anatomical and developmental consequences of the selection
procedure used by Hetzer and his colleagues suggest that some other
pig genotypes will curtail their deposition of fat in the same
manner as the Pietrain.   Standal and his co-workers in Norway
have shown that Norwegian Landrace pigs selected in the same
manner as the American Yorkshires also modify their fat metabolism
as they become increasingly lean to one similar to that we have
described for the Pietrain.   Lipid mobilisation from the fatty
tissue of lean pigs can be induced more readily in vitro (45);
the fatty acid composition of depot fats is altered (46), and the
fasting levels of plasma free fatty acids are increased (47).

There are additional fundamental issues involved.   Sensitivity to stress may not simply be a feature associated with mesomorphism but a direct cause of it and a regulatory mechanism for
body form generally.   The role of catecholamines and of the
sympathetic nervous system are of key importance in the more
bizarre metabolic responses of the Pietrain.   The interrelationships of these with thyroid and pancreatic hormones in lipolysis

Table 1

Performance of Duroc and Yorkshire boars and gilts selected over >10 generations for backfat thickness*

Trait		Duroc High-fat	Duroc Control	Duroc Low-fat	Yorkshire High-fat	Yorkshire Control	Yorkshire Low-fat
Backfat thickness (cm)	♂	4.9±0.05	3.7±0.05	3.0±0.04	3.6±0.05	3.1±0.04	2.6±0.04
	♀	5.1±0.05	4.0±0.04	3.2±0.04	3.9±0.05	3.2±0.03	2.7±0.03
Days on test from 56 days of age	♂	98.8±2.0	95.1±1.5	92.0±2.6	97.3±1.2	98.9±1.4	111.5±2.0
	♀	98.4±2.3	94.7±1.2	92.7±2.1	110.8±1.5	109.2±1.5	118.0±1.9
†Age at end of test (80 kg)	♂	154.8	151.1	148.0	153.3	154.9	167.5
	♀	154.4	150.7	148.7	166.8	165.2	174.1
†Length of body (cm)		91.6±0.1	94.9±0.1	96.4±0.1	101.6±0.2	101.9±0.2	102.8±0.3
†Height of body (cm)		57.8±0.1	59.3±0.1	60.3±0.1	59.8±0.1	59.9±0.2	62.1±0.3

* After (43)

† Mean of ♂ and ♀ (after (44))

and lipogenesis are well documented (see also Turner & Munday: **this** volume), but there are other associations of importance.   Catecholamines play an important part, via the hypothalamus, in the control of eating and satiety (48) and tend to be anorexic (49). Such effects together with the resultant suppression of insulin secretion (50) may collectively be responsible for the small appetite of animals of the Pietrain type.   We have observed that the voluntary food intake of Pietrains is of the order of 100 g dry matter/kg$^{0.75}$ compared with >120g/kg$^{0.75}$ for Large Whites of about 30–50 kg liveweight.   Thus the reduced level of fat in Pietrains can be explained in part by their lower voluntary food intake.

The discussion so far has emphasised the part played by neuro-hormonal mechanisms in energy balance but they can also have important consequences in connection with protein deposition.   It is now considered that protein deposition can only occur in the presence of adequate circulating levels of insulin which might not be found in animals such as the Pietrain or those animals selected to develop an exaggerated capacity not to deposit fat.   There is certain evidence that the fattest Durocs and the leanest Yorkshires of Hetzer's experiments responded to energy and protein in the diet in different ways;   the fat Durocs deposit more protein in response to increased dietary protein but not to increased dietary energy whereas the reverse applied for the leanest Yorkshires (51,52). Thus not only is excessive leanness likely to prove energetically inefficient but it may also prejudice the efficiency with which protein is deposited.   For maximum efficiency of food use it appears necessary for the deposition of protein to be combined with the deposition of a fixed proportion of fat.   It is interesting in this connection that Gregory (53) in our laboratory has observed a close correlation (>0:98) between the fasting level of plasma insulin of pigs and the quotient obtained by dividing the weight of subcutaneous fat by that of the longissimus dorsi muscle in the dissected carcass.

Although it might have been fortuitous that the rates of lean deposition which we observed in Pietrain and Large White pigs were similar, it was, nevertheless, useful for comparisons of the pattern and efficiency of growth that they were.   The observed differences in the efficiency of food utilisation were almost entirely attributable to the differing capacities of the two breeds to deposit fat.   Further, because the animals were fed matched quantities of food, the contributions to the efficiency of food use which were attributable to the proportions of deposited fat or lean, resting heat production or substrate used for energy purposes, could be assessed.   The Large Whites deposited more fat more efficiently than the Pietrains failed to deposit it and in a comparison of the efficiency of food use over a fixed weight range this would have led to even greater economy in the rapidly growing

Large Whites because of their reduced net daily energy cost of
maintenance.   Reduced net daily energy costs of maintenance may
also conceal energetic inefficiency in animals with a Pietrain-
like metabolism which grow particularly fast.   It is possible
that some strains of Landrace pigs fall into this category.
Thus it is not reasonable to expect that all animals with Pietrain-
like characteristics would automatically be identified during
performance testing by their inherent energetic inefficiency.

The evidence that the usual methods employed in selection
programmes to encourage leanness by reducing fat are more likely
to discourage fat deposition in animals which already demonstrate
superiority of performance rather than stimulate the deposition of
lean is now no longer circumstantial.   The physiological mechanisms responsible for the particular fat metabolism induced in
pigs selected for leanness are demonstrably those employed by the
Pietrain.   Thus we can expect that these mechanisms will ultimately induce, if they have not done so already, the energetic
inefficiency and stress related problems which are so apparent in
the Pietrain.

In our search for the ideal pig we have assumed that leanness
in a carcass can be equated with efficiency of food use, in the
belief that all pigs behave like Hetzer's Durocs.   Clearly all
pigs do not!   The production of increasingly lean pigs is not,
perhaps, the objective we should aspire to if, as we have seen,
overall efficiency and productivity are prejudiced.   We need to
reassess the criteria we use for selection purposes.   Do we wish
to produce meat in the fastest and most efficient way or to custom-
build carcasses for specific purposes?   The choice ultimately
depends on the market requirements and pricing policy for the
commodity and these are controlled by neither the farmer nor the
animal scientist.   The considerations presented in this paper
suggest that the most efficient animals will always produce more
fat than the leanest animals but if the Pietrain/Large White
comparison holds, the extra fat is a bonus accruing from the
consumption of the same amount or even less food.   The appropriate
combination of lean and fat in the body should also maintain
flexibility for increasing the absolute rates of lean deposition
which are probably brought about via increases in mature lean body
size (30).   The benefits of doing this, however, must be carefully
weighed, for enormity of size does not of itself lead to efficiency
especially when the maintenance of the parent stock is taken into
account.

There is one overriding feature which emerges from this
discussion.   The identification of future breeding stock to meet
new production criteria can only be achieved by the use of test
procedures which make use of the physiological principles governing
growth, development and metabolism which are now emerging.   The
animal breeders' rule of thumb can no longer suffice.

REFERENCES

1. Ludvigsen, J. (1954). Investigations on what is called
    'Muscular Degeneration' in Hogs. 272 beretning fra forsøgs-
    laboratoriet, København

2. Lister, D. (1970). In: "Physiology and Biochemistry of Muscle
    as a Food II" (Ed. E. J. Briskey, R. G. Cassens and
    B. B. Marsh), p. 705 (Univ. Wisconsin Press, Madison)

3. McLoughlin, J. V. (1963). In: "Proceedings of 9th Conference
    of European Meat Research Workers", Budapest

4. Bendall, J. R. (1966). J. Sci. Fd Agric. 17, 333

5. Lister, D., Scopes, R. K. and Bendall, J. R. (1969). Anim.
    Prod. 11, 288

6. Sair, R. A., Lister, D., Moody, W. G., Cassens, R. G.,
    Hoekstra, W. G. and Briskey, E. J. (1970). Am. J. Physiol.
    218, 108

7. Zaimis, E. (1969). In: "Disorders of Voluntary Muscle"
    (Ed. J. N. Walton) p. 57 (J. & A. Churchill Ltd, London)

8. Swatland, H. J. and Cassens, R. G. (1972). J. Comp. Path.
    82, 229

9. McArdle, B. (1969). In: "Disorders of Voluntary Muscle"
    (Ed. J. N. Walton) p. 607 (J. & A. Churchill Ltd, London)

10. Lister, D. (1971). In: "Proceedings of the 2nd International
    Symposium on Condition and Meat Quality of Pigs" p. 235
    (Pudoc, Wageningen)

11. Passbach, F. L., Mullins, A. M., Wipf, V. K. and Paul, B. A.
    (1970). J. Anim. Sci. 30, 507

12. Monin, G., Lacourt, A. and Henry, M. (1974). In: "Proceedings
    of 20th European Meeting of Meat Research Workers", p. 24,
    Dublin

13. Greaser, M. L., Cassens, R. G., Briskey, E. J. and Hoekstra,
    W. G. (1969). J. Fd Sci. 34, 120

14. Greaser, M. L., Cassens, R. G., Hoekstra, W. G. & Briskey,
    E. J. (1969). J. Fd Sci. 34, 633

15. Cheah, K. S. (1973). J. Sci. Fd Agric. <u>24</u>, 51

16. Nelson, T. E., Jones, E., Henrickson, R., Falk, S. and Kerr,
     D. (1974). Am. J. Vet. Res. <u>35</u>, 347

17. Lister, D., Hall, G. M. and Lucke, J. N. (1975). Lancet <u>i</u>,
     519

18. Hall, G. M., Lucke, J. N. and Lister, D. (1975). Anaesthesia
     <u>30</u>, 308

19. Murphy, B. P. (1965). J. Lab. Clin. Med. <u>66</u>, 161

20. Herbert, V., Gottlieb, C. W., Kam-Seng Lau, Gilbert, P. and
     Silver, S. (1965). J. Lab. Clin. Med. <u>66</u>, 814

21. Lister, D. (1973). Br. med. J. <u>1</u>, 208

22. Lister, D., Hall, G. M. and Lucke, J. N. (1974). Br. J.
     Anaesth. <u>46</u>, 803

23. Lister, D. (1974). In: "Proceedings of 20th European Meeting
     of Meat Research Workers", Rapporteurs' Papers, p. 17, Dublin

24. Lister, D. and Wood, J. D. (1975) - in preparation

25. Hall, G. M., Lucke, J. N. and Lister, D. (1975) - in
     preparation

26. Bowman, W. C. and Nott, M. W. (1969). Pharmacol. Rev. <u>27</u>, 21

27. Judge, M. D. (1972). In: "Proceedings of Pork Quality
     Symposium" (Ed. R. G. Cassens, F. Giesler and Q. Kolb),
     p. 68 (Univ. Wisconsin Press, Madison)

28. Vos, M. P. M. and Sybesma, W. (1971). In: Proceedings of the
     2nd International Symposium on Condition and Meat Quality of
     Pigs p. 278 (Pudoc, Wageningen)

29. Wood, J. D. (1973). Anim. Prod. <u>17</u>, 281.

30. Lister, D., Wood, J. D. and Berry, B. N. (1974). In:
     "Proceedings of 25th Meeting of European Association for
     Animal Production - Pigs Commission", Copenhagen

31. Lean, I. J., Curran, M. K., Duckworth, J. E. and Holmes, W.
     (1972). Anim. Prod. <u>15</u>, 1

32. Lister, D., Lucke, J. N. and Perry, B. N. (1972). J. Endocr.
     <u>53</u>, 505

33. Marple, D. N. and Cassens, R. G. (1973). J. Anim. Sci. 36, 1139

34. Judge, M. D., Briskey, E. J., Cassens, R. G., Forrest, J. C. and Meyer, R. K. (1968). Am. J. Physiol. 214, 146

35. Sorensen, P. H. (1962). In: "Use of Radioisotopes in Animal Biology and the Medical Sciences" Vol. 1, p. 455 (Academic Press, New York)

36. Romack, F. E., Turner, C. W., Lasley, J. F. and Day, B. N. (1964). J. Anim. Sci. 23, 1143

37. Palludan, B. (1972). In: "Isotope Studies on the Physiology of Domestic Animals" p. 199 (International Atomic Energy Agency, Vienna)

38. Werner, S. C. and Ingbar, S. H. (1971). Ed. "The Thyroid", 3rd Ed. (Harper and Row, New York)

39. Nicoloff, J. T. (1970). J. Clin. Invest. 49, 267

40. Moss, B. W. (1975). Ph.D. Thesis, University of Bristol

41. Boyd, G. S. (1963). In: "The Control of Lipid Metabolism" (Ed. J. K. Grant), p. 79 (Academic Press, New York)

42. Wood, J. D., Gregory, N. G., Hall, G. M. and Lister, D. (1975) - in preparation

43. Hetzer, H. O. and Miller, R. H. (1973a). J. Anim. Sci. 35, 730

44. Hetzer, H. O. and Miller, R. H. (1973b). ibid, 743

45. Standal, N., Vold, E., Trygstad, O. and Foss, I. (1973). Anim. Prod. 16, 37.

46. Vold, E. (1974). In: "Proceedings of the 20th European Meeting of Meat Research Workers" p. 176, Dublin

47. Bakke, H. (1975). Acta Agric. Scand. - in press

48. Bray, G. A. (1974). Fed. Proc. 33, 1140

49. Porte, D. and Robertson, R. P. (1973). Fed. Proc. 32, 1792

50. Cahill, G. F. Jr., Aoki, T. T. and Marliss, E. G. (1972). In: "Handbook of Physiology" Section 7, Vol. 1, p. 563 (Williams & Wilkins, Baltimore)

51.   Davey, R. J., Morgan, D. P. and Kincaid, C. M. (1969).   J.
      Anim. Sci. $\underline{28}$, 197

52.   Davey, R. J. and Morgan, D. P. (1969).   J. Anim. Sci. $\underline{28}$, 831

53.   Gregory, N. G. (1975) - in preparation

OBSERVATIONS OF THE APPARENT ANTAGONISM BETWEEN MEAT

PRODUCING CAPACITY AND MEAT QUALITY IN PIGS

D. Steinhauf, J.H. Weniger, H.-P. Mäder

Institut für Tierproduktion der Technischen

Universität Berlin

It is a well-known fact that in meat production, especially pork production, negative associations exist between meat producing capacity or muscularity on the one hand and the traits of meat quality on the other. This relationship manifests itself in different ways. Negative correlations have been observed between muscularity and meat quality in many investigations, and during the course of the selection for higher muscularity one can notice the deterioration of meat quality. This process can be traced back even to the last century (1,2). The increasing frequency of pale, soft, exudative (PSE) muscle is not limited to certain climatic regions nor continents but pertains, as far as is known, to all populations of domestic pigs. It is more pronounced the higher the meat productive capacity.

It is also well established that just as meat quality has decreased consistently with the improvement of muscularity, so too has adaptability, stress resistance and endurance of the domestic pig. This lower adaptability shows itself most distinctly in the dramatic increase in the number of pigs dying during transport from the farm to the slaughterhouse. During the last 15 years this figure has increased in Germany from 0.2% up to 1.5% in some regions (3). The growing frequency of cardio-respiratory disorders, increased sensitivity to physically and psychologically stressful situations of all kinds and the greater needs for special housing and management are further symptoms of impaired adaptability. As in the case of meat quality, negative correlations have been calculated between adaptability and meatiness, or muscularity of the animal, though the number of investigations into adaptability is far lower than those concerning meat quality(4,5).

The view that muscularity and meat producing capacity of the animal, meat quality and adaptability are causally associated is not new and seems to be beyond doubt. But there is still uncertainty about the nature of this association. The various hypotheses suggested may be simply summarized into three groups (6,7, 8,9):

1.   hormonal disorders

2.   deficiencies of the circulatory or oxygen-transport systems

3.   disorders of carbohydrate metabolism of the muscle.

    1.   It is believed that selection for fast muscular growth involves selection for the increased production of anabolic hormones and a decreased ability to produce ACTH. There is evidence for this in that the adrenal cortex of modern meat type pigs in general seems to be less developed than that of more primitive types.

The result of this is the general tendency for increased sensitivity to stress in the better muscled pigs. Thus, even limited stress before slaughter can lead to a raised body temperature, changes in the amount of glucose, and the acid-base balance of the blood and an increased incidence of PSE muscle.

    2.   Impaired oxygen transport capacity suggests that, in pigs with the best productive performance and desirable carcass conformation, the capacity for oxygen supply to the tissues is insufficient under stressful conditions. In support of this it has been reported that stroke volume, hemoglobin concentration, viscosity, and oxygen tension in blood, systolic time/diastolic time and the capillarity of the muscular tissue are inadequate in animals with the highest performance, especially in the Landrace and Pietrain breeds.

Consequently meat type pigs develop hypoxia and anaerobic metabolism even under minimal stress such as heat excitement or even slight physical effort. If these features develop at slaughter, the carcass will become PSE.

    3.   Pigs, cattle, chicken or sheep with a rapid growth rate and increased muscularity probably achieve this through an increased number of muscle fibers, a larger proportion of which are of the fast, white, anaerobic type (10,11,12,13). Thus in animals with the best productive performance the tendency to anaerobic metabolism and acidosis under stress is increased as is the tendency to develop PSE muscle.

    Apart from these there are obviously many other potential

causes such as disorders of the sympathetic/parasympathetic system and the adrenal medulla, alterations to the permeability of mitochondria or muscle fiber membranes and thermoregulatory ability. It is obvious, too, that all these are by no means independent of each other. They are all closely linked to, or caused by, fast growth and high muscularity which means that in the course of further selection for leanness and performance an increasing frequency of unsatisfactory meat quality can be expected and, with that, decreasing adaptability (see Fig. 1 and Table 1).

Three questions emerge from the previous discussion:

1.   Is it possible to improve both muscularity and meat quality by selection in spite of the obvious causal connexion between meatiness and the deterioration of meat quality in pigs?

2.   Which parameters are appropriate for assessing stress reactions quantitatively and associations with impaired meat quality?

3.   What factors exist which are likely to limit selection for meatiness and what is the nature of these factors?

Meat quality and meatiness are genetically influenced by a great number of factors many of which are interrelated. "Simple" relationships between these traits expressed as rigid phenotypic and genetic correlation-coefficients have not so far been found nor does it seem likely that they ever will be. Therefore, simultanous selection for both meatiness and meat quality seems possible although progress is likely to be slow (see Tables 2 and 3).

Several investigations furnished evidence that there are many ways of reliably measuring the reactions of pigs to stress. These measurements include meat quality (brightness, pH-value, water-binding capacity etc.), hormonal status, especially corticosteroids, enzyme activities (CPK, LDH, LDH-isoenzymes, ATP, GOT etc.), traits of the cardiorespiratory system (respiration rate, heart frequency, blood pressure, body temperature etc.), the acid-base, or the O_2-CO_2 status of blood and concentrations of microelements eg. Zn, in blood or tissues (see Figs. 2 and 3).

It should also be possible by means of carefully controlled tests to determine the degree of adaptability of endurance (see Figs. 3, 4,5,6,7). The expenditure of money, manpower and time would, however, be so high that it seems doubtful whether such a test is feasible for routine use in animal production.

On the other hand it is not so easy to devise tests which

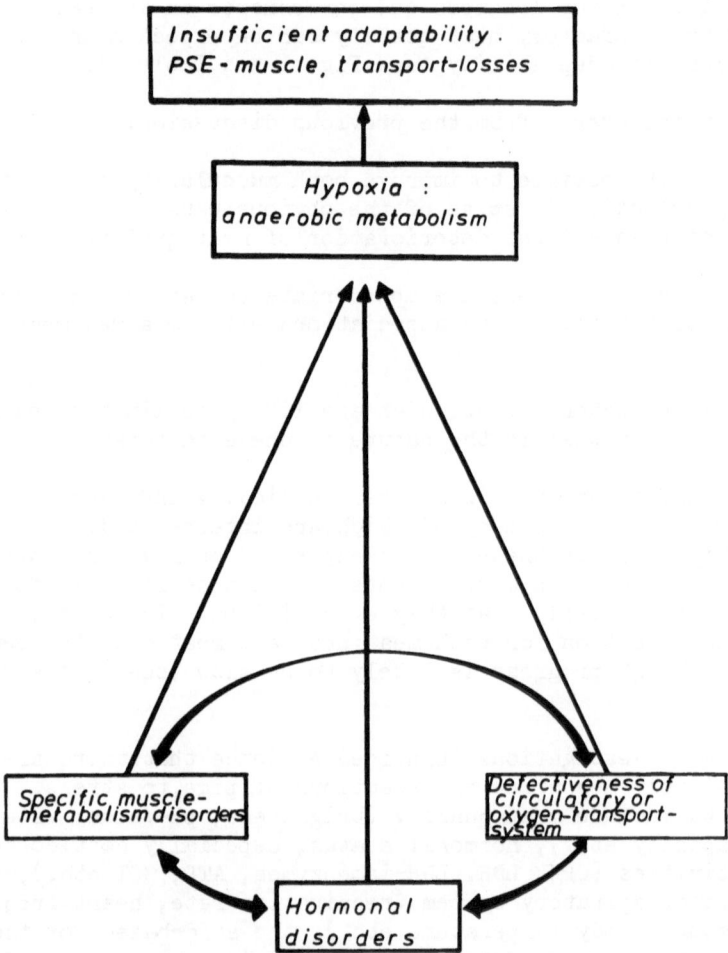

Fig. 1.   Diagrammatic representation of the stress/metabolism/
meat quality pathway (after 9).

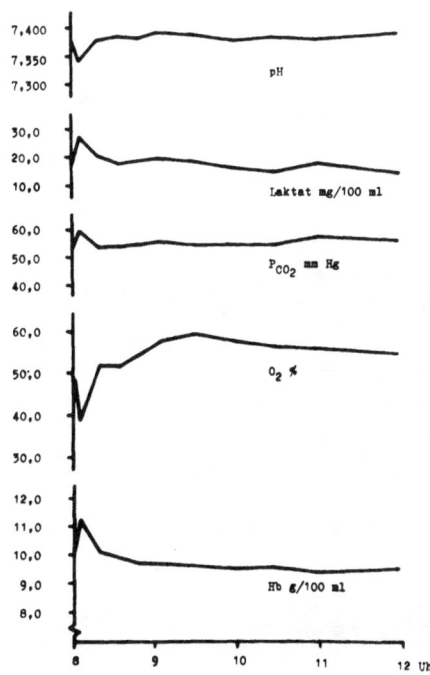

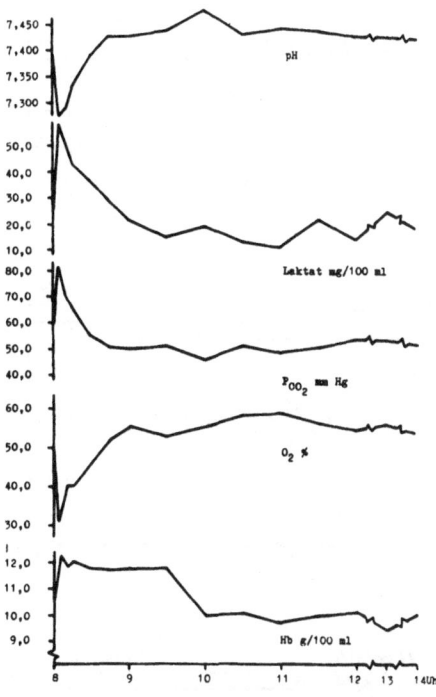

Fig 2.   Effect of stress by
electric stimulation (cattle
goad) for 15 sec on five parameters of oxygen metabolism in
venous blood. Mean values of
5 boars, tested 3 times each
(after 5).

Fig. 3.   Effect of stress by
electric stimulation (cattle
goad) for 1 min on five parameters of oxygen metabolism in
venous blood. Mean values of 5
boars, tested 3 times each
(after 5).

identify both potentially poor adaptability and poor meat quality.
The parameters described above are often seen after only the
slightest, unavoidable, and often almost imperceptible environmental influences. Therefore, accurate baseline or resting values
are difficult to obtain. This becomes evident in the low repeatability coefficients of the parameters which undoubtedly can not
be attributed to the methods of measurement (see Tables 4 and 5).
Irrespective of the methodological difficulties, the benefit from
the use of so called "physiological parameters" in selection
seems to be doubtful, especially if these parameters are enzyme
or hormone activities, heart frequency, pH-value or others, which
are elements of closed-loop systems. Even if one succeeds in
varying such an element by means of selection and breeding, this

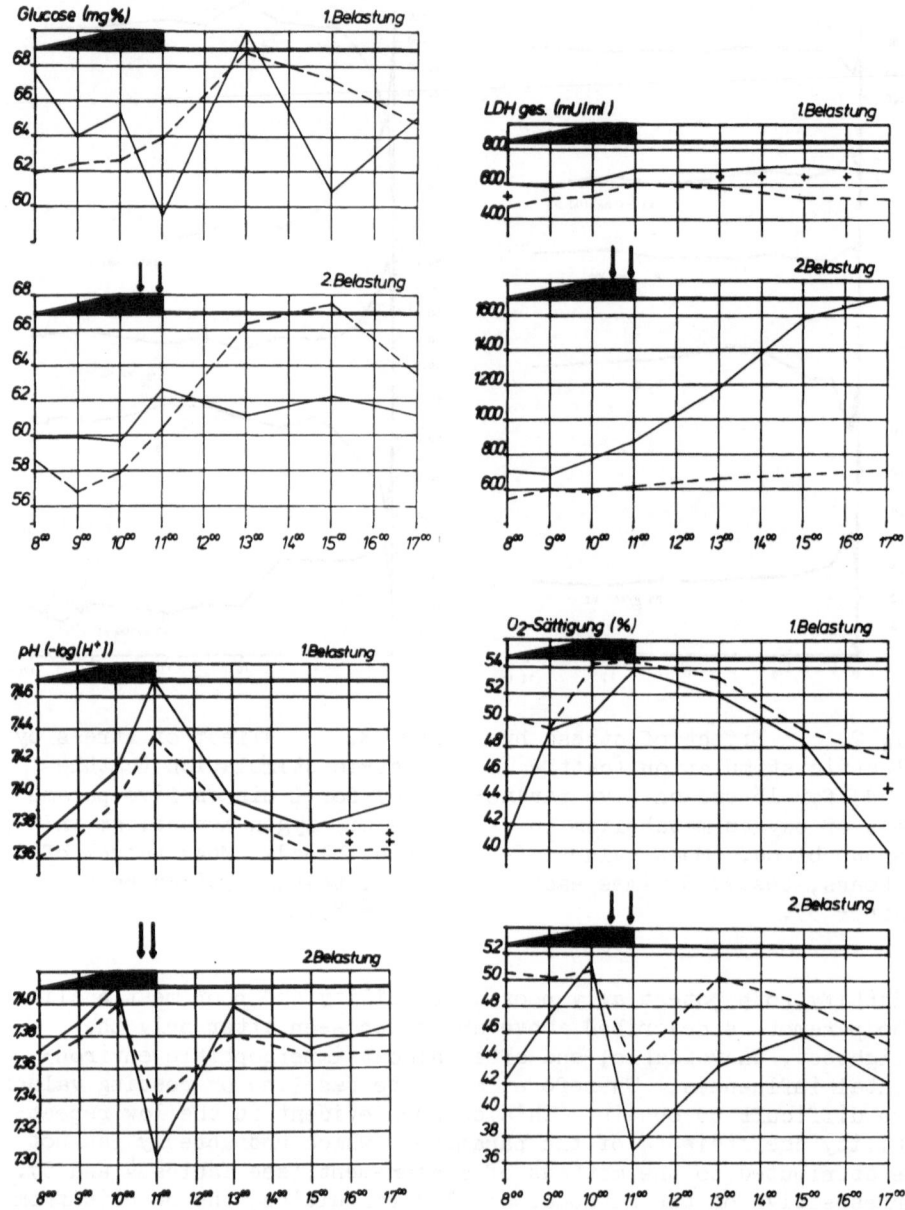

Effect of heat stress (37°C), relative humidity (95%) and exercise
(walking 180 m) on glucose and LDH (Figs. 4 and 5) and pH and O_2
saturation (Figs. 6 and 7) in venous blood in two tests.
Black marks: Temperature and humidity, ↓↓: walking,

        – – 7 "PSE" animals, —10 normal animals, + p<5%, ‡ p<1%

Table 1

Summary of the metabolic features associated
with problems of meat quality

Symptoms	Insufficient adaptability: PSE musculature, transport-losses		
General obser- vations	Acidosis, high lactate concentration in the serum, low oxygen saturation in venous blood, tachycardia, hyper- thermia, dyspnea, cyanosis. Generalized observation: Hypoxia, anaerobic metabolism		
Special obser- vations	Large muscle fibres, shift from red to intermediate muscle fibres, insufficient oxygen diffu- sion, low mitochondrial respiration rates, low activities of anaerobic muscle enzymes, elevated serum enzyme activi- ties especially after stress	Low 17-Keto- steroid pro- duction, eosinopenia hyperglycemia and increased histaminia after stress. Disturbed oxidation of protein and fat.	Low capillarisation of musculature, low relative heart weight, uneconomic heart action, high blood viscosity, low blood volume, low Hb-values, diameter of muscle fibres increased, shift from red to intermediate muscle fibres, insufficient oxygen diffusion.
Causes	Specific dis- orders in the muscular metabolism, mitochondrial defects, deficiency in mitochondrial pacemaker enzymes	Low produc- tion of adap- tation hor- mones, endo- crine in- sufficiency, shift from the produc- tion of cata- bolic hormones in the pituary ant- erior lobe.	Defectiveness of the circulatory or oxygen transport system, impaired oxygen supply of the musculature.

Table 2

Phenotypic correlations between some traits of meat quality and fattening performance (n = 863) (after 14)

	brightness low=poor high=good	Waterbinding capacity		pH_{45} low=poor high=good
		centrifuge value low=good high=poor	press value low=good high=poor	
Ribeye muscle area cm²	$-0,18^{++}$	$0,22^{+++}$	$0,26^{+++}$	$-0,21^{+++}$
Fat/lean ratio	$0,11^{+}$	$-0,14^{+++}$	$-0,17^{+++}$	$0,16^{+++}$
Valuable % of carcass weight cuts	$-0,06$	$0,15^{+++}$	$0,18^{+++}$	$-0,16^{+++}$
Fat cuts % of carcass weight	$0,03$	$-0,10^{++}$	$-0,15^{+++}$	$0,10^{++}$
Feed conversion ratio kg feed per kg live weight	$0,04$	$-0,17^{+++}$	$-0,19^{+++}$	$0,05$

Table 3

Genetic correlations between some traits of meat quality and fattening performance (n = 616) (after 14)

	brightness low=poor high=good	Waterbinding capacity centrifuge value low=good high=poor	press value low=good high=poor	pH_{45} low=poor high=good
Ribeye muscle area cm^2	-0,45	0,80	1,11	-0,59
Fat/lean ratio	0,05	-0,18	-0,36	0,27
Valuable cuts % of carcass weight	-0,40	0,32	0,33	0,04
Fat cuts % of carcass weight	0,12	-0,32	-0,27	-0,24
Feed conversion ratio kg feed per kg live weight	0,06	-0,47	-0,27	-0,33

Table 4

Repeatability coefficients[1] of some parameters of
adaptability tested under strict resting conditions
of the animal (after 15,16)

(The coefficients refer to 8 measurements per
day of test and different numbers of days)

Trait		1 day	2 days	3 days
pH	in blood	0,27	0,43	0,53
P_{CO_2}	"	0,59	0,75	0,81
Base excess	"	0,57	0,72	0,80
Buffer base	"	0,47	0,64	0,73
Stand. Bicarb.	"	0,45	0,62	0,71
Act. Bicarb.	"	0,63	0,77	0,84
Total CO_2	"	0,62	0,77	0,84
O_2 %	"	0,83	0,91	0,93
Hemoglobin	"	0,60	0,75	0,82
P_{O_2}	"	0,26	0,41	0,51
Glucose	"	0,40	0,57	0,66
Lactate	"	0,87	0,92	0,95
Body temperature		0,70	0,82	0,87
Heart frequency		0,23	0,37	0,47

1) $$W_t = \frac{s_R^2}{s_I^2 + s_{IT}^2/t + s_R^2/t \; n}$$ where:-

s_I^2 = effect of animal

s_{IT}^2 = effect of interaction animal x day of test

s_R^2 = remainder

t   = number of days tested = 1 or 2 or 3

n   = number of measurements per animal and day = 8

Table 5

Repeatability coefficients[1] of some parameters of
adaptability tested with slightly disturbed
animals (after 15,16)

(The coefficients refer to 11 measurements per
day of test and different numbers of days)

Trait		1 day	2 days	3 days
pH	in blood	0,35	0,52	0,62
P_{CO_2}	"	0,01	0,11	0,15
Base excess	"	0,17	0,29	0,38
Stand. Bicarb	"	0,15	0,26	0,35
Act. Bicarb.	"	0,09	0,16	0,22
Total CO_2	"	0,09	0,13	0,19
O_2 %	"	0,52	0,68	0,76
Hemoglobin	"	0,77	0,87	0,91
P_{O_2}	"	0,45	0,62	0,71
Glucose	"	0,43	0,59	0,69
Lactate	"	0,56	0,72	0,82
Body temperature		0,54	0,70	0,78
Heart frequency		0,64	0,78	0,84

1) calculated as in Table 4.

variation may be very quickly compensated for by other elements
and the desired improvement in performance of the animal may not
materialise. Selection parameters seem to be more useful and
meaningful if they combine several single components in an index.

At the present state of knowledge the question of limiting
factors in selection for performance remains of general interest.
These limiting factors may pertain to either the physiology of
the animal or the economy of animal production. Contrary to the
opinions voiced within the last 10 or 20 years, it is our belief
that physiological factors such as the existence of an upper limit to
nitrogen retention will not be important for some time to come.
It seems, however, that we have to expect a limit to the economy
of pig production, since with increasing performance either the
quality of the product will decrease or greater efforts will be
needed to keep the animals healthy and productive.

It seems, therefore, to be not so much a question of biology
but of economics in how to avoid the apparent negative relationships between muscularity and meat quality. One can envisage three
ways to do this:

1.   By improving techniques of management, housing, transport and
slaughter by health control, but even these may ultimately be
limited by economic considerations.

2.   By cross breeding programmes designed to make use of hybrid
vigour, although this seems unlikely to be as effective in
relation to meat quality and adaptability as was expected a few
years ago.

3.   By breeding for both meatiness and meat quality but employing
selection indices. This is both time consuming and expensive and
the choice of selection parameters is not as simple as has often
been claimed.

REFERENCES

1.   Herter, M. and Wilsdorf, G. (1914) "Die Bedeutung des Schwei-
      nes fur die Fleischversorgung." Arbeiten der Deutschen
      Landwirtschaftsgesellschaft, Berlin.

2.   Bourrier, T. (1897) "Les industries des abattoires."
      Baillère, Paris.

3.   Löhr, J. (1972) Vieh- und Fleisch-Handelsz., 16, 3.

4.   Sybesma, W., Van der Wal, P.G., Walstra, P. (1968) "Recent
      points of view on the condition and meat quality of pigs for
      slaughter." Research Institute for Animal Husbandry "Schoon-

oord", Zeist.

5. Anon.   (1971) "Proceedings of the 2nd international
    symposium on condition and meat quality of pigs", Pudoc,
    Wageningen.

6. Bentler, W. (1972) Fleischwirtschaft, $\underline{52}$; 861, 1015, 1149,
    1321, 1591.

7. Bickhardt, K. (1971) "Pathophysiologische Zusammenhange
    zwischen Belastungsreaktionen und Fleischqualitat."
    Hülsenberger Gesprache 1971, Verlagsgesellschaft fur Tier-
    zuchterische Nachrichten, Hamburg.

8. Charpentier, J. and Goutefongea, R. (1972) World Rev.
    Anim. Prod., $\underline{8}$; 44.

9. Mader, H.-P. (1974) "Beziehungen zwischen Belastungsreaktionen
    und der Fleischbeschaffenheit beim Hausschwin."
    Dissertation Berlin.

10. Ashmore, C. R., Addis, P. B., Doerr, L. (1973) J. Anim.
    Sci., $\underline{36}$, 1088.

11. Hendricks, H. B., Aberle, E.D., Jones, D.J., Martin, T.G.,
    J. Anim. Sci., $\underline{37}$, 1305.

12. Ashmore, C.R., Robinson, D.W., Rattray, P., Doerr, L. (1972)
    Exp. Neurol., $\underline{37}$; 241.

13. Ashmore, C. R., Tompkins, G., Doerr, L. (1972) Exp. Neurol.,
    $\underline{35}$, 413.

14. Weiss, F. K. (1967). "Merkmale der Fleischbeshaffenheit beim
    Schwein, ihre Erblichkeit und ihre Beziehungen zur Mastlei-
    stung und Schlachtkorperwert." Dissertation Göttingen.

15. Haase, W., Petersen, J., Steinhauf, D., Weniger, J. H. (1973)
    Z. Tierz. Zuchtungsbiol., $\underline{90}$, 180.

16. Steinhauf, D., Haase, S., Petersen, J., Weniger, J. H. (1973)
    Z. Tierz. Zuchtungsbiol., $\underline{90}$, 186.

DISCUSSION

Dr. Turner asked about the mechanism which caused the apparent association between body fatness and menarche. He referred to the theories of Prof. Hervey on the regulation of feed intake which, if accepted, would suggest that Dr. Frisch's observations might be explained by the coincidence that both appetite and sexual maturity were perhaps controlled by similar steroid hormones feeding back on the hypothalamus. In reply, *Dr. Frisch* felt that the teleological argument was very clear, and believed that a specific mechanism located outside the hypothalamus sensed that the body was of an appropriate weight and composition for menarche. *Prof. Burch* suggested that a comparison of differentially-nourished monozygotic twins might provide a critical test of Dr. Frisch's hypothesis. *Dr. Frisch* said that such a pair, which were included in her data, did in fact support her view. *Dr. Bichard* pointed out that selection for increased leanness had not necessarily resulted in PSE problems; he wondered whether selection for increased muscularity or reduced backfat thickness might have different consequences for meat quality. *Dr. Steinhauß* agreed that body type was probably more important than the lean to fat ratio which he had found to have no correlation with meat quality. In reply to Dr. Dickerson, *Dr. Steinhauß* said that the most reliable measure of PSE was the brightness or reflectance of the cut surface of the meat. In the live animal he felt that any single method used as a basis for selection was likely to identify only one aspect of a complex syndrome. *Dr. Braude* commented that it was possible with highly specialized diets to produce extremely fast growing and lean pigs with excellent meat quality. *Dr. Steinhauß* agreed that in certain populations this was true, but that in Denmark and the Netherlands correlations between meatiness and meat quality of up to -0.3 had been found. *Dr. Lister* pointed out that genetic selection for high rate of lean tissue deposition would not of itself result in problems of meat quality. If a lower rate of fat deposition is to be achieved by genetic, rather than by nutritional, means, there appears to be an inevitable association between the β-action of catecholamines responsible for fat mobilization and their   — adrenergic action which leads to a higher sensitivity to stress, with consequently increased transport deaths and poorer meat quality. The Pietrain is the extreme example of this type. The Large White represents the opposite extreme. The time has come to decide, in our selection programmes, how far we are prepared to accept energetic inefficiency as the price of producing a carcass with reduced fat. *Dr. Rhodes* observed that in his experience Large white boars, though leaner than females and castrates, were not subject to PSE meat. *Dr. Steinhauß* said that there was a significantly higher incidence of PSE meat in boars of the German Landrace breed. *Dr. Webster,* in agreeing with the main propo-

sitions of Dr. Lister, suggested that the pig was, in any case,
peculiarly sensitive to stresses such as those of heat and
halothane anaesthesia; it developed a malignant hyperpyrexia due
to its poor capacity for evaporative heat loss. This susceptibility was also seen in the association between transport mortality and season. *Dr. Fowler* questioned the inevitability of the
association between low fat content and meat quality in boars.
He asked if Dr. Lister had made any measurements of stress-
induced thyroid metabolism in boars. *Dr. Lister* replied that he
had not, but that he believed that boars would always tend to have
a thyroid metabolism tending towards the Pietrain type. *Dr. Fuller*
noted that much of the carbohydrate fed to pigs is converted to
fat. This energy is not needed to meet the energy demands of
maintenance or protein synthesis, but is necessary to obtain
efficient utilization of dietary protein. Selection of animals
which utilized dietary protein efficiently with a low intake of
dietary carbohydrate could give higher rates of lean tissue
deposition with less food and hence less body fat.

The Technology of Producing Meat Animals

MEAT PRODUCTION FROM RUMINANTS

A. J. H. van Es
Institute for animal feeding and nutrition research
Hoorn, and Department of animal physiology
Wageningen
The Netherlands

INTRODUCTION

Ruminants produce meat in several systems of animal husbandry. In this paper attention will be paid to veal calves and beef cattle and occasionally to dairy cattle and lambs. For each of these animals the economy of meat production depends on many factors: costs of producing the newborn animals, costs of investments, housing, feeding and care for the animals until slaughter. Important too are the choice of breed, type, feeding regime and age at slaughter which affects carcass quality and the quantity of meat produced. The complexity of these branches of animal husbandry is the greater when meat production is not the single purpose, but both milk and meat are being produced. This became clear in my country when in a long-term comparison of American Holstein-Friesians with Dutch Friesians the formers' higher milk yield was offset economically by their lower carcass quality.

Even when there is only one main purpose the complexity of the enterprise may be great, e.g. in the case of lamb production where the choice of the breeds of the sire and dam can considerably influence the number of offspring and the quantity and the quality of the meat produced. Multiple births, moreover, may lower the growth potential of some of the lambs (1).

The complexity is increased further by the difficulty of predicting by simple means available on the farm the genetic meat potential of the individual animal and the feeding regimes it needs for optimal feed conversion and the meat quality the consumer prefers. Even with advanced techniques available at research institutes such predictions are thought far from easy. Partly this

is due to the lack of sufficient information on the basic aspects
of growth and development. For another part the integration of
available knowledge on the growing animal – its genetics, endocrin-
ology and biochemistry, and on nutrition and meat technology is not
ideal. Finally, consumer preferences are variable and it is often
not clear what are the qualities desired in meat. Even if these
preferences were made clearer, they would, in the rather traditional
system of meat marketing, be transmitted only slowly to the farmer.
Special preferences of consumers for meat from young animals sometimes also limit the full development of their growth potential.

     In this paper an attempt will first be made to describe the
essentials of the process of development of the animal body as far
as it is important for meat production. Next, attention will be
paid to nutrition, for feed costs play an important part in the
economy of meat production. Finally, some remarks will be made on
technology.

                GROWTH AND DEVELOPMENT OF THE ANIMAL BODY

            Protein and fat synthesis, live weight gain and
                 preferred degree of carcass fat content.

     Initially in the growing animal muscle synthesis is quanti-
tatively most important; at a later stage synthesis of fat in the
tissues may become predominant, especially at high feeding levels
(2). Muscle synthesis results in considerable live weight gain as
a great part of muscles consists of water. Fat is mostly laid down
in the fat cells and there is a partial replacement of water by fat.
In terms of energy deposition per kg weight gain, the difference
between these two kinds of synthesis is remarkable. The synthesis
in muscle of 0.21 g of protein containing 1.2 kcal (5 KJ) is
usually accompanied by the retention of about 0.78 g water,
resulting in a total weight gain of 1 g. A fat deposition of
1.4 g results in some 0.4 g water being replaced by fat in the
growing animal, resulting again in a weight gain of 1 g and increasing body energy content by 13.3 kcal (56 kJ), 11 times as
much as in the case of muscle. The data on weight gain are from a
study (3,4) with veal calves in which live weight gain was regressed
on protein and fat deposition, and from chemical analysis (5) of
bulls weighing 150–580 kg; they do not apply to very young or full-
grown animals.

     It will be clear from this that in general it is not profitable
to allow the animal to deposit more fat than is needed for obtaining
the desired meat quality. To do so it would be important to have a
good knowledge of the total quantities of protein and energy which
have to be produced, so that rations can be fed which suit the
requirements of the animal. Here problems arise. Non-carcass
parts of the body have little value, but in ruminating animals may

amount to 40 to 50% of their weight. The N content of these non-carcass parts is not very much below that of the whole body (5): their energy content also cannot be neglected; it increases with the fatness of the animal. Obviously this involves considerable amounts of protein and energy but information on the composition of non-carcass parts is scarce.

Information on the composition of the carcass is less scarce although most of it comes from dissection studies and specific gravity measurements rather than from direct determinations of fat, protein and energy content. The ideal carcass should have a high muscle content and a high ratio of muscle to bone. However, with regard to its fat content - from the point of view of animal nutrition a very important item - preferences in the various countries differ. In the continent of Europe leaner meat is preferred than in the United Kingdom and the USA; besides this there is a general tendency in the world toward leaner carcasses. Low fat contents make the meat less tender, especially unappreciated in older animals as tenderness of the muscles decreases with age. Thus, opinions differ, and are changing, on the desired composition of the carcass and information on preferred fat, protein and energy contents is far from abundant.

Measures to reduce maintenance costs

To improve the animal's gross feed efficiency the growth period should usually be as short as possible (less maintenance feed). With rising feeding level daily gain will be enhanced, so this might serve the purpose. However, since fat synthesis is far more susceptible to an increase in feeding level than protein synthesis, and in view of the optimal fat content of the carcass mentioned above, the improvement cannot be found by increase of feeding level. The improvement obviously should be found by using animals with a high genetic potential for muscle synthesis. I doubt whether extremely high energy levels or protein levels which are above protein requirements (provided that the energy supply is sufficient) enhance muscle synthesis. The first tends to produce over-fat animals; the impression of a beneficial effect on muscle synthesis may be due to the fact that increased fat deposition in the muscles has been taken for protein synthesis.
The second measure may only be effective if during parts of the day shortages of limiting amino acids at the cell level may be prevented.

Protein synthesis can also be increased by using anabolic agents. With veal calves Berende et al.(6) and van Weerden et al. (7) showed that considerable improvement of live weight gain and N deposition could be obtained by anabolic agents. However, the time of slaughter should not be too remote from the time of treatment, otherwise net gain is small. My own measurements (8)

on their calves showed that there was no fall in the efficiency of
utilisation of the energy of the feed. Treatment too early
resulted in a temporary weight increase followed by a decrease of
similar magnitude. Moreover, from their figures it appears that
the increase was greater when the treatment was applied to older
calves, i.e. when they are in a stage in which protein synthesis
makes up a decreasing part of total synthesis. Also in view of
the greater effect of diethylstilboestrol on the growth of steers
than of bulls (6) it may be postulated that the effect of anabolic
compounds is a temporary speeding up of protein synthesis which is
especially effective when the body's own production of anabolic
hormones is low. Because of the possible presence of undesired
residues of anabolic substances or their degradation products in
the meat it seems better to use dehorned bulls rather than steers
for beef production; this also leads to rapid growth and low
maintenance costs. However, in view of the world food shortage
the advantages and the disadvantages of the use of anabolic agents
in animal husbandry should be very carefully weighed. Human
health organizations tend to reject this use, often more for
psychological reasons than because of scientific evidence. FAO
and WHO intend to organize a conference on this topic; agreement on
such subjects, of course, should be reached at the international
level.

Methods of selection at an early age

Within-breed variation of genetic potential for muscle
synthesis is considerable even within single-purpose beef breeds as
shown by Geay et al (9). The variation is also great in the case
of dual purpose breeds for which selection for meat production is
of secondary importance. For such animals it would be very useful
if at an early age they could be divided in groups with higher and
lower muscle growth potentials. This would facilitate appropriate
feeding measures, e.g. allowing continuous ad lib. feeding to the
former group and restricted feeding - especially near slaughter
time - of the other group, resulting in animals which are not too
fat. For single purpose beef breeds also the possibility of
distinguishing promising animals at an early age would greatly
help selection.

The cause of the higher daily rate of lean deposition is not
quite clear. Lister et al. (10) use McCance and Widdowson's (1)
metabolic clock theory and believe that the lean mass in animals
with a higher mature weight has to be synthesized in about the
same time period as the smaller lean mass of animals with a smaller
mature size. Bergström (11) is of the opinion that the time in
years needed to reach maturity in cattle is equal to the mature
weight raised to the 0.3 power. This also makes it necessary for
the animal which has a higher mature weight to have a higher
absolute and relative daily synthesis of lean mass although not to

such a degree as according to the other theory. Bergström, however, also mentions between-breed and within-breed differences in muscle to bone ratios.

It seems not illogical to assume that mature bone size determines mature age rather than mature lean mass. Regardless of which of these theories is correct it would be useful to predict the animal's genetic potential for muscle synthesis at an early age. One wonders if it would be sufficient for this purpose to follow the weight increase and feed intake of the animals over a given age or weight interval while on a ration not too low in protein (to exclude compensatory growth). Poor feed conversion figures would point to a higher proportion of fat being synthesized. The preference for group feeding, probably, would interfere with this method of testing. Enzyme (12,2) or hormone assays of biopsy samples of fatty tissues or blood might give information on fat synthesis. Protein turnover studies might give more direct information on the genetic potential for protein synthesis but these techniques are very difficult to perform with large animals (13) even at a research institute. Similar studies on muscle biopsies could be considered. The observation of Jentsch et al (14) that high propionic acid levels in the rumen of beef cattle are correlated with higher levels of protein synthesis is interesting. This could be an indication of the importance of gluconeogenesis in ruminating cattle for protein synthesis so that assay of the enzymes involved might be useful. However, the high rate of gluconeogenesis might also be due to the high level of fat synthesis. This possible effect of propionic acid on protein synthesis might be related to a similar positive effect of fatty acids with a medium chain length in veal calves and pigs, a theme of research in France.

Lister et al. (10) suggested that the increasing leanness of pigs, with only a small variation in mature weight, might be due to selection for low fat synthesis rather than for high protein synthesis. In cattle with a high mature weight the high level of protein synthesis would automatically mean a low level of fat synthesis unless protein synthesis would stimulate feed intake or would lower maintenance needs. High intake and low physical activity, however, are usually found in animals which tend to fatness.

ENERGY REQUIREMENTS

Due to the uniform and highly digestible rations of <u>veal calves</u> energy requirements for these animals can be simply expressed in metabolisable energy (M_E) measured with nonruminating calves (3,4). The animals need about 110 kcal M_E (460 kJ) per unit of metabolic weight ($kg^{\frac{3}{4}}$) for maintenance and nearly 70% of the M_E

present above the amount needed for maintenance is converted into
energy in fat and protein. The amount of energy lost as methane
in these animals is very small and can be neglected; urinary energy
losses increase with age from 2 to 5% of the gross energy. Thus,
it is clear that digestibility is the main determinant of the M_E-
content of the feed. Lower digestibilities are usually met when
the expensive milk proteins are replaced by plant proteins; the
presence of partly or completely undigested plant proteins in the
gut appears to reduce growth rate still more. Exchange of milk
lactose by cheaper starch products is possible only to a limited
extent, especially in young calves (15).

Protein turnover rate is thought to play an important part
with regard to the feed energy required for protein deposition in
young calves. Biochemically, linking the amino acids to make a
protein molecule requires little energy. Measurements with growing
animals suggest a considerably higher energy requirement (16).
Hoffmann et al. (17) (quoted by Kielanowski (18)), and Kielanowski
(18), are of the opinion that all kinds of protein synthesis
require large amounts of energy. From my own balance data (16)
on milk and egg protein production I derived a lower energy
requirement per gram of protein synthesized. This might be due to
the fact that the body tissues of the mature animals involved had
a low rate of protein turnover. In the rapidly growing animal
this rate is thought to be higher (19), resulting in a greater
energy requirement per g net production of protein. The discrepancy between these views is due to the limited amount of experimental information, to difficulties of interpretation because of
the lack of precision of estimates of the maintenance needs of
growing animals, and to the minor contribution of protein to total
energy deposition.

A model was made of the relation between live weight, feed
intake and feed composition and growth rate which for Dutch
Friesian bull-calves approximately fits the experimental data (3,4).

Energy utilisation in ruminating beef cattle is considerably
more complicated due to the fermentation in the forestomachs, the
higher slaughterweight and the variety of feeding stuffs which may
be used. Recent investigations by Jentsch et al. (14) have shown
that information on the M_E content of feeding stuffs used in
rations for beef production can easily be derived from data on
their digestibility when fed to sheep near the maintenance feeding
level. On average, the feeding level of beef cattle seldom
exceeds twice maintenance; at this level there is a small depression
of digestibility which, however, is nearly compensated by lower
energy losses in methane and urine.

The effect of ration composition, especially the ratio of M_E

to gross energy, on the utilisation of M_E for maintenance and for production is of considerable importance. Thus, for a correct energetic evaluation of the ration, the ratio of maintenance to production metabolism should be known (20). This means that the same ration or feedstuff may have different net energy values for high compared to low daily gains. Obviously, net energy varies with the animal production level (21). However, according to a proposal to account for it presented by Alderman (22), and slightly modified by me, the differences are not very great: at daily live-weight gains of 0.75 (moderate) and 1.25 kg (high) the computed net energy contents for barley, hay with 20, and hay with 40% crude fibre in dry matter are 2109 and 2037, 1521 and 1408, 867 and 753 cal/g dry matter, respectively or, relative to the barley values, 100 and 100, 72 and 69, 41 and 37. In view of this, instead of working with the whole range of net energy contents (= feeding values), calculated by computer for each production level, it seems sufficient, while accepting a slight inaccuracy, to work with only two feeding values, one for high and one for moderate production levels. It should be clear that in that case the requirements should be expressed in the appropriate feeding values.

PROTEIN REQUIREMENTS

Experiments (3,23) have shown that in <u>veal calves</u> the protein content of the diet may be reduced with increasing liveweight. Homb (24) found a similar effect in pigs. It is explained by the fact that with advancing body weight, protein deposition becomes less important relative to total metabolism. In practice use is made of this principle either by a steady increase of a component in the diet which is low in protein or by changing to a diet with less protein after some weeks. Another method is to use the same ration throughout the whole growth period but with a protein content which is somewhat low relative to the animal's needs in the first weeks and rather high during the final period.

In this case obviously use is made of compensatory growth.

Protein or N standards for (ruminating) <u>beef cattle</u> used in practice appear to be rather high. Both Jentsch et al. (14) and Schulz et al. (5) derived lower N requirements in their countries. De Boer (25) came to a similar conclusion from results of feeding trials performed over several years with Friesian and Meuse-Rhine-Yssel bulls fed primarily on beet pulp and 1 kg hay daily. He considered 7 g protein (N x 6.25) per unit of metabolic weight to be about the minimum amount required by these bulls in the weight range of 250-500 kg for a satisfactory daily gain.

The discrepancy between these research results and the recommended standards can be only partly explained by the necessary safety margins included in the latter figures. It is true that the

former are from experiments with dual purpose breeds so that for
breeds with higher potentials for muscle synthesis the minimum
requirements may be higher. For ruminant animals the composition
of the N-free components of the ration is, of course, important
for the availability of N.

TECHNOLOGY

In view of the world food situation and the high prices of
those feedstuffs which are suitable for monogastrics, forages and
concentrates with higher levels of cellulose seem most suited for
ruminant feeding, including beef production. However, the intake
of bulky feeds with high lignin contents, given as single feeds,
is often too low to raise the production level enough to make beef
production economically attractive. Obviously, feeds of higher
nutritive value should be given along with such forages. Feeds
like fresh and artificially dried grass, corn silage and beet- and
citrus-pulp suit the purpose very well.

Berner (26) recently wrote excellent papers on the production
of beef by bulls kept at pasture in the northern part of Germany.
Due attention was given to the quality of the grass as related to
its digestibility and intake when the pasture was used permanently
or in a rotation system in the various months of the year. In the
rotation system digestibilities of organic matter ingested change
only slightly from April to October (from 80 to 75%, similar to
values found by me (27) for Dutch grass fed ad lib. to lactating
cows). Intake was estimated to be 1.8-2.0 kg dry matter per 100
kg body weight. In the permanent pasture considerable reductions
were assumed to occur with regard to digestibility in the course
of time, as this system involves older grass being eaten. This
was thought to be accompanied also by some decrease of intake.
The net energy intakes in April to August in the rotation system
are sufficient to allow daily gains of about one kg; thereafter,
and also on permanent pasture, they are lower. Additional
feeding of appropriate quantities of a low protein concentrate
(1-3.5 kg per head daily) in this and other such unfavourable
situations is very beneficial as it restores the rate of growth to
normal, allowing a weight gain during the pasture season of some
200 kg. Similar studies at the "Hoorn" institute by Weide are in
progress with lambs to see if it is necessary, in Dutch circumstances and using the rotation system, to feed additional concentrates. An attempt is also being made, by total collection of faeces,
to obtain some information on the replacement of grass by concentrates.

The use of pelleted artificially dried grass for beef production is favoured by the high intakes but hampered by their high
prices when they have a high M_E- content. Grass or hay pellets
with lower M_E-contents should be used together with some concentrates to assure a sufficient rate of weight gain. As is the case

with the rations based on beet- or citrus-pulp some long forage should be fed or given as bedding to prevent digestive disturbances.

Corn silage has come increasingly into use in Europe thanks to the development of better corn varieties and of harvesting machines which chop to a length of about 6 mm; this favours the ensiling process. Very mature corn contains whole grain which cattle do not chew to a sufficient degree, resulting in lower digestibilities. Similar results have been found (28,29) when comparing whole grain with cornmeal in dairy cows. Sheep do chew the hard grains so that the digestibility data found with them for mature corn silage may be higher than for cattle. The development of equipment for tractors suited for taking from the silo some 500-100 kg quantities as a whole facilitates feeding and allows storage for a few days without heating.

REFERENCES

1. McCance, R. A. and Widdowson, E. M. (1974) Proc. R. Soc. Lond. B. 185, 1.

2. Bergen, W. G. (1974) J. Anim. Sci., 38, 1079.

3. Es, A. J. H. van (1970) In: "Energy Metabolism of Farm Animals: Proceedings of the 5th Symposium of the European Association for Animal Production" (Eds. A. Schürch and C. Wenk) p. 97 (Juris Druck Verlag, Zurich).

4. Es, A. J. H. van and Weerden, E. J. van (1970) Landbouwk. Tijdschr. 82, 109.

5. Schulz, E., Oslage, H. J. and Daenicke, R. (1974) Fortschr. Tierphysiol. Tierenähr. 4.

6. Berende, P. L. M., Wal, P. van der and Sprietsma, J. E. (1973) Landbouwk. Tijdschr., 85, 395.

7. Weerden, E. J. van, Wal, P. van der, Berende, P. L. M., Huisman, J. and Sprietsma, J. E. (1973) Landbouwk. Tijdschr. 84, 409.

8. Es, A. J. H. van, unpublished.

9. Geay, Y., Robelin, J. and Jarrige, R. (1974) In: "Energy Metabolism of Farm Animals: Proceedings of the 6th Symposium of the European Association for Animal Production" (Eds. K. H. Menke, H.-J. Lantzsch and J. R. Reichl) p. 129 (University Documentation Department, Hohenheim).

10. Lister, D., Wood, J. D. and Perry, B. N. (1974) Ann. Meeting EAAP, Copenhagen.

11. Bergström, P. L. (1974) "Slachtkwaliteit bij runderen", Pudoc, Wageningen.

12. Rogdakis, E. (1974) Z. Tierphysiol. Tierernähr. Futtermittelk. 33, 329.

13. Perry, B. N. (1974) Br. J. Nutr. 31, 35.

14. Jentsch, W., Hoffman, L., Schiemann, R. and Wittenburg, H. Arch. Tierernähr - In Press.

15. Es, A. J. H. van, Dommerholt, J., Nijkamp, H. J. and Vogt, J. E. (1970) Z Tierphysiol. Tierernähr. Futtermittelk. 27, 71.

16. Es, A. J. H. van (1971) In: "Hulsenberger Gespräche 1971", p. 10. Verlag Gesellsch. tierzücht. Nachrichten, Hamburg.

17. Hoffman, L., Schiemann, R., Jentsch, W. and Henseler, G. (1974) Arch. Tierernähr, 24, 245.

18. Kielanowski, J. Proc. 1st Symp. Protein Metab. - In Press.

19. Millward, D. J., Garlick, P. J., James, W. P. T., Sender, P. and Waterlow, J. C. Proc. 1st Symp. Protein Metab. - In Press.

20. A.R.C. (1965) Nutrient requirements of farm livestock, no. 2. Ruminants. Agricultural Research Council, London.

21. Harkins, J., Edwards, R. A. and McDonald, P. (1974) Anim. Prod. 19, 141.

22. Alderman, G. et al. Proc. 7th Nutr. Conf. Feed MFRS., Univ. Nottm. - In Press.

23. Weerden, E. J. van, Es., A. J. H. van and Hellemond, K. H. van (1970) Landbouwk. Tijdschr. 82, 115.

24. Homb, Th. (1972). In: "Festskrift til Knut Breirem" (Ed. L. S. Spildo, Th. Homb, H. Hvidsten) p. 61(Mariendals Boktrykkeri As., Gjøvik).

25. Boer, F. de (1974) Ann. Meeting EAAP, Copenhagen.

26. Berner, W. D. (1974) Lohmann Information August, 1; September, 18.

27. Es, A. J. H. van, et al. (1974) Z. Tierphysiol, Tierernähr,

Futtermittelk. <u>33</u>, 193.

28. Cottign, B. G., Boucque, R., Aertsen, J. V. and Buysse, F. X.
(1973) In: "Deegrijpe mais in de rundveehouderij" p. 44
(Rijkscentrum landbouwk.onderzoek Gent).

29. Moe, P. W., Tyrrell, H. F. and Hooven, N. W. (1973) J. Dairy
Sci. <u>56</u>, 1298.

ADVANCES IN PIG TECHNOLOGY

A. Rérat

Laboratoire de Physiologie de la Nutrition - C.N.R.Z.

78350 Jouy-en-Josas, France
Translated from the French by Kirsten Rérat

The problems of pig production have been changing in recent years because of modifications in agricultural practice and in human eating habits. Whilst in the past the main concern of farmers was with increased productivity, this is now associated with high quality production.

Productivity depends on the performance of the animals during both the rearing and the growing-finishing periods. The annual production of piglets per sow is much below the real production potential. This is due to very high pre-weaning mortality (commonly 20 to 30%), to morbidity, to fertility problems in the breeding animals and to deficient farm management. As an example, let us mention that prevention of half of the losses of piglets would allow the whole stock of sows of the E.E.C. (8 million) to be reduced by about 1 million with no change in output, allowing a saving of more than 1 million tons of feed.

As regards growing-finishing animals, the problem does not lie in mortality, which is very low during this phase, but in errors of feeding (bad adjustment of nutrient supplies to the requirements) and hygiene (latent morbidity) bringing about a lowering of the growth rate and excessive expenditures for maintenance. For instance the feed intake of the 100 million pigs slaughtered each year in the E.E.C. being about 35 million tons, an improvement of only 10% would lead to a saving of feed 3 times greater than that mentioned for the rearing period. From an examination of the best and worst herds of our countries, this improvement seems to be easy to obtain.

The quality of the final product constitutes another important problem for present day pig production. Two aspects will be considered; the first concerns body composition. The carcasses of pigs produced according to traditional methods are too fat for present taste. It therefore becomes necessary to produce lean pigs which requires precise control of the methods of production. The other aspect is related to the increasing frequency of exudative myopathies causing so-called "pale soft exudative" meats.

Confronted with changing human eating habits, new production systems have been developed in the last twenty years. Formerly, almost all pigs came from small farms with one or two breeding animals and their offspring, whereas at present 50 to 90% of European pig production is located in semi-industrial units with stocks exceeding 100 pigs and 20 sows. These units are characterized by specialization ("pig rearers" and "pig feeders") and intensification. This results in an increase in the size of units, and a more rapid diffusion of new techniques. The trends towards the disappearance of small farms because of their low profit-earning capacity, arising from the relatively high investment and labour costs, will become more and more marked during the coming years.

Modification of performance during the different periods of the life of the pig requires knowledge and control of several factors: we shall attempt to consider recent advances in nutrition and feeding practice and changes in rearing and breeding methods. Problems concerning the quality of meat will not be considered.

NUTRITION OF GROWING PIGS

The advances made in the last few years in the field of both nutrition and feeding concern the reduction of feeding costs relative to total production costs. The animal required is lean; its growth, though not necessarily the most rapid, is the most advantageous from an economic point of view, i.e. in terms of feed efficiency and carcass quality. In most cases feed efficiency and the resulting body composition are related. This is clearly shown by analysis of the feed conversion ratio which is the ratio of the amount of feed, I, eaten by a growing animal during a given period to the live weight gain, G, during the same period. As the amount of feed eaten is intended to meet the requirements for maintenance, E, and for production, PG (P being the energy cost per unit of gain) the formula can be expressed as follows.

$$FCR = \frac{I}{G} = \frac{E + PG}{G} = \frac{E}{G} + P \qquad (1)$$

These formulae show that the feed efficiency depends primarily on the nutrient content of the diet, but also on the relationship between the requirement for maintenance and the growth rate as

well as on the expenditures for each unit of tissue formed. This
expenditure is not constant but depends on the composition of the
tissues formed, each type of tissue being synthesized with a
specific efficiency (2,3,4) and having different energy contents.
Because of the water content of the muscular tissue (about 80%)
and the adipose tissue (about 10%), the energy expended in the
synthesis of unit weight of muscular tissue is only a quarter of
that used in the synthesis of adipose tissue. The feed conversion
ratio is therefore lower as the diet is better adapted to the
requirements, ensuring their satisfaction with the lowest amount
of feed, and as the fatness of the animals is reduced and their
growth more rapid.

Recent reviews (5,6,7,8) have considered the effects of
various dietary factors amongst which protein and energy are
especially important.

Energy

Many studies have been carried out with a view to controlling
the energy intake in order to obtain lean carcasses at the lowest
price. The methods used may be divided into three main groups:
- variations of the feeding level
- variations of the energy concentration of the diet
- feeding methods and physical form of the feeds

These three points have been analysed in detail in a recent
review (9); in this paper we are only going to comment briefly upon
the first.

General effects of changing feed intake. When the animal
receives a feed mixture ad libitum, its dry matter intake depends
upon the composition of the diet, its energy density, protein amino
acid, mineral and vitamin contents. As the pig is a gluttonous
animal, the consumption of a well balanced diet may be excessive,
in particular the amount of energy.

On the basis of various experiments, Vanschoubroek (10) has
studied the influence of feed restriction on the performance of
the pig, the magnitude of restriction being calculated from an
ad libitum reference level (between 30 and 90 kg liveweight, an
average daily intake of 2.7 kg of a diet containing 3,000
digestible kcal/kg and bringing about a mean daily gain of 750 g
and a backfat thickness of 37 mm). The results of these calculations are reported in table 4.

As the intensity of the restriction increases, the following
phenomena can be observed:
- a reduction of the growth rate, the greater as the intensity
  of the restriction increases

- a linear decrease in the backfat thickness
- an improvement of feed conversion ratio, which falls to the minimum value with about 25% restriction and increases again with greater restriction.

These variations depend on two phenomena acting in opposite directions. On the one hand, the extension of the fattening period causes an increase in the part of the total requirement for growth which is due to maintenance. On the other, the production cost per unit of gain decreases as the animal produces less fat.

From this, it would be tempting to conclude that there is an optimum reduction of the feeding level of about 20 to 25% compared to ad libitum feeding. Such a restriction, corresponding to a mean intake of 2.0 to 2 2 kg of feed between 50 and 90 kg live-weight, results in a 15 to 20% lowering of the growth rate (i.e. to 600-630 g/day), an improvement of the feed conversion ratio by about 6% (i.e. a saving of 0.2 kg feed per kg weight gain) and a decrease of backfat thickness of about 8% (i.e. about 3 mm). Variation in fatness is accompanied by variation in the composition of the depot fats; their content of unsaturated fatty acids increases as fatness is reduced by feed restriction (11,12, 13,14). This might be explained by a decrease in the de novo synthesis of saturated fatty acids.

The calculations of (1) do not take into account a certain number of variables, each playing its part. The optimum feed restriction varies with the type of carcass wanted; it also varies with other factors, among which the most important are the genetic origin, the sex, and the age of the pigs, as well as the composition of the diet and the environment (especially the temperature) (15).

Genetic origin. Some strains or breeds are less fat than others. For a given carcass quality, the restriction required will be smaller for animals which exhibit greater ability to produce muscular tissue. This point can be illustrated by an example taken from (16). Animals from a lean line, receiving a diet according to a relatively liberal feeding level were as lean as animals from a fat line receiving the same diet according to a feeding level limited by 25%. (Table 1). The use of restricted feeding of a diet with a low energy concentration might therefore prevent the appearance of the potential differences between strains (17,18).

Sex. The feed intake of the female is lower than that of the castrated male (19). Therefore, restriction according to weight may mean that the castrated male is really restricted while the female receives an amount of feed almost corresponding to its voluntary intake (Table 2).

Table 1

Variations in the effects of a feed restriction according to
the genetic potential of the line in the Yorkshire breed (16)

Line Feeding level	Fat normal	fat restricted	lean Normal	lean restricted
Mean daily gain (g/d) Feed conversion ratio	604 3.79	494 3.56	640 3.77	542 3.51
carcass weight (kg) lean cuts (kg) Fat cuts (kg)	66.7 23.7 34.4	55.9 24.8 22.7	68.1 30.2 28.2	54.1 26.6 18.5
lean cuts % fat cuts %	35.5 51.6	44.4 40.6	44.3 41.4	49.2 34.2

In the female, growth and body composition are only slightly
modified by this type of restriction, whereas in the castrated
male it brings about a reduction of the growth rate and a marked
improvement of body composition (21).

The castrated male seems to be much less able than the female
to compensate for the lengthening of its fattening period by
improvement of feed efficiency. The interaction between feed
restriction and sex has been emphasized by some authors (14,18,17)
but others deny its existence (22,23).
Age. It is generally admitted that continuous restriction
during the whole growth period leads to leaner carcasses and a
better utilization of the feed than the method of ad libitum
feeding followed by a feed restriction from 50 kg liveweight (15).
However, the adaptation of restriction according to age and sex may
be profitable (21,24). Thus, Walker et al. (25) comparing various
growth profiles between 20 and 87 kg liveweight, considered that
the best performance can be obtained with a moderate restriction
up to 55 kg liveweight followed by liberal feeding.

Attention must be drawn to the phenomenon of feed intake and
compensatory growth when the animals have been restricted during
their growing period (26,27,28). This compensatory growth only
lasts a short time when the restriction is applied to very young

Table 2

Variations in the effects of food restriction
according to sex (20)

	Semi ad libitum		Feeding scale	
	Castrated male	females	Castrated male	females
Daily food intake (kg)	2.50	2.12	2.03	2.06
Mean daily gain (g)	686	600	544	602
Feed conversion ratio	3.18	3.10	3.26	2.98
Lean cuts %	48.6	50.9	50.0	52.1
Fat cuts %	22.8	17.7	19.9	18.1

piglets between the age of 5 and 42 days (27) but lasts longer
when applied to animals between 23 and 50 kg (26) or more (28).
The feed efficiency is not improved, which seems to prove that
compensatory growth reflects only an increased appetite.

Protein

Studies have been made on pigs to estimate the protein and
amino acid requirements, and of diets, to determine the best
combinations of protein sources. Only the aspects concerning the
animal will be considered here.
Dietary protein concentration. The influence of dietary
protein concentration has been the subject of many studies (9).

These investigations show that when the nitrogen level is low,
the nitrogen requirement for synthesis of the muscular tissue is
not satisfied; growth rate is reduced and a great portion of the
energy, in excess relative to the protein, is deposited in the
form of lipids: the animals are then fatter and the feed conversion
ratio is higher (19). Before 50 kg liveweight, it is possible to
increase the growth rate and improve the feed conversion ratio by
raising the crude protein level up to a plateau of 16-17.5% of the

dietary dry matter. The same result can be obtained with a level
of 13-14.5% between 50 and 90 kg liveweight. The body composition
is improved even more if the crude protein level reaches 17.5 -
20.5% of the diet (29).

It may be asked whether performance can be improved by
increasing the nitrogen level even more : apparently this is not
possible (30,31). Trials using a very large excess of protein (32),
appear to be of great interest. The supplementary supply of protein has an unfavourable action on feed intake and growth, but the
feed conversion ratio is unchanged. On the other hand, the fatness
of the animals is greatly reduced when the protein level becomes
higher (33). This may depend on the decrease of feed intake as
well as on preferential utilization of the energy from the carbon
chains of the amino acids in excess for protein synthesis (34).

Protein Quality. The necessary level of dietary protein
naturally depends on its essential amino acid content. The mere
addition of an amino acid to a deficient protein will modify the
level of this protein necessary in the diet (35,36).

Introducing a synthetic amino acid into the diet modifies the
total protein requirement. At each protein level, the addition
of lysine gives diets of greater efficiency than those containing
more non-supplemented protein. As regards body composition it
appears that balancing the diet with lysine reduces the fatness of
the animals (37,38,39), but adding an excess of the limiting
amino acid brings about less improvement, particularly when the
nitrogen levels are low (40,41,42,43,44,45). This is associated
with a decreased feed intake, which can be used to reduce the
fatness of the carcass (37,46).

Interaction between protein supply and sex. When the
nitrogen or amino acid content of the diet varies, the performance
of castrated males and females is affected in different ways.
Generally, castrated males, whose feed intake is higher than that
of the females, adapt themselves much better to low-protein diets,
their growth is more rapid than that of the females (47, 48,49,50).
which react better to the addition of the limiting amino acid (51)
or to an increase in protein (52,53,30).   Baker et al. (54)
consider that the optimum protein levels for growth and feed
efficiency are 14% for the females and 12% for the castrated males,
the lowest fatness of the carcasses being obtained with a level of
16% for the females and 14% for the castrated males. The requirement for sulphur amino acids (35) and for lysine (38) are higher
in the female than in the castrated male.

Relations between protein and energy. The relations between
the levels of protein or amino acids and the energy supply are
different with ad libitum and restricted feeding.

It is generally admitted that increases in the energy content
of the diet by addition of lipids reduce the feed intake, whatever

the level of protein; this reduction is not sufficient to decrease
the growth rate as a result of lower protein intake; however, the
protein/energy imbalance tends to be greater and results in
greater fatness of the carcasses. Clawson (55) thinks that decrease
in the feed intake only occurs when the protein supply is insufficient in quality or quantity.

Cellulose level also affects the performance of growing pigs,
in the opposite manner to the lipid level. This phenomenon is
well illustrated by an experiment (56,57) in which the animals
were given different proportions of protein and of cellulose.
Increasing cellulose content of the diet caused an increase in the
dry matter intake, but that was insufficient to compensate for the
decrease in energy value of the diet. The increased protein intake
(+ 3%) is counterbalanced by the decrease in nitrogen digestibility
(- 4%) so that growth is a little slower; fatness is reduced
because of the much lower energy intake (-7%).

The necessary level of amino acids in the diet increases
with a higher dietary lipid content because of the accompanying
reduction in feed intake (58,59). The opposite happens when the
energy level is reduced by the incorporation of an inert diluent
into the diet (35,60).

The relation between feed restriction and the supply of
necessary crude protein for optimum performance are of great
importance, but have only been little studied. It may be asked
whether limiting the feed supply may not lead to an insufficient
supply of other nutrients necessary for growth.

From the recommendations of the N.R.C. (61), it appears that
the suggested protein level of the diets restricted at the end of
the growing period is maintained relatively high. No experimental
proof was available to support these recommendations, but recently
it has been shown (18,62) that during the finishing period, limiting the daily amount of a diet containing 11-12% protein to 5
pounds brings about an insufficiency in the supply of protein
leading to poor performance.   When the protein level is increased,
growth is more rapid and the feed conversion ratio as well as the
body composition are improved.

When feed restriction is applied to the animals during the
whole growth period (63), it appears that the supply of protein
is much less limiting than the supply of energy, which confirms
results mentioned above.

In an experiment to study the necessity to make compensatory
adjustments to the protein supply when restricting feed, it was
shown (64) that a 20% reduction of the energy supply during the
growing period results, in all cases, in a decrease of growth rate

without significant change in feed efficiency or body composition.
When the nitrogen supply is partly or totally restored, the
decrease of growth rate is less pronounced; the feed conversion
ratio is improved and the fatness reduced as compared to the non
restricted animals, but the expenditure of protein per kg gain is
higher. It appears that the protein added to the diet is used
only partly for protein synthesis. It follows that a reduction
of feeding from near the ad libitum level must be accompanied by
an increase in the proportion of protein (and of amino acids) in
the diet, the amount of this increase being less than that of the
energy restriction. However, some studies (65) are in contra-
diction to this.

  Method of feeding and protein utilization. The method of
feeding may also affect the feed intake, and consequently the
protein required for optimum growth and body composition in the
pig. It may be asked whether alternate feeding high and low nitrogen
diets can modify the nitrogen utilization.   Menke et al. (67) as
well as Yeo and Chamberlain (68) showed that when this alternate
administration took place within a 24 hour cycle, neither the
growth performance nor the nitrogen retention was modified. This
phenomenon was recently shown in animals subjected to restricted
feeding (66) (Table 3).

  According to our results it seems that the hyper-nitrogenous
diet, offered in the morning or in the evening, efficiently
supplements the crude protein of the cereals, provided that the
interval between the meals does not exceed 10-12h. If, on the
contrary, the protein concentrate is only given every two days, a
decrease of the growth rate and feed efficiency is observed. On
the basis of these results, a feeding schedule can be imagined
according to which the cereals could be offered once a day in
alternation with the protein supplements, thus saving much labour.

  Diets high or low in nitrogen, alternating during longer
periods (18-35 days), result in a decreased growth rate but also
in an improvement of the utilization of nitrogen for growth during
the depletion periods, followed by a decrease during the repletion
periods (69).

Conclusions

  It is now possible to control the growth impulse and the
fatness of the pig by varying the composition of the diet and the
feed intake. The most striking advances in this field concern
determination of the requirements for energy and amino acids.
However, as it has been demonstrated that the requirements differ
according to the genetic origin and sex of the animals, the
problems are complex and further studies are required. At the
present time, it may be concluded that if we want to obtain the
best productivity with our current breeds, we have to separate

Table 3

Influence of alternating high and low protein diets
on the performance of growing pigs (66)

GROUP	1		2		3		4			
Protein level (%) sequences	16	16	23	9	9	23	37	9	9	9
Feeding time	M	E	M	E	M	E	M	E	M	E
Growing period (20–60 kg)										
Mean daily gain (g)	540		527		527		483			
Feed conversion ratio	2.85		2.99		2.95		3.26			
Finishing period (60–90 kg)										
Mean daily gain (g)	778		750		804		804			
Feed conversion ratio	3.60		3.87		3.53		3.72			
Growing and finishing periods (20–90 kg)										
Mean daily gain (g)	620		601		613		573			
Feed conversion ratio	3.14		3.28		3.18		3.45			
Body composition										
Yield (%)	71.22		71.15		70.77		71.06			
Lean cuts (%)	52.88		52.60		52.29		52.06			
Fat cuts (%)	17.27		17.64		17.25		18.83			
Backfat thickness (mean of loin and back; mm)	23.7		24.2		24.6		27.1			

castrated males from the females during their growing period and
supply different diets and amounts of feed. The protein content
of the diet can be decreased if a better balance of amino acids
is provided.

NUTRITION OF THE BREEDING SOW

The efficiency of a female breeding pig mainly depends on the number and viability of the piglets produced during its life and on the amount of feed supplied. In the reproductive life of the sow a number of factors are of importance : age at sexual maturity, weight and size of the litter at birth and weaning, longevity, regularity of the cyles and resistance to diseases. According to the data of Legault (70) concerning more than 10,000 litters, the heritability of rearing characteristics is very low. Feeding may therefore be very important. Yet our knowledge of feeding was rather limited until the last few years. Twelve years ago, the feeding standards recommended in the different European countries for the sow during pregnancy varied considerably (E.A.A.P, 1962). It is only recently that very important studies have been made in France (71,72,73) Great Britain (74-84) and in the United States (44,49, 85-89). These studies have elucidated part of the problems concerning the feeding of the sow.

Feeding during the Oestrus Cycle

The numerous studies of the last few years have been summarised in recent reviews (90,91). They show that the feeding level may affect ovulation rate in the gilts and sometimes the litter size in multiparous sows. For the moment the practice of flushing the sow does not seem to be justified.

Nutrition and pregnancy

The principal characteristics of pregnancy which lasts 114 days can be divided into two main phenomena, the development of the contents of the uterus and the synthesis of maternal tissues (which has been neglected for a long time). The uterine contents, composed of the concepta and the foetal fluids and membranes have been analysed and growth curves of the various components have been described. The maternal weight however has long been mistaken for that of the concepta; it is only recently that accurate date on the anabolism of pregnancy were supplied (73,80). The study of the development of all these new tissues (from the foetus or the mother) has led to a better determination of the nutritional expenditures and consequently to the calculation of requirements (29, 92).

Classical feeding methods lead to weight gains in the pregnant sow, followed by losses of the same magnitude during lactation. The question is to know if such weight variations are required for of maximum reproductive performance. Studies on this subject have been reviewed (79,91,93).

The influence of energy nutrition and in particular of

feeding level on liveweight variations and reproductive performance in the sow is rather important. On the one hand, an excess
of energy in early pregnancy may lead to an increase in embryonic
mortality (94,95,96), but this does not seem to be a general rule
(97,98). If the excess of energy is maintained during the whole
pregnancy, it may result in slightly heavier piglets (78,86,99).
On the other hand, an energy deficiency may reduce the weight of
the piglets at birth as well as the maternal tissue deposition
(78,100,101,102). According to Lodge et al. (75) the optimum
gain would be 70 lb for 3 pregnancies. In addition, it seems that
a constant feeding level during the whole of pregnancy represents
a practical solution (74, 83,103,104).

The specific effect of the nitrogen supply and especially of
a protein deficiency has only been recently established. As a
matter of fact, the requirement for protein in the pregnant sow
was long considered as identical to that of the animal in late
growth. In fact, the protein metabolism of the pregnant animal is
not very similar to that of the non pregnant one as the nitrogen
retention efficiency is increased under the influence of the
hormonal complex specific to pregnancy (105) especially during
late pregnancy (73,80). All the recent experiments show that
quantitative variations in the supply of protein only very slightly
affect the reproductive performance of the sow (73,106,107,108,109).
Only a very severe restriction (protein free diet during the whole
pregnancy) leads to reduction in the weight of the piglets at
birth without changing the litter size (25,110-112). Amino acid
deficiency seems to have the same effect as a drastic protein
restriction (113).

Quantitative and qualitative modifications of the protein
supply markedly affect pregnancy anabolism, the weight gain of the
sow being lower as the amounts of crude protein supplied are
reduced in quantity and quality (87,88,89,114,115) and this is
clearly shown by the nitrogen balance (73,116).

As a consequence, it is possible in practice to decrease the
protein supply during pregnancy within certain limits, so as to
reduce the weight gain of the sow to a minimum. A protein level
of 10 to 12% can be recommended during pregnancy.

Nutrition and lactation

Under natural conditions lactation in the sow lasts between
57 and 77 days. In practice, piglets have for a long time been
weaned after 8 weeks of lactation. For about ten years, through
more precise knowledge of the physiology of the piglet, the length
of the suckling period was reduced for economic   reasons first to
5 weeks and then to 3 weeks. At present, frequent and successful
attempts have been made to wean the piglets 6 to 10 days after

birth. It is therefore necessary to reconsider the problems of
nutritional requirements in lactation according to the shortening
of the suckling period. Since catabolism of tissue gained in
pregnancy meets part of the expenditures for lactation (73), it is
obvious that these requirements - which are imprecisely known -
have to be considered according to nutrition during pregnancy.

Studies concerning the relations between lactation and
nutrition have been reported in various reviews (9,79,93). These
studies show that very large variations in the supply of nutrients
result in lesser but still important variations in the quantity and
quality of milk. Thus, reduction of the feeding level during
lactation brings about a decrease of the milk yield (73,117), the
relation between the two being linear (77). Likewise, raising the
dietary protein level from 10 to 14% increases the amount of milk
produced (118), but over 14% has no further effect (72,118,119).
Variations in the dietary supply of energy do not cause any significant alteration of either the number of piglets or their weight
at weaning (84,120). On the other hand, increasing the dietary
protein level of the lactating sow leads to an acceleration of the
growth of the piglets during the first weeks of their life (89,
118,121), but this benefit is not increased by a further rise of
the protein level (122). This difference in the effects of
protein and energy supply on the performance of the piglets may
be explained either by the compensatory intake of dry feed by the
young suckling animals (74,117,120), or by the influence of the
protein level in the diet of the sow on the composition of its
milk (72,118,122). The large variations in the weight of the
lactating sow, directly related to the variations in the dietary
supply of protein and energy (73,74,84,117,120) have also to be
taken into account. A reduced supply of nutrients, a decreased
protein level (122,123,124), and use of protein with a low biological value (71,125) all bring about increased weight losses
during lactation.

It may be concluded that variations in milk production are
relatively small because of the very large contribution of the
maternal tissues when the dietary supply of nutrients is insufficient. It is at this level that nutrition in pregnancy and
lactation interact.

Thus, the lower the feed intake during pregnancy, the higher
the voluntary feed intake during lactation (73,126).

Also the higher the weight gain during pregnancy, the greater
the weight loss during lactation. The reserves deposited during
pregnancy appear to be easily mobilized afterwards (81,82).

When considering both the milk production and the variations
in the weight of the mother, the feed efficiency becomes optimum

when a severe feed restriction during pregnancy is followed by
free access to food during lactation (73,127). Feed restriction
during pregnancy does not cause any unfavourable long term effects
on reproduction or nutritional status of the sow (128) except if the
the animal is subjected to severe environmental conditions (81).

Conclusion

It can be concluded that protein and energy may be greatly
reduced during pregnancy without affecting very much the reproductive performance of the sow. Only a very low energy or protein
supply may cause reduction of the weight of the piglets at birth,
without altering their number. Conversely, the levels of protein
and energy alloted affect pregnancy anabolism and the appetite of
the sow during subsequent lactation. In this way, the feeding
level during pregnancy indirectly affects milk production and the
magnitude of the weight losses during lactation. However, the
feeding level during lactation is the main factor affecting the
milk production and   the weight change of the sow during
lactation.

In terms of feed efficiency, the L-H feeding sequence (low
during pregnancy, high during lactation) is the most efficient one
(73). In this way, a noticeable sparing of feed can be obtained
as compared to the traditional method (61). This sparing is around
20% (i.e. 100 kg feed) for energy and may reach 25% for protein
(i.e. 25 kg protein). It will probably be possible to increase
this saving further when the requirements for protein, and
especially for amino acids, are more accurately defined.

IMPROVEMENT OF REPRODUCTIVE EFFICIENCY

As regards the female breeding pig, the age at first mating,
the length of lactation or the induction of a new pregnancy during
lactation have been the subject of many recent studies. In the
case of the growing pig, the practice of castrating the males is
contested. Each of these techniques has a noticeable influence on
pig production efficiency, and their modification may lead to
important advances.

The efforts made in this field aim at reducing the length of
the unproductive periods of life of the female (shortening the
growing period to puberty, and the interval from weaning to oestrus,
the rapid culling of infertile sows) and by shortening the interval
between two successive farrowings (early weaning, induction of
pregnancy during lactation) or by stimulating ovulation.

Early weaning

The use of early or very early weaning presents two very

important aspects: the rearing and feeding of the very early
weaned piglets and the reproductive rhythm of the sow.

Environment and feeding of the piglet before early weaning

The development of very early weaning methods has long been
hindered by insufficient knowledge of the nutrition of the young
animal. Research on digestion has classified the chronology of
appearance of the digestive enzymes (129,130) as well as the
great faculty of adaptation of the young animal (131,132). In
addition, the palatability of feeds and their feeding value are
better known. It is now possible to prepare diets for early
weaning. Some authors recommended liquid diets for the very
young animals (133,134,135) as was proposed twenty years ago in
the United States, as well as automatic feeding systems (136,137,
138). Other authors developed dry weaning diets and studied in
particular various protein sources including whole egg (139),
skim-milk (140,141,142), soybean and compound feeds (135,143,144,
145). However, it must be mentioned that these diets were mainly
experimental, as they were intended for very young animals. It
is only from the studies carried out recently   (146,147) that
one may expect development of a method for the weaning of animals
at the age of ten days.

Beside the "natural" weaning at 56 days, various weaning
techniques are now available for the pig producer : very early
weaning (7-15 days) using artificial milk offered in liquid or
dry form, or early weaning (15-35 days) based on the use of a
pre-starter feed of a composition similar to that of the sow's
milk followed by that of a feed more rich in cereals.

Aumaitre (148) gives examples of the diets used in the
various types of weaning. These diets permit satisfactory growth,
though the weight at 8 weeks is higher in the case of weaning at
21 or 35 days (Table 4). In addition the mortality of the piglets
to 8 weeks is slightly lower in the case of weaning at 21 days
(149).

As regards environmental conditions, temperature, in
particular, has been the subject of systematic investigations (150).
It is well established that because of its small energy reserves at
birth and its poor thermal insulation, the piglet is particularly
susceptible to cold. Keeping the young animals in a cold environment leads to early and high mortality, due to hypoglycemia (151),
or to crushing by their mother when they try to get warm. Those
that survive grow slowly. The feed conversion ratio is poor
because of the large energy expenditure for thermogenesis and the
morbodity is high, mainly because of digestive disturbance. Thus,
very low temperatures are disastrous for the piglet, but very high
temperatures are not to be advised either as they lead to reduced

Table 4

Effects of weaning three successive litters
at different ages (149)

Weaning age (d)	10	21	35
Number of litters	113	95	162
PIGLETS Mean weight at 8 weeks (kg)	15.2	17.0	16.8
Mortality 0-8 weeks %	20	17	21
Mean number in the litter born alive still born	9.13 0.87	9.75 0.95	9.26 0.79

feed intake and a consequently lower growth rate (152). The room
temperature must be about 25°C (146) between birth and three
weeks of age and the relative humidity must be about 50-60%.
Thus, very early weaning is technically possible for a great
number of pig rearers.

Early weaning of the piglet and reproductive performance of
the sow. Regarding the dam, two types of problem exist (153).
What is the influence of shortening the lactation period on the
interval between farrowings? What is its influence on the size of
the subsequent litter? According to many authors the interval
from weaning to oestrus increases when the length of lactation
decreases, and the inverse relation between these two variables
appears to be very significant (154,155). This lengthening of the
weaning-oestrus period is particularly striking when weaning takes
place before 10 days (148,156,157).

However, some authors (158,159) do not find any significant
differences between the mean values recorded in very early and
late weaning. In fact, this discordance may be attributed to the
period when the study is made: there is actually a different
chronological distribution of heats for these two types of weaning
(159).

Analysis of the data shows that a larger number of sows
exhibit oestrus 8-9 days after weaning at 7 d and 5-6 days after

late weaning (42 d). This was verified in practice by Aumaitre
et al. (160). Even though a tendency towards earlier onset of a
post-weaning oestrus exists when weaning takes place at 35 days
(Table 5), the percentage of sows having exhibited oestrus is the
same, 10 days, after early (13 d) or late (35 d) weaning.
Analogous facts were found with respect to the weaning - conception
interval (Table 6).

The percentage of non-fertilized sows within the two months
following weaning also appears to be higher in the case of very
early weaning than after late weaning (Table 7).

The prolificacy of the sow seems to be reduced by early
weaning (149,156,158,159,161). The number of piglets per litter
is reduced by 0.7 piglets per litter when changing from weaning at
35 days to weaning at 13 days (160); this difference in the number
of piglets born is still found two months later and persists during
the course of successive farrowings. This decrease due to early
weaning is however negligible and very largely compensated by the
acceleration of the reproductive cycle.

According to the observations of Aumaitre et al. (160) the
prolificacy of the sow is higher, the less is the age at weaning,
(Table 8), however there seem not to be great differences when
lactation varies between 13 and 21 days. As the mean weight of
the piglets at the age of 2 months is slightly lower in the case
of early weaning (13 d), the total 63-day weight of the piglets
produced per year under these conditions is only slightly higher
(about 10 kg) than that obtained after later weaning. This might
be corrected by more sophisticated feeding.

Table 5

Percentage of sows returning to oestrus
within 5, 10 or 15 days of weaning
at different ages (160)

Interval (d) between weaning and oestrus	<5	<10	<15
Weaning at 13 days	5	71.8	85.6
Weaning at 35 days	17.0	71.5	81.1
Significance of the differences	p<0.05	NS	NS

Table 6

Effect of weaning at 13 or 35 days on reproduction in the sow (160)

Sample of sows	Weaning age (d)	Weaning oestrus interval (d)	Weaning fertilization interval (d)	Culled sows (%)	Farrowing interval (d)	Litters/ year/ sow
All sows	13	12.5	19.9	9	147	2.48
	35	13.5	20.6	13	173	2.11
Sows showing Oestrus within 60 d. of weaning	13	10.9	13.5	17	141	2.59
	35	11.5	14.0	21	167	2.27
Sows showing Oestrus within 30 d. of weaning	13	9.7	10.1	25	137	2.66
	35	9.5	10.0	29	163	2.24

However that may be, the increased prolificacy of the sow
spreads the costs of the breeding herd over a greater number of
offspring, as well as reducing the food eaten by the sow, and by
its piglets in the two months after birth.

Induction of Pregnancy during lactation. An acceleration of
the reproductive cycle of the sow can also be obtained if conception takes place before late weaning. It is known that the
oestrus cycles of the sow are interrupted by lactation, the ovary
being then inactive (162,163). The onset of ovarian activity,
leading to oestrus accompanied by ovulation, can be induced by
means of intramuscular injection of 2000 I.U. of P.M.S.G.
(Pregnant mare serum gonadotrophin) in 72% (164) to 82% (165) of
the females treated. However, the percentage of induced pregnancies followed by farrowings is lower the shorter the time
between farrowing and injection, and the greater the litter size
(166). It is not significantly changed if, in addition, the mother
is separated from the piglets for 12 h over three successive days
(169). The size of the litters obtained in this way is not
significantly altered. Thus, it appears that the treatment results
in satisfactory reproductive performance provided it is applied
during the 4th week of lactation (169). However, the very large
delay in the onset of post-weaning oestrus in the case of
unsuccessful treatment has to be noted : 3 out of 4 of the sows
in which the treatment did not succeed exhibited their post-
weaning oestrus more than 10 days after the end of lactation (166).
At present, this represents one of the main disadvantages of the
method. On account of the failures of the P.M.S.G. treatment,
even when applied at 4 weeks, the interval between farrowings is
154-157 days (169) whereas in the case of weaning at 13 d it is
reduced to 147 days (160). However, this treatment offers the
advantage of ensuring regularity in the onset of oestrus, allowing
insemination 4 to 5 days after the injection.

Table 7

Effect of weaning at 10, 21 & 35 days on
return to oestrus and on fertility of sow (149)

Age at weaning (d)	10	21	35
Number of sows	194	178	74
Sows returning to oestrus (%)	15.5	9.6	10.8
Sows not pregnant within 2 months (%)	10-14	5-8	2-3.3

Table 8

Productivity of the sow according to the time
of weaning (149, 16)

Age at weaning (d)	10	13	21	35	39
Theoretical production					
Number of litters (1)	2.80	2.72	2.55	2.32	2.28
Number of piglets born alive (2)	28.0	27.2	25.5	23.2	22.8
Observed production					
Number of litters	2.65	2.49	2.52	2.22	2.10
Number of piglets born alive	26.4	24.6	25.7	23.1	21.8

(1) Assuming an interval of 7 d from weaning to conception.
(2) Assuming 10 piglets born alive per litter.

154-157 days (169) whereas in the case of weaning at 13 d it is
reduced to 147 days (160). However, this treatment offers the
advantage of ensuring regularity in the onset of oestrus, allowing
insemination 4 to 5 days after the injection.

Reduction of the Length of Unproductive
periods during the life of the female

Unproductive periods include the duration of pre-pubertal
growth in the gilt, the intervals between weaning and oestrus in
the sow, and lastly the time necessary to diagnose reproductive
inadequacy and cull sows which have become unfit for further
breeding.

Age at puberty. The age at puberty determines the age at
first farrowing. According to a retrospective analysis by
Legault and Dagoorn of 8500 sows, the prolificacy of the sows
decreases by 0.2-0.3 piglets per year of their presence in the
herd when first farrowing is delayed by 10 days; the age at
culling is delayed by 15 days and the interval between generations
is increased by 12 days. On the other hand, the prolificacy of
the sow throughout its active life is not modified by its age at
first farrowing. Age at puberty generally varies both from one
herd to another, and within the same herd (172). Various factors
seem to be involved in determining the appearance of puberty and
the age at first farrowing, especially the housing conditions
(173), the feeding during growth and during the pre-oestrus period

(174) and the transport of the animals (172). However, the action of these various factors has to be defined more accurately.

In the late maturing sow oestrus can be induced by injection of P.M.S.G. However, this possibility is of only limited application since it cannot be used in the immature sow far from sexual maturity. Generally, this treatment cannot induce an oestrus cycle; although it sometimes leads to ovulation and fertilization, pregnancy is interrupted (172) since the corpora lutea regress probably because of a deficient endogenous hypophyseal support (175).

The most promising solution seems to be in selection or crossbreeding. Sexual precocity appears to vary according to breed; sows of the Pietrain breed for instance exhibit earlier maturity than those of the Belgian Landrace (176). The high heritability (0.46) of the age at puberty is a very promising factor in the selection of young females. Furthermore, young crossbred Large White x Landrace gilts gain almost one month in sexual precocity as compared to purebred animals because of a large effect of heterosis (177); this is seen in Table 9.

<u>Shortening the mean interval between weaning and conception;</u>
<u>Synchronization of heat.</u> From an analysis of 100,000 litters Legault (153) found that the interval between weaning and conception varies between 20 and 25 days according to the season. The causes of this very long delay appears to depend on variations in the interval from weaning to oestrus and on a high rate of return to heat.

The interval from weaning to oestrus is itself very variable.

Table 9

Effect of crossbreeding on age and weight at puberty,
ovulation rate and number of embryos (177)

Breed	Weight at puberty (kg)	Age at puberty (d)	Number of corpora lutea	Number of embryos alive
Large White	$108.5^+_-18.2$	$208.3^+_-29.1$	$13.93^+_-1.97$	$9.15^+_-3.25$
Large White X Landrace	$98.1^+_-17.1$	$182.0^+_-21.4$	$13.69^+_-2.09$	$9.69^+_-2.94$
Landrace	$86.8^+_-15.5$	$194.7^+_-28.1$	$12.53^+_-2.34$	$8.78^+_-3.02$

The onset of oestrus more than 10 days after weaning leads to a
lower farrowing rate (178) and consequently to an increase in the
returns to heat. Early oestrus is therefore particularly desirable. According to Martinat et al. (179) this may be achieved
by treatment with gonadotrophic hormone (2000 I.U. of P.M.S.G.):
91% of the sows treated exhibit oestrus within five days after
weaning and 2/3 of these sows become pregnant after insemination. However, the percentage of sows pregnant one month after
such treatment is no greater than when no treatment is applied.
In addition, the non-pregnant sows, after treatment and insemination, show a great irregularity in their returns to heat, and
the mean interval between weaning and conception is greater in
the treated sows than in the controls. However, the hormonal
treatment, as well as the grouping of weaning, allows synchronization of heat between the 3rd and the 5th day in the majority
of the sows and consequently permits grouping of matings and
farrowings with all its advantages. The use of methallibure to
synchronize heats in cycling gilts (180,181) is very efficient,
but has now been prohibited because of the malformations it may
cause in the young piglets (182).

Finally, the decrease in the percentage of returns to heat
depends on the use of a highly fertile boar and on a precise
detection of oestrus allowing artificial insemination or mating
at the most favourable moment. This is 22-23 h after the
beginning of heats which according to Boender (183) is 10 to 20 h
before ovulation. The results of artificial insemination are
improved when using crossbred sows because of a large effect of
heterosis (18%). The increased fertility of these animals ranges
between 0.1 and 0.5 piglets per year (184); this can be added to
the effect of heterosis on fertility (see below).

Early diagnosis of sows for culling. A certain percentage
of both nulliparous gilts and multiparous sows may be classified
provisionally or definitely unsuitable for reproduction. These
animals may significantly affect the total production costs; it
is therefore important to detect and eliminate them as soon as
possible. The early elimination of non-pregnant sows can be
achieved through early diagnosis of pregnancy by means of
vaginal biopsies (185). Unfortunately, at the present time, no
test is available to forecast the reproductive value of gilts
before puberty.

Increased prolificacy

Apart from increasing the number of litters during the life
of the sow, an increase in the number of piglets per litter also
improves prolificacy. It is well established that this characteristic is very variable and this fact is clearly shown by the
statistical analyses made by Legault (186) of 16,000 litters of
Large White pigs tested in France. Some effects seem to be

unchangeable, the litter size for example increases until a
plateau reached at the 3rd litter at about 2 years. Winter
litters are generally larger at birth and smaller at weaning than
summer litters (153).

In the sow, it is easy to obtain superovulation by means of
exogenous gonadotrophins. Thus, Hunter showed (187) that ovulation rate is directly related to the level of P.M.S.G. used
(1.89 corpora lutea for 100 I.U. P.M.S.G.), the optimum period for
the treatment being around days 15 and 16 of the oestrus cycle.
However, there is no corresponding increase in the number of
surviving embryos. Thus according to Anderson and Melampy (188),
injections of 500 to 1500 I.U. of P.M.S.G. in 13 different trials
led to 4.8 additional ovulations, but only one embryo after 30
days of pregnancy. Furthermore, the survival rate of the embryos
may be very slightly increased by treatment with progestagens and
oestrogens in very early pregnancy. Lastly, the transfer of
young embryos into an already gravid uterus (189,190) even though
it is followed by the survival of some of the embryos up to the
25th day of pregnancy, does not lead to a systematic increase in
the number of foetuses alive after the 105th day of pregnancy. In
the present state of knowledge, all these physiological techniques
seem to be too hazardous to be currently applied to improving
prolificacy.

The heritability of prolificacy in the female is low.
According to Legault (70) on the basis of the bibliographical data
available at that time it amounts to 0.23 for the litter size at
birth and 0.12 for the same characteristic at weaning. In further
studies Legault (177) reported that the heritability of ovulation
rate is 0.10, that of the number of embryos alive 0.09. This
implies that selection will lead to a real, but slow improvement
of reproductive performance in the sow. If the pig producer
chooses his breeding animals from the best half of sows classified
according to their prolificacy, the expected improvement may
represent 0.05 piglets per litter per year. On the basis of a
generation interval of 2 years, 20 years of selection will be
necessary to increase the litter size by one piglet. If selection
for this character is practised simultaneously on the breeding
animals of both sexes, a more marked improvement can be obtained.
Thus, Ollivier (191) recorded an annual genetic change of 0.15
piglet per litter over five generations of selection. Judicious
use of crossbreeding would probably lead to a more rapid genetic
change (192).

Sellier (192) has given mean values of the improvement obtain-
able by cross-breeding in the principal criteria of economic
interest. The increase of prolificacy is very important and
especially for crossings of 3 breeds: 8% in the case of piglets
born alive, 16% in the case of weaned piglets, this improvement

being in particular related to a decrease of 8% in embryonic mortality
and of 5 to 8% in the mortality between birth and weaning.
According to Legault (70) an increase of one pig per litter can
thus be obtained within 2 generations by means of crossbreeding
compared with 10 generations of selection. However, the efficiency
of crossbreeding in the improvement of prolificacy must not
exclude the selection. If intense selection is not practised
within the lines and pure breeds, the spectacular effect of crossbreeding will only be temporary. Long term improvement of the
productivity of the sows therefore depends on the simultaneous
application of both methods.

    It is known that when used in artificial insemination the
boar has a significant effect on the size of the litter (193);
this effect has not been noted in the case of natural mating (70),
probably because of the much greater number (20 to 30 times) of
spermatozoa involved. Thus, the direct effect of the boar is
responsible for almost 5% of the variation of the litter size at
birth in the case of artificial insemination (193) compared with
only 1% for natural mating (70). In artificial insemination, the
direct influence of the boar on the size of the litter is
noticeably greater than the genetic influence of the father of the
sow which has produced the litter. It has to be noted, with
respect to artificial insemination, that prolificacy and number of
spermatozoa per ejaculate are moderately heritable characteristics
in the boar (0.35) which can therefore be improved by selection.
Because of the high correlation (0.52) between fertility and prolificacy in the boar, selection for fertility practised in the
Artificial Insemination Centres automatically leads to improvement of prolificacy (194). In an inquiry by Ollivier and Legault
(193) including more than 1000 litters from 30 boars, the
difference between the boar with the highest and the boar the
lowest prolificacy represented 5.10 piglets at birth and 2.92 at
weaning. As each boar was able, in these conditions. to produce
500 litters per year, it is clear that the boar may have a
considerable effect on prolificacy.

MEAT PRODUCTION BY THE ENTIRE MALE PIG

    Traditionally, the male pig used in meat production is
castrated in order to avoid the problems of "boar taint". This
was probably justified at a time when the animals grew more slowly
and reached sexual maturity at a low weight as can still be
noticed in some breeds, in particular the Corsican breed. But
since the breeding value of the animals has considerably increased
and since the age at slaughter is generally less than that of
sexual maturity, is castration still necessary? Many studies have
been carried out in recent years not only in order to measure such
factors as feed intake, growth and body composition, but also to
determine the conditions of appearance of disagreeable odours in

the tissues. These studies were reported in several reviews (195, 196,197,198,199,200).

The entire male certainly presents a number of advantages as compared to the castrate male and the female. Its growth rate is generally higher, its fatness lower and its feed conversion ratio better than in the castrate (201,202). However, it appears that these differences depend on the genetic origin. In the Duroc breed no difference can be recorded between the two categories (203). In the Pietrain, there is no difference in N retention before 60 kg liveweight and the difference of 18% is only reached at 80 kg (204). The proportion of edible tissues is higher in the male than in the female, and higher in the female than in the castrated male (205). The daily nitrogen retention of these animals, which were of Landrace origin, decreased in the same order : between 30 and 110 kg liveweight, it was 21.4 g in the boar, 18.4 g in the female and 16.5 g in the castrated male. Unlike the other categories of animals, the nitrogen retention of the boar is not uniform between 30 and 110 kg liveweight, and its advantage is principally apparent at the end of the growing period (206). These facts concerning nitrogen retention were confirmed recently by Desmoulin et al. (207) (Table 10).

Thus, the capacity of muscular synthesis in boars is undeniably higher than that of the castrated males. This leads both to increased requirements for protein and to a better response to feed restriction. The response of boars to increasing protein level is better than that of the castrated males (208) or of females (53). According to Hays et al. (209) the most rapid growth can be obtained in the entire male with a dietary protein content of 18% between 22 and 57 kg liveweight and 16% between 57 kg and slaughter; the best body composition was obtained with a 20-18 sequence. These conclusions have been confirmed by Walstra (210), Fowler et al. (211), Newell and Bowland (212) and by Desmoulin et al. (213), even though some of the levels indicated by these authors are slightly lower than those determined by Hays et al. (209) and do not exceed 14-16% during the finishing period (210,213). Such variations in the required level of protein may correspond to variation in the composition of the dietary proteins used. With respect to protein quality, the requirement of the boar is also higher and the response to the addition of the limiting factor is more marked in the boar than in the castrate (53).

Furthermore, these higher anabolic abilities are particularly marked in feed restriction. Under these conditions, the growth difference between entire and castrated males is increased (210) and the feed efficiency is excellent in spite of the growth reduction. This is clearly shown by the results of Desmoulin (196); (Table 11) during which male or female pigs, castrated or

Table 10

Influence of castration on nitrogen retention (207)

Weight range (kg)	30 - 33		64 - 71	
Sex	Male	Castrated male	Male	Castrated male
Apparent digestibility of N (%)	85.6	88.6	86.3	96.9
N retention (% of N ingested)	49.3	49.9	44.5	34.7
N retention (% of N app. digested)	57.6	56.3	51.6	39.9
Protein deposited (g/d)	118	119	136	119

not, were fed according to two restriction schedules (semi ad libitum and 25% reduction).

In pigs of the Large White breed subjected to relatively severe feed restriction, the growth rate of the boar is 15% higher than that of the castrated male, and the feed efficiency is 10% better. At slaughter, 60% of the male carcasses and 80% of the female carcasses had a specific gravity above 1.050 and 90% of the carcasses of the castrates a specific gravity below 1.050. This corresponds to a 25-30% greater amount of lean tissue in the male. Desmoulin (196) consider that the fatness of the castrated male slaughtered at 100 kg is equivalent to that of the boar slaughtered at 140 kg. However, it must be noted that the dressing percentage of the carcass (carcass weight as % of liveweight) is 2 to 3% units lower in the male, this fact being related to a higher weight of the viscera and to the presence of testicles (204,214). In addition, the distribution of lean mass is slightly different since in the boar the weight of the shoulder is increased by 5-6% units, whereas that of the ham is unchanged (214). In spite of these disadvantages, the boar is definitely superior to the castrated animal.

Unfortunately, the carcasses produced may have more or less marked defects. One of these defects, which is relatively minor, concerns the composition of body lipids. The percentage of unsaturated fatty acids is higher in the boar than in the castrated male (215). From a technological point of view, this leads to a higher susceptibility to oxidation and to insufficient

Table 11

Influence of sex and feed intake on growth and feed
efficiency (T = control; R = 25% restriction)(196)

Sex	Males		Castrated males		Females		Castrated Females	
Feeding level	T	R	T	R	T	R	T	R
Mean daily feed intake (kg)	2.22:1.77		2.30:1.80		2.12:1.68		2.13:1.67	
Total feed intake (kg)	254a:252a		295b:308b		280b:297b		297b:327c	
Growth rate (g/d)	682a:555b		627a:470c		610 :440c		568b:408c	
Fattening length (d)	115 :143		129 :171		132 :176		137 :196	
Feed conversion ratio (kg feed/kg gain)	3.26:3.18 a	a	3.69:3.84 b	b	3.49:3.85 ab	b	3.74:4.19 b	c

Values with the same subscript are not significantly different.

firmness of the meat. In addition, boar meat in certain conditions
has a disagreeable odour (196,216), for which androsterone is
supposed to be partly responsible (217). This "sexual" odour,
mainly located in the fats of the male, generally does not appear
before 60 kg liveweight in the different breeds studied (217).
From 90 kg onwards the proportion of animals with a marked sexual
odour increases with age, but it is not possible to establish a
relation between the age of the animal and the risks of odours
(218). In a large proportion of young males of 90 kg, which have
reached sexual maturity, the intensity of the odour from their
meat after cooking does not distinguish them from the castrated
animals (76,219). This fact was confirmed by Desmoulin et al.
(213); under their experimental conditions the risk of unpleasant
meat odours involves 10% of the Large White males slaughtered at
80 kg (155 days of age) and 15% of the males slaughtered at 100
kg (175 days of age). This relatively low percentage cannot,
however, be considered applicable to all breeds and all production
conditions. It is increased when feed is restricted (78). It is
also noticeably higher in the Landrace pig (218), this being
connected with a presumed difference in sexual maturation.

According to Jonsson and Wismer-Pedersen (220), the heritability
of this trait seems to be rather high (0.54); selection to
eliminate this defect can therefore be imagined. The risk of
supplying consumers with poorly edible meats can be reduced by
detection at the slaughterhouse of meat with boar taint. Such
meat can be used in the meat processing industry, as it seems not
to be possible to distinguish it from the other meats when used
in this way. However, for some products it seems to be necessary
not to exceed a certain level of incorporation (221). Sorting of
meat according to its odour requires a rapid olfactory test such
as the "soldering iron" technique proposed by Jarmoluck et al.
(222). In these conditions, it should be possible to detect the
70-75% of pigs with no defects. Doubtful carcasses (25%) should
be subjected to supplementary tests in the laboratory, but the
disadvantage of the latter is that they are long and laborious.

IMPROVEMENT OF HEALTH

The economic losses related to pathology include the losses
due to mortality, morbidity, and therapeutic expenses, whether
preventative or curative, and the condemnations at the slaughterhouse. The loss arising from mortality is the easiest to define
in current animal production. It is very high and varies between
15 and 25% (223). The estimation of the loss due to morbidity is
much more difficult; it corresponds to the delay in growth and
consequently to the increased maintenance of the animal for the
same production and to the cost of veterinary interventions. It
has been the subject of theoretical calculations by Labouche (223).
The cost of infectious swine pneumonia in Great Britain is very
great (225).

One piglet in four does not reach weaning. Mortality, even
at birth, represents 6-8% (226,227); almost all these losses can
be attributed to anoxia during farrowing (228).

The causes of mortality after birth have also been analysed
(226,229) and change with age (230).

During the first week of life, more than half the mortality
can be attributed to physical or developmental failures. Thus,
more than 10% of piglets die during this period. The loss of
animals is greater when their weight at birth is lower (231) and
the litter larger (232).

Accordingly, the resistance of young animals has to be
increased by genetic techniques, especially crossbreeding.
Accidents during the first days must be eliminated. This can be
achieved by controlling the environment of the piglets (temperature, and in particular the housing conditions) (232,233).
Bacterial and viral infections must be prevented. The use of

specific pathogen free (S.P.F.) animals in herds, maintained in isolation represents an excellent means of achieving this aim.

What is the effect of protection from specific diseases on production efficiency? Before answering this question, it must be noted that, in most cases, the comparisons between S.P.F. and conventional animals are not strictly rigorous particularly because of accompanying genetic differences. With this reservation, and provided that no reinfection occurs, the disappearance of specific diseases leads to decreased mortality and improved performance (234,235,236,237,238).

Inoculation of a group of pigs with a suspension of lungs presenting infectious pneumonia leads to irregular growth and a 25% increase in feed conversion ratio. The savings in this case are of the same magnitude as the estimates of Betts and Beveridge (225) of the influence of respiratory diseases on feed conversion ratio.

All these results show that, by reducing mortality and improving performance, the S.P.F. technique is of great economic interest. However, it has to be emphazised that this improvement can only be maintained by rigorous precautions to prevent further infection.

REFERENCES

1. Fevrier, R., 1952. Ann. Zootech. $\underline{1}$ (1), 175-184.

2. Kielanowski, J. and Kotarbinska, M., (1970). Further studies on energy metabolism in the pig. in "Energy Metabolism of Farm Animals". Proc. 5th Symposium Vitznau, Sept. 1970, E.A.A.P., publ. No 13, 129-132, Juris Verlag, Zurich.

3. Thorbek, G., 1970. The utilization of metabolisable energy for protein and fat gain in growing pigs in "Energy meta- bolism of farm animals". Proc. Vth symposium, Vitznau, E.Q.A.P. publ. No 13, 129-132, Juris Verlag, Zurich.

4. Oslage, H. J., Gädeken, D. and Fliegel H., (1970) in Schürch and Wenk C. "Energy metabolism of farm animals" Proc. 5th Symp. Vitznau, E.A.A.P., publ. No. 13, 133-136, Juris Druck, Verlag, Zurich.

5. Kielanowski, J. (1972) in "Pig Production" (COLE, D.J.A. ed.), 183-201, Butterworths, London.

6. Chamberlain, A.G., (1972) in "Pig Production" (Cole, D.J.A. ed.), 203-223, Butterworths, London.

7. Cunha, T.J., (1972) in "Pig Production" (Cole, D.J.A. ed.),

225-242 Butterworths, London.

8.   Cole, D.J.A., Hardy, B. and Lewis, D., (1972) in "Pig Production" (Cole, D.J.A. ed), 243-257, Butterworths, London.

9.   Rerat, A., (1972).   2eme Congrès Mondial d'alimentation animale, Madrid, 4, 39-155.

10. Vanschoubroek, F., De Wilde, R. and Lampo, P.H., (1967). Anim. Prod. 9, 67.

11. Greer, S.A.N., Hays, V.W., Speer, V.C., McCall, J.T. and Hammond, E.G., (1965). J. Anim. Sci. 24, 1008-1013.

12. Babatunde, G.M., Pond, W.G., Van Vleck, L.D., Kroening, G.H., Reid, J.T., Stouffer, J. R. and Wellington, G.H., (1966). J. Anim. Sci. 25, 526-531.

13. Babatunde, G.M., Pond, W.G., Van Vleck, L.D., Kroenig, G.H. and Reid, J.T., (1967). J. Anim. Sci. 26, 718.

14. Friend, D.W. and Mc.Intyre, R.M., (1970).   J. Anim. Sci. 30, 931.

15. Rerat, A., (1970). Rec. Med. Vet. Alfort, 146, 1243-1295.

16. Davey, R.J., Morgan, D.P. and Kincaid, C.M., (1969). J. Anim. Sci. 28, 197-203.

17. Skitsko, P.J., Bowland, J.P. and Elliot, I.I., (1969). Performance of Duroc. Hampshire and Yorkshire sired pigs fed two energy levels and slaughtered at 150, 200 and 250 pounds. 1. Growth feed conversion and carcass merit. (1969) Alberta Feeders' Day Rep., p.32.

18. Wallace, H.D., Palmer, A.Z., Carpenter, J. and Combs, G.E. (1966). Bulletin 706 Florida Agricultural Experiment Station, University of Florida, Gainesville, Florida.

19. Rerat, A. and Henry, Y., (1964). Ann. Zootech. 13, 5-33.

20. Desmoulin, B., (1969). Journées Rech. porcine en France, 67-76 I.N.R.A., I.T.P. ed. Paris.

21. Desmoulin, B. and Bourdon, D., (1971). Journées Rech. Porcine en France, 73-90, I.N.R.A., I.T.P. ed., Paris.

22. Hines, R.H., Hoefer, J.A., Miller, E.R. and Luecke, R.W., (1966). J. Anim. Sci. 25 1245 (Abstr.).

23. Hines, R.H., Hoefer, J.A., Miller, E.R. and Luecke, R.W.,
     (1966). J. Anim. Sci. 25, 1277 (Abstr.).

24. Moal, J., Gaye, A. and Desmoulin, B., (1972). Journee Rech.
     Porcine en France, Paris, 121-126, I.N.R.A.-I.T.P. ed.

25. Walker, N., Holme, D.W. and Forbes, T.J. (1968). J. Agric.
     Sci. (Camb.) 71, 311-326.

26. Cole, D.J.A., Duckworth, J.E., Holmes, W. and Cuthbertson, A.,
     (1968). Animal Prod. 10, 345.

27. Fowler, V.R., (1966). Anim. Prod. 8, 354 (Abstr.)

28. Owen, J.B., Ridgman, W.J. and Wyllie, D., (1971). Anim.
     Prod. 13, 537.

29. Agricultural Research Council (A.R.C.), (1967). Agric. Res.
     Counc., London, pp Xi + 278.

30. Rerat, A. and Henry, Y., (1967). Ann. Zootech. 16, 203-211.

31. Clausen, H., (1965). World Rev. Anim. Prod. No. 1, 28-42.

32. Sugahara, M., Baker, D.H., Harmon, B.G. and Jensen, A.H.,
     (1969). J. Animal Sci. 29, 598.

33. Lewis, D. and Hardy, B., (1970). The effect of dietary energy
     and protein in carcass composition in the growing pig.
     R.I.K.E.N.A., Mallorca, 1-3 Mai.

34. Allee, C.L., Baker, D.H. and Leveille, G.A., (1971).
     J. Anim. Sci. 33, 1248-1254.

35. Rerat, A. and Henry, Y., (1970). Journées Rech. Porcine en
     France, Paris, 61-66, I.N.R.A-I.T.P. ed.

36. Kropf, D.H., Bray, R.W., Phillips, P.H. and Grummer, R.H.,
     (1959). J. Anim. Sci. 18, 755.

37. Rerat, A. and Lougnon, J., (1965). Ann. Zootech. 14, 247-260.

38. Henry, Y., Rerat, A. and Tomassone, R., (1971). Ann. Zootech.
     20, 521-550.

39. Chamberlain, A.G. and Cooke, B.C., (1970). Animal Prod. 12,
     125-137.

40. Rerat, A., Lougnon, J. and Pion, R., (1962). Ann. Zootech.
     11, 159-172.

41. Jarov, I.I., (1968). Svinovodstvo No. 9, 17-18.

42. Long, J.I., (1966). Dissertation Abstr. (B), 27, 10-B.

43. Holck, G.L., (1966). Dissertation Abstr. (B) 27, 1007B-
     1008B.

44. Baker, D.H., Clausing, W.W., Harmon, B.G., Jensen, A.H. and
     Becker, D.E., (1969). J. Anim. Sci. 29, 581-584.

45. Berry, T.H., Combs, G.E., Wallace, H.D. and Robbins, R.C.,
     (1966). J. Anim. Sci. 25, 722-728.

46. Rerat, A. and Henry, Y., (1972). Unpublished results.

47. Bell, J.M., (1965). Can.   J. Anim. Sci. 45, 105-110.

48. Smith, J.S., Jr., Clawson, A.J. and Barrick, E.R. (1967).
     J. Anim. Sci. 26, 752.

49. Baker, D.H., Jordan, C.E., Waitt, W.P. and Gouwens, D.W.,
     (1967). J. Animal Sci. 26, 1059.

50. Wallace, H.D., Lucas, E.W., Palmer, A.Z. and Combs, G.E.,
     (1970). Fla. Agric. Exp. Sta. Mimeo, series No. AN 71-1.

51. Robinson, D.W., (1966). J. Sci. Fd Agric. 17, 1-6.

52. McBee, J.L., Jr., Horvath, D.J., Lee, C. and Scherer, C.W.,
     (1969). Relationship of dietary protein level to carcass
     characteristics and growth performances of swine. W. Virginia
     Univ. Agric. Exp. Stat. Bull. No. 576 1, March 1969, p.21.

53. Bayley, H.S. and Summers, J.D., (1968). Can.   J. Anim. Sci.,
     48, 181-188.

54. Baker, D.H., Becker, D.W., Jensen, A.H. & Harman, B.G. (1970).
     J. Anim. Sci. 30, 364-367.

55. Clawson, A.J. (1967). J. Anim. Sci. 26, 328-334.

56. Henry, Y., (1966). 9th Int. Congr. Anim. Prod., Edinburgh,
     Oliver and Boyd L.T.D., 9-10.

57. Henry, Y. and Etienne, M., (1969). Ann. Zootech. 18, 337-357.

58. Mitchell, J.R., Becker, D.E., Jensen, A.H., Norton, H.W. and
     Harmon, B.G., (1965). J. Anim. Sci. 24, 977-980.

59. Anderson, G.H. and Bowland, J.P., (1967). Can.   J. Anim.

Sci. 47, 47-55.

60. Lerner, J.T., (1968). These Doc. 3ème Cycle, Paris, Fac.
     Sciences.

61. National Research Council (N.R.C.), (1968). Nutrient require-
     ments of swine, publ. 1599, National Academy Sciences,
     Washington D.C.

62. Cliplef, R.L., Hanson, L.E., Meade, R.J. and Wass, D.F.,
     (1966). J. Anim. Sci. 25, 1243 (Abstr.).

63. Baird, D.M., McCampbell, H.C. and Allison, J.R., (1971).
     J. Anim. Sci. 33, 390-393.

64. Rerat A., Henry, Y. and Desmoulin, B., (1971). Journees Rech.
     Porcine en France, Paris, 65-72, I.N.R.A-L.T.P. ed.

65. Henk, G. and Laube, W., (1967). Arch. Tierernähr, 17, 393-
     407.

66. Rerat, A. and Bourdon, D., (1972). Journees Rech. Porcine en
     France, Paris, 215-224, I.N.R.A.-I.T.P. ed.

67. Menke, K.H., Lantzsch, H.J., Ehrensvard, U. and Schneider, W.,
     (1969). Landwirtsch. Forsch. 22, 173-181.

68. Yeo, M.L. and Chamberlain, A.G., (1966). Proc. Nutr. Soc.
     25, XL1.

69. Penzes, L. and Mentler, L., (1968). Acta Agronom. Hung. 17,
     195-204.

70. Legault, C., (1970). Ann. Genet. Sel. Anim. 2, 209-227.

71. Salmon-Legagneur, E., (1964). Ann. Zootech. 13, 51-61.

72. Salmon-Legagneur, E., (1964). Ann. Biol. Anim. Bioch.
     Biophys. 4, 49-62.

73. Salmon-Legagneur, E., (1955). Ann. Zootech., 14, No. 1, H.S.,
     pp. 1-137.

74. Lodge, G.A., (1969). Anim. Prod. 11, 133-143.

75. Lodge, G.A., Elsley, F.W.H. and McPherson, R.M. (1966).
     Anim. Prod. 8, 499-506.

76. Elsley, R.W.H., (1968). E.A.A.P. pig commission, Dublin.

77. Elsley, F.W.H., (1970). Nutrition and lactation in the sow
     in Proc. 17th East. School. Agr. Sci. Univ. Nottingham.
     (Falconer, I. R., ed.), 393. Butterworths, London.

78. Elsley, F.W.H. and Livingstone, R.M., (1969) in "Meat
     production from entire male animals" (Rhodes, D.N. ed.), 273,
     Churchill, London.

79. Elsley, F.W.H., and MacPherson, R.M., (1972) in "Pig
     Production" (Cole, D.J.A., ed.), 417-434, Butterworths,
     London.

80. Elsley, F.W.H., Anderson, D.M., McDonald, J., McPherson, R.M.
     and Smart, R., (1966). Anim. Prod. 8, 391-400.

81. Elsley, F.W.H., McPherson, R.M. and Lodge, G.A., (1968).
     Anim. Prod. 10, 149-156.

82. Elsley, F.W.H., Bannerman, M., Bathurst, E.V.J., Bracewell,
     A.G., Cunningham, J.M.M., Dodworth, T.L., Dodds, P.A.,
     Forbes, T.J. and Laird, R., (1969). Anim. Prod. 11, 225-241.

83. Elsley, F.W.H., Bathurst, E.V.J., Bracewell, A.G.,
     Cunningham, J.M.M., Dent, J.B., Dodsworth, T.L., McPherson,
     R.M. and Walker, N., (1971). Anim. Prod. 13, 257-270.

84. Elsley, F.W.H., Kneale, W.A., Lightfoot, A.L., Petchey, A.M.,
     Saul, D.W., Taylor, A.G., Walker, N., Willmot, W.E. and Yeo,
     M.L., (1971). 53th Meet. Br.   Soc. Anim. Prod. 13, 386
     (Abstr.).

85. Baker, D.H., Becker, D.E., Jensen, A.H. and Harmon, B.G.,
     (1968). J. Anim. Sci. 27, 1332.

86. Baker, D.H., Becker, D.E., Norton, H.W., Sasse, C., Jensen,
     A.H., and Harmon, B.G., (1969). J. Nutr. 97, 489-495.

87. Baker, D.H., Becker, D.E., Jensen, A.H. and Harmon, B.G.
     (1970). J. Anim. Sci. 30, 364-367.

88. Baker, D.H., Becker, D.E., Jensen, A.H. and Harmon, B.G.,
     (1970). J. Anim. Sci. 31, 526-530.

89. Rippel, R.H., Harmon, B.G., Jensen, A.H., Norton, H.W. and
     Becker, D.E., (1965). J. Anim. Sci. 24, 203-208.

90. Brooks, P.H. and Cooper, K.J., (1972) in "Pig Production"
     (Cole, D.J.A. ed). 385-398, Butterworths, London.

91. Rerat, A. and Duee, P.H., (1972). Nutrition et reproduction

chez la Truie. Journee du Porc, Pau. Synd. nat. Vet. fr.,
4-25.

92. Salmon-Legagneur, E., (1968). B.O.C.M. Refresher Course
Shuttleworth College.

93. Lodge, G.A.,(1972).in "Pig Production" (Cole, D.J.A., ed.)
399-416, Butterworths, London.

94. Goode, L., Warnick, A.C., Wallace, H.D., (1965). J. Anim.
Sci. 24, 959-963.

95. Bazer, F.W., Clawson, A.J., Robison, O.W., Vincent, C.K. and
Uuberg, L.C., (1968). J. Anim. Sci. 27, 1021-1026.

96. Frobish, L.T., (1970). J. Anim. Sci. 31, 486-490.

97. Schultz, J.R., Speer, V.C., Hays, V.W. and Melampy, R.M.,
(1966). J. Anim. Sci. 25, 157-16.

98. Heap, F.C., Lodge, G.A., and Lamming, G.E., (1967). J. Reprod.
Fert. 13, 269-279.

99. Clawson, A.J., (1969). Proc. Georgia. Nutr. Conf., p.45.

100. Frobish, L.T., Steele, N.C. and Davey, R.J., (1971). J. Anim.
Sci. 33, 1148 (Abstr.)

101. Vermedahl, L.D., Meade, R.J., Hanke, H.E. and Rust, J.W. (1969).
(1969). J. Anim. Sci. 28, 465-472.

102. Buitrago, J., Maner, J.H. and Gallo, J.T., (1970).
J. Anim. Sci. 31, 197 (Abstr.)

103. Frape, D.L. and Hocken, R.W., (1967). Anim. Prod. 9, 547-552.

104. O'Grady, J.F., (1967). Ir. J. Agric. Res. 6, 57-71.

105. Rombauts, P. and Febre, J., (1966). Zeitschr. Tierphysiol.
Tierernähr. Futtermittelk. 21, 91-102.

106. Clawson, A.J., Richards, H.L., Matrone, G. and Barrick, E. R.,
(1963). J. Anim. Sci. 22, 662-669.

107. Frobish, L.T., Hays, V.W., Speer, V.C. and Ewan, R.C., (1966).
J. Anim. Sci. 25, 1249 and 1250 (Abstr.).

108. Nielsen, A.J., (197o). Beretn. Forsogslab. Kobenhavn No. 381,
p. 212.

109. Boaz, T. G., (1962). Vet. Rec. 74, 1482-1493.

110. Pond, W.G., Wagner, W.C., Dunn, J.A. and Walker, J.F.,
     (1968). J. Nutr. 94, 309-316.

111. Pond, W.G., Dunn, J.A., Wellington, G.H., Stouffer, J.R. and
     Van-Vleck, L.D., (1968). J. Anim. Sci. 27, 1583-1586.

112. Strachan, D.N., Walker, E.F., Pond, W.G., O'Connor, J.R.,
     Dunn, J.A. and Barnes, R.H., (1968). J. Anim. Sci. 27, 1157
     (Abstr.).

113. Duee, P.H. and Rerat, A., (1974). Journées Rech. Porcine en
     France, 49-56, I.N.R.A.-I.T.P. ed., Paris.

114. Pike, I.H. and Boaz, T.G., (1969). J. Agric. Sci., Camb. 73, 301
     309.

115. Hawton, J.D. and Meade, R.J., (1971). J. Anim. Sci., Camb. 32,
     88-95.

116. Pike, I.H., (1970). J. Agric. Sci., Camb. 74, 209-215.

117. O'Grady, J.F., Elsley, F.W.H. and McPherson, A.M., (1970).
     Anim. Prod. 12, 374 (Abstr.)

118. Mahan, D.C., Becker, D.E., Harmon, B.G. and Jensen, A.H.,
     (1971). J. Anim. Sci. 32, 482-486.

119. McPherson, R.M., Elsley, F.W.H. and Smart, R.I., (1969).
     Anim. Prod. 11, 443-451.

120. Hitchcock, J.P., Sherritt, G.W., Gobble, J.L. and Hazlett,
     V.E., (1971). J. Anim. Sci. 33, 30-34.

121. Elliott, R.F., Van der Noot, G.W., Gilbreath, R.L. and
     Fisher, H., (1971). J. Anim. Sci. 32, 1128-1137.

122. Holden, P.J., Lucas, E.W., Speer, V.C. and Hays, V.W., (1968).
     J. Anim. Sci. 27, 1587-1590.

123. O'Grady, J.F., (1971). Ir. J. Agric. Res. 10, 17-29.

124. Mahan, D.C., Becker, D.E. and Jensen, A.H., (1971).
     J. Anim. Sci. 32, 470-475.

125. Mahan, D.C., Becker, D.E. and Jensen, A.H., (1971).
     J. Anim. Sci. 32, 476-481.

126. Friend, D.W., (1971). J. Anim. Sci. 32, 658-666.

127. Bowland, J.P., (1967). J. Anim. Sci. 26, 533-539.

128.   Salmon-Lagneur, E., (1969). Journees Rech. Porcine en France, Paris, 77-81, I.N.R.A.-I.T.P. ed.

129.   Aumatre, A., (1969). Communication au Congres de la F.E.Z., Commission de production porcine, Helsinki.

130.   Aumaitre A., (1971). Ann. Zootech, 20, 551-575.

131.   Corring, T. and Saucier, Rlk (1972). Ann. Biol. Anim. Bioch. Biophys. 12, 233-241.

132.   Corring, T., Aumaitre, A. and Rerat, A., (1972). Ann. Biol. anim. Bioch. Biophys. 12, 109-124.

133.   Manners, M.J., (1970). J. Sci. Fd Agric. 21, 333-340.

134.   Moody, N.W., Speer, V.C. and Hays, V.W., (1966). J. Anim. Sci. 25, 1250 (Abstr.)

135.   Lennon, A.M., Ramsey, H.A., Alsmeyer, W.L., Clawson, A.J. and Barrick, E.R., (1968). J.Anim.Sci. 27, 1176 (Abstr.).

136.   Perry, G.C. and Lecce, J.G.,(168). Anim. Prod. 10, 433.

137.   Danielson, D.M., (1968). J. Anim. Sci. 27, 1132 (Abstr.).

138.   Lecce, J.C., (1969). J. Anim. Sci. 28, 27.

139.   Peo, E.R., Jr., Everison, D.M., Wehrbein, G.F., Vipperman, P.E., Jr. and Cunningham, P.J. (1969). J. Anim. Sci. 29, 141 (Abstr.).

140.   Drews, J.E., Hays, V.W., Speer, V.C. and Ewan, R.C., (1967). J. Anim. Sci. 26, 1472 (Abstr.).

141.   Young, L.G., (1967). J. Anim. Sci. 26, 912 (Abstr.).

142.   Hawton, J.D. and Meade, R.J., (1968). J. Anim. Sci. 17, 1775 (Abstr.).

143.   Schneider, D.L. and Sarett, H.P., (1969). J. Nutr. 98, 279.

144.   Hines, R.H. and Koch, B.A., (1969). Simple and complex pig starters compared. Kans. Agr. Exp. Stat. Rep. 151, 23.

145.   Bayley, H.S. and Carlson, W. E., (1970). J. Anim. Sci. 30, 394.

146.   Van der Heyde, M., (1969). Revue Agric. 22, 1419-1428.

147. Aumaitre, A., Du Mesnil Du Buisson, F. and Renoux, E., (1971)
      Bull. Tech. Inf., Paris, 257, 197-204.

148. Aumaitre, A., (1972). Pig Farming, 8, 68-77.

149. Aumaitre, A. and Rettagliatt, J., (1972). Journées Rech.
      Porcine en France, 273-286, I.N.R.A. I.T.P. ed., Paris.

150. Mount, L.E., (1972) in "Pig Production" (Cole, D.J.A. ed.),
      71-89, Butterworths, London.

151. Harvey, D.G., (1967). Metabolic diseases in pigs, in
      management and diseases of pigs, the Royal Veterinary College
      (University of London) and the British Council, 87-94.

152. Heitman, H., Jr., Kelly, C. F. and Bond, T.E., (1958).
      J. Anim. Sci. 17, 62-67.

153. Legault, C., Aumaitre, A. and Du Mesnil du Buisson, F.,
      (1974). 25th E.A.A.P. Pig Commission, Copenhagen 17-21
      August.

154. Lynch, G., (1965). Norges Lanbrukshøgskole Melding No. 192,
      1-6.

155. Moody, N.W., Baker, D.S., Hays, V.W. and Speer, V.C., (1969).
      J. Anim. Sci. 28, 76-79.

156. Smidt, D., Scheven, B. and Steinbach, J., (1965).
      Züchtungskunde 37, 23-35.

157. Self, H.L., and Grummer, R.H., (1958). J. Anim. Sci. 17,
      862-868.

158. Van der Heyde, M., (1971). Intern. Milk Replacer Symposium.
      Zurich, 26-27 May.

159. Te Brake, J.H.A., (1972). E.A.A.P. Pig Commission, Verona,
      Italy 5-9th October.

160. Aumaitre A., Le Pan, J., Rettagliati, J., Bina L. and
      Rousseau, P., (1974). Journées Rech. Porcine en France,
      77-91, I.N.R.A., I.T.P. ed., Paris.

161. Moody, N.W., Speer, V.C., (1971). J. Anim. Sci. 32, 510-514.

162. Palmer, W.M., (1965). Macroscopic and microscopic changes
      in the reproductive tract of the lactating sow. The Ohio
      State Univ., Ph. D. Thesis, 87 pp.

163. Graves, W.E., Lauderdale, J.W., Kirkpatrick, R.L., First, N. L. and Casida, L. E., (1967). J. Anim. Sci. 26, 365-371.

164. Martinat, F., Legault, C. and Du Mesnil du Buisson, F., (1972). Journées Rech. Porcine en France, Paris, 37-43, I.N.R.A.-I.T.P. ed.

165. Epstein, H. and Kadmon, S., (1969). J. Agric. Sci. (Cambridge) 72, 365-370.

166. Martinat-Botte, F., Du Mesnil Du Buisson, F., Bariteau, F. and Mauleon, P., (1974). Journees Rec h. Porcine en France, Paris, 37-41, I.N.R.A.-I.T.P. ed.

167. Cole, H.H. and Hughes, E.H., (1946). J. Anim. Sci. 5, 25-29.

168. Kudlac, E., (1962). Vet. Med. Praha 35, 507-512.

169. Crighton, D.G., (1970). Anim. Prod. 12, 611-617.

171. Legault, C. and Dagorn, J., (1973). Journees Rech. Porcine en France, Paris, 227-237, I.N.R.A.-I.T.P. ed.

172. Martinat, F., Legault C. du Mesnil du Buisson F., Olivier, L. and Signoret, J.P., (1970). Journees Rech. Porcine en France, Paris, 47-54, I.N.R.A.-I.T.P. ed.

173. Salmon-Legagneur, E., (1970). Journees Rech. Porcine en France, Paris, 41-46, I.N.R.A.-I.T.P. ed.

174. Duee, P.H. and Etienne, M., (1974). Journées Rech. Porcine en France, 43-47, I.N.R.A.-I.T.P. ed., Paris.

175. Polge, E.J.C. (1972) in "Pig Production" (Cole, D. J. A. ed) 315-327, Butterworths, London.

176. Etienne, M. and Legault, C., (1974). Journees Rech. Porcine en France, I.N.R.A., I.T.P. ed., Paris 57-62.

177. Legault, C., (1973). Journées Rech. Porcine en France, Paris, 147-154, I.N.R.A.-I.T.P. ed.

178. Du Mesnil du Buisson, F., and Signoret, J.P., (1969). Journées Rech. Porcine en France, Paris; 53-55, I.N.R.A.-I.T.P. ed.

179. Martinat, F., Du Mesnil du Buisson, F. and Bariteau, F., (1972). Journées Rech. Porcine en France, Paris, 45-50, I.N.R.A.-I.T.P. ed.

180. Polge, C., Day, B.N. and Groves, T.W., (1968).
      Vet. Rec. 83, 136-142.

181. Du Mesnil du Buisson, F. and Mauleon, P., (1970). Journées
      Rech. Porcine en France, Paris, 17-27, I.N.R.A.-I.T.P. ed.

182. King, G.J., (1969). J. Reprod. Fert. 20, 551.

183. Boender, J., (1966). World Rev. Anim. Prod. (Special issue
      2), 29.

184. Legault, C., Gruand, J. and Provost, J.P., (1973).
      Journées Rech. Porcine en France, Paris 155-158, I.N.R.A.-
      I.T.P. ed.

185. Girardot, J., Bosc, M. J. and Bariteau, F., (1972).
      Journées Rech. Porcine en France, I.N.R.A., I.T.P. ed,
      Paris, 59-67.

186. Legault, C., (1969). Ann. Gent. Sel. Anim., 1, 281-298.

187. Hunter, R.H.F., (1964). Anim. Prod. 6, 189-194.

188. Anderson, L. L. and Melampy, R. M., (1972) in "Pig
      Production" (ed. Cole, D. J. A.). 329-366, Butterworths,
      London.

189. Bazer, F. W., Clawson, A. J., Robison, O. W., and Ulberg,
      L. C., (1969). J. Reprod. Fert. 18, 121.

190. Dziuk, D. J., (1968). J. Anim. Sci. 27, 673-676.

191. Ollivier, L., (1973). Proc. 13th Intern. Cong. Genet. S 202
      (Abstr.) Berkeley 20-29 Aout.

192. Sellier, P., (1970). Ann. Genet. Sel. Anim. 2, 145-207.

193. Ollivier, L., and Legault, C., (1967). Ann. Zootch. 16,
      247-254.

194. Du Mesnil du Buisson, F., Millanvoye, B., Bariteau, F. and
      Legault, C., 1974. Journées Rech. Porcine en France,
      Paris, 63-70, I.N.R.A.-I.T.P. ed.

195. Desmoulin, B., (1973). Journées Rech. porcine en France,
      189-199, I.N.R.A.-I.T.P. ed. Paris.

196. Desmoulin, B., (1973). E.A.A.P. Pig Commission, Vienne
      (Autriche).

197. Turton, J. D., (1969). In "Meat production from entire male animals".(Rhodes, D. N., ed.), 5. Churchill, London.

198. Vold, E., (1969). Meld. Fra. Norges Landbruksk\u00f8gskole 48, No. 280, Norway.

199. Vold, E., (1970). Meld. Fra. Norges Landbruksk\u00f8gskole, 49, No. 283, Norway.

200. Wismer-Pedersen, J., (1968). Report at the Pig Commission, E.A.A.P., Dublin.

201. Blair, R. and English, P.R., (1965). J. Agric. Sci., Camb. 64, 169-176.

202. Texier, C., Desmoulin, B. and Dumont, B.L., (1970). Journ\u00e9es Rech. porcine en France, Paris, 209-216, I.N.R.A.-I.T.P. ed.

203. Hetzer, H. O. and Miller, R. H., (1972). J. Anim. Sci. 35, 730-742.

204. Eeckhout, W., Bekaert, H. and Casteels, M., (1971). Revue Agric. 24 (1), 41 and (10). 13-63.

205. Jung, H. and Piatkowski, B., (1967). Arch. Tierzucht 9, 399-408.

206. Piatkowski, B. and Jung, H., (1966). Arch. Tierzucht. 9, 307-319.

207. Desmoulin, B., Donneau, M. and Bourdon, D., (1974). Journ\u00e9es Rech. Porcine en France 247-255, I.N.R.A.-I.T.P. ed., Paris.

208. Prescott, J. H. D. and Lamming, G. E., (1967). Anim. Prod. 9, 535-545.

209. Hays, V. W., Speer, V. C., Frobish, L. T. and Ewan, R. C., (1966). J. Anim. Sci. 25, 1278, (Abst.)

210. Walstra, P., (1969) in "Meat Production from entire male animals". (Rhodes, D. N., ed.), 129. Churchill, London.

211. Fowler, V. R., Taylor, A. G. and Livingstone, R. M., (1969) in "Meat production from entire male animals" (Rhodes, D. N. ed.) 51-62, Churchill, J. A., London.

212. Newell, J. A. and Bowland, J. P., (1972). Can J. Anim. Sci. 52, 543-551.

213.   Desmoulin, B., Dumont, B. L. and Jacquet, B., (1971).
       Journées Rech. Porc ine en France, 187-192, I.N.R.A.-I.T.P.,
       ed. Paris.

214.   Horst, P. and Bader, J., (1969). Zuchtungskunde 41, 226 et
       248.

215.   Koch, D. E., Parr, A. R. and Merkel, R. A., (1968).
       J. Food Sci. 33, 176.

216.   Rhodes, D. N., (1969) in "Meat production from entire male
       animals" (Rhodes, D. N. ed.) 189, Churchill, London.

217.   Patterson, R. L. S., (1968). J. Sci. Fd Agric. 19, 31,
       38 and 434.

218.   Patterson, R. L. S. and Stinson, G. G., (1971). 17th Europ.
       Meeting of Meat Res. Workers, Bristol, England, B12, 148.

219.   Martin, A. H., Fredeen, H. T. and Stothardt, J., (1968).
       Can.   J. Anim. Sci. 48, 171.

220.   Jonsson Per and Wismer-Pedersen, J., (1974). Livestock Prod.
       Sci. (1), 53-66.

221.   Maarse, H., Moerman, P. C. and Walstra, P., (1972).
       I.V.O., Rapport C. 180.

222.   Jarmoluck, L., Martin, A. H., and Fredeen, H. T., (1970).
       Can.   J. Anim. Sci. 50, 750.

223.   Arbuckle, J. B. R., (1967). British Council Course, Royal
       Vet. College, London, 117.

224.   Labouche, C., (1972) in "Les maladies animales, leur incid-
       ence sur l'economie agricole (Regards sur la France), 171-179,
       S.P.E.I. ed, Paris.

225.   Betts, A. O. and Beveridge, W. I. B., (1953). Vet. Rec. 65,
       515.

226.   Sharpe, H. B. A., (1966). Brit. Vet. J., 122, 99-211.

227.   Sovljanski, B. and Milosavljevic, S., (1968).
       Vet. Glas, 22, 469-474.

228.   Randall, G. L. B. and Penny, R. H. C.,(1967). Vet. Rec. 81,
       359-360.

229.   Preston, K. S. and Mayrose, V. B., (1963). Modern Veterinary

Practice 44, 48.

230. Ministry of Agriculture, 1959. Vet. Rec. 71, 777-786.

231. Korkman, N., (1947). Act. Agric. Suecana 2, 253-310.

232. Akkermans, J. P. W. M., (1964). Veeteelt Zuivel Berichten 7, 540.

233. Robertson, J. B., Laird, R., Hall, J. K. S., Forsyth, R. J., Thomson, J. M. and Walker-Love J., (1966). Anim. Prod. 8, 171-178.

234. Keller, H., (1971). Schweizer Arch. Tierheilk T. 113, 130-138.

235. Sickel, E., (1971). Deutsche Landwirtschafts Gesellschaft, 880-882, August.

236. Peo, E. R., (1965). Veterinarian 26, 10-13.

237. Sharman, G., Jones, A. V., Deherley, A. and Elsley, F. W. H., (1971). Rec. Vet. Sci. 12, 65.

238. Young, G. A., Caldwell, J. D. and Underdahl, N. R., (1959). J. A. V. M. A., 134, 231.

239. Yeo, M. L. and Chamberlain, A. G., (1966). Proc. Nutr. Soc. 25, XL1.

The Use of Genetic Potential

THE CHOICE OF SELECTION OBJECTIVES IN MEAT PRODUCING ANIMALS

Gordon E. Dickerson

U.S. Meat Animal Research Center
Agricultural Research Service
U.S. Department of Agriculture

INTRODUCTION

In the breeding of animals used for meat production, the choice of biological objectives presumably should be guided primarily by their expected contributions to efficiency (E) in terms of lower total production cost per unit of animal product, C/U. Cost per unit seems more realistic than profit per breeding female per year, $(U^{\cdot}V-C)/YI$, because selling prices (V) for products tend toward a fluctuating margin above production costs, so that lower costs benefit consumers more than producers. The cost and the profit definitions of efficiency lead to the same performance rankings when product price (V) is constant, but selection programs deal with future reductions in relative cost per unit product when price per unit may decline correspondingly. Also note that discounting the value of future gains in efficiency for cumulative interest charges on capital used in breeding programs is justified only for the _excess_ of interest rates over (steady) rates of currency inflation, since rates of inflation affect scales of both the interest and the future cost savings from improved efficiency.

The cost per unit of animal products depends largely upon the efficiency of three basic biological functions: (1) reproduction, (2) female production and (3) growth of the young. Total costs can be separated into those for (1) maintaining the female population and (2) growing progeny to market size. Volume of animal product also arises from these same two sources: (1) directly from the breeding females, e.g., as milk, wool or eggs and (2) from progeny, e.g., as meat. Net or life-cycle economic efficiency is the ratio of total costs to total animal product (in economic equivalent units) from breeding females and their progeny over a representative period of time.

A difficult problem for animal breeders is to determine which biological components offer greatest opportunity for genetic improvement in net efficiency of production. Too often, in the history of animal breeding, objectives have been matters of fashion or ease of producing genetic change, without sufficient examination of net effects on life-cycle economic efficiency of production. This has led to fads and cyles, particularly in size and conformation. Even in pilot selection experiments with laboratory species, effects on life-cycle efficiency of meat production have not been examined. Our present purpose is to re-examine a rationale (1) for choice of selection objectives in meat producing animals and to propose some tentative choices.

RATIONALE FOR BREEDING OBJECTIVES

What kinds of biological and economic information are needed to choose the combination of breeding objectives most likely to improve net efficiency of meat production? Because maximum improvement in breeding value for net-efficiency (G_E) is the objective, ultimate information desired is relative "scorecard" importance of each potential objective (P_i) in predicting genetic differences (G_E) in efficiency (technically, the standard partial regressions of G_E on the P_i; see Figure 1). Of course, actual choice of selection criteria would need to consider also relative costs of measuring and using each objective in selection. Several kinds of information are required for such evaluation of relative importance of components (P_i) in predicting G_E, i.e., in choosing meaningful selection objectives.

Economic Importance

Basic information from production accounting analysis is the direct effect on efficiency ($E=C/U$) per unit of change in each component, when other components remain unchanged (b_{EP}). These are partial regressions and permit definition of efficiency in terms of its components as –

$$E = \sum_1^k (b_{EP_i} \cdot P_i) + \text{Constant}$$

These same economic weightings apply in defining breeding value for efficiency as –

$$G_E \sum_1^k (b_{EP_i} \cdot G_i) + \text{Constant}$$

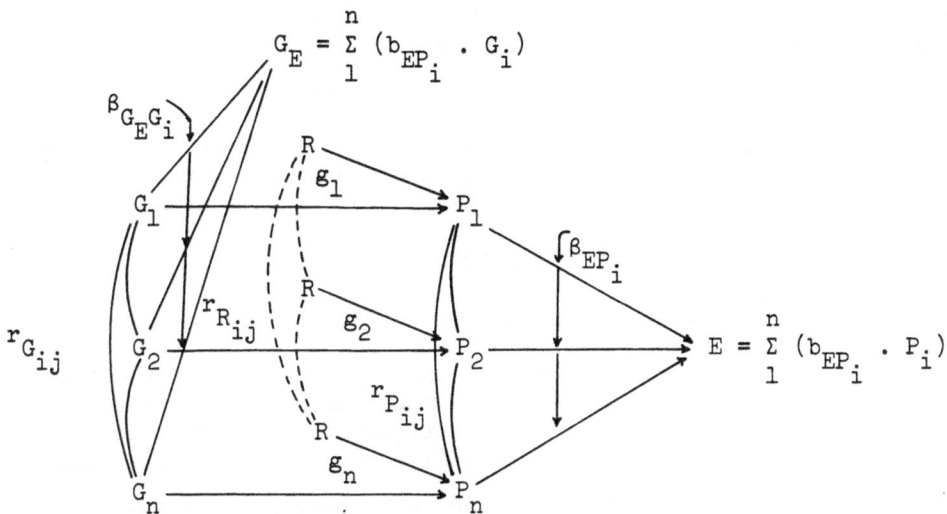

Figure 1. Path coefficient diagram of associations among
phenotypes (P_i) and genotypes (G_i) for several (n) components
of phenotype (E) and genotype (G_E^i) for productive efficiency.

Even for components which have curvilinear effects on efficiency
(e.g., size of litters reared), the linear weightings appropriate
for the population mean can be used and adjusted as the mean
changes.

Variation

The direct and indirect importance of phenotypic variation
in any component (P_i) in predicting phenotypic efficiency (E) is
proportional to the product of its standard deviation and its
unit economic effect on efficiency, $\sigma p \cdot b_{EP}$. Similarly,
importance of genotypic variation in each component in predicting
breeding value for efficiency (G_E) is proportional to the genotypic
standard deviation and unit effect, $\sigma G_i \cdot b_{EP_i} = g_i \cdot \sigma P_i \cdot b_{EP_i}$.
The latter is certainly a major consideration in choosing selection
objectives but the ratio of genetic to phenotypic variation
($g_i = \sigma_{G_i} / p_i$) also directly affects the breeding value importance
of phenotypic variation in each component, $g_i \cdot \sigma_{G_i} \cdot b_{EP_i} =$
$g_i^2 \cdot \sigma P_i \cdot b_{EP_i}$. This applies rigorously only for components
which are uncorrelated. It says that usefulness of phenotypic

biological components of efficiency in breeding programs is proportional to the product of variability (p_i), heritability (g_i^2) and unit economic importance (b_{EP_i}). Potential usefulness of physiological and biochemical indicators of efficiency components or of genotypes for blood groups, proteins or enzymes can be quantitatively and dispassionately evaluated in this manner.

Correlation

Correlation among phenotypes (rp_{ij}) and genotypes ($r_{G_{ij}}$) of different components of efficiency can modify not only the optimum weighting of the components but also the over-all effectiveness of selection for efficiency. Genetic antagonisms between major components of efficiency (e.g., negative rG_{ij} of growth rate with maintenance cost of breeding females) can sharply limit the genetic variation in net efficiency and hence the opportunity for significant improvement. Positive genetic correlation among components indicates some degree of duplicating or joint effect on efficiency and hence leads to appropriately reduced independent emphasis on each (i.e., proportional to $\beta_{G_E \cdot P_i}$).

Environments

Environments include differences among production and marketing regimes as they influence relative economic importance of components (b_{EP_i}) and the variation in and correlations among components of efficiency. Examples would include climate, housing, feed resources, early weaning of young, market preferences.

Breeding Systems

Selection operates between breeds or breed crosses as well as among individuals or families within such populations. In selection among means for different breeds or breed crosses, accuracy of selection can be much higher than among individuals within breeds (i.e., $g_{\bar{P}}^2 \gg g_{P_i}^2$) and the relative attention to different components will be determined automatically by magnitude of breed differences in each component ($\sigma_{\bar{P}_i}$) and their economic weightings (b_{EP_i}). Here the primary concern is having correct economic weig ings and direct comparison of economic efficiency (E) may be the best alternative.

Commercial breeding systems can have a marked effect on

choice of selection objectives. In breeds used primarily to supply males for terminal-sire crossing, from which all progeny are slaughtered, much less attention to reproductive and maternal components is warranted than in breeds used to produce commercial crossbred females. However, in existing or new breeds to be used primarily as purebreds or in sire-breed rotation cross-breeding, a balanced combination of individual and maternal components of efficiency is desired since the same genotype will be represented in both breeders and market animals.

Cost of Information

Alternative measures of efficiency components may differ greatly in cost or in the intensity of selection permitted as well as in accuracy for prediction of breeding value for efficiency. For example, carcass yield or meat quality information may be more accurate than live animal indicators but requires slaughter of animals, reduces intensity of selection and can only be used in family or progeny test selection. Such alternatives can be compared in terms of expected genetic change in efficiency per unit of time.

Usefulness of endocrine, other biochemical or immunological indicators can also be examined in terms of additional economic progress expected in efficiency relative to added costs for indicator information.

TENTATIVE CHOICES

Much of the information necessary to make effective choices of selection objectives for improving efficiency of meat production is still incomplete. This is especially true concerning the extent of (1) genetic antagonisms among components of efficiency and of (2) genetic variation in maintenance requirements per unit of metabolic size and in nutrient utilization for synthesis of fat and protein tissues. However, rough initial definition of selection objectives for meat producing animals may stimulate their refinement.

Bio-Economic Model

A much oversimplified model (1) will be used as a starting point (Formula 1). The items of cost and of product represented in this formula illustrate how cost per unit of product value (Efficiency, E) is influenced by each variable. The partial regression of such a measure of efficiency on each component indicates its direct linear effect on efficiency (b_{EP_i}), as outlined earlier.

$$\frac{\text{Expense/year}}{\text{Product year}} = \frac{(A/Y)+(I_d+B_d\cdot F_{md}+F_{pd})+N}{P_d\cdot V_d} + \frac{D(I_o+B_o\cdot F_{mo}+F_{po})+S_o}{N\cdot P_o\cdot V_o}$$

For breeding female   For her progeny

Where:

A/Y = (Cost, young female – value, old female)/years in production.

I_d = Yearly fixed costs/female, for labor, housing, etc.

B_d = Metabolic body size of female, relative to population mean.

F_{md} = Average maintenance feed costs/female/year for population.

F_{pd} = Feed cost above maintenance/female/year.

N = Number progeny reared/female/year.

D = Days from weaning to market weight for individual.

I_o = Average fixed costs/animal-day.

B_o = Average postweaning metabolic body size for individual, relative to population mean.

F_{mo} = Average maintenance feed cost/animal-day for population.

F_{po} = Average feed costs above maintenance/day for individual.

S_o = Fixed costs/animal for slaughter, marketing, vaccines, etc.

P_d = Yearly volume of product/female.

V_d = Value per unit of female product.

P_o = Live weight of meat animal when marketed.

V_o = Value per unit of live weight.

Reproductive rate. Increasing the number of progeny reared per female per year (N) reduces curvilinearly all breeding female costs per offspring marketed or kept for replacement in proportion to $1/N$, except as it increases female feed costs above maintenance (F_{pd}) for gestation and suckling and lengthens the postweaning feeding period (D) due to smaller size at weaning. The relative economic gain in changing N from 1 to 2 in cattle or sheep is 7 or 8 times greater than in changing N from 14 to 15 in pigs and nearly 40 times greater than in changing N from 79 to 80 in meat chickens. There is no doubt of the potential economic gains from increasing reproductive rate, particularly in cattle and sheep. The difficulties are in the limited initial genetic variation in cattle plus the low heritability of variation and the dependence upon improved nutrition and protection for mother and young in both species. In swine and poultry variability in reproductive rate is no problem, but heritability of litter size is low in swine unless there is adjustment for varying competition (litter size) effects on development of breeding females.

Growth of young. Efficiency of growth to market weight (P_o) in meat animals is represented by days fed for fixed per day costs ($D \cdot I_o$) plus total feed required for maintenance $D(B_o \cdot F_{mo})$ and weight gain $D \cdot F_{po}$, divided by market weight adjusted for composition or value per unit ($P_o \cdot V_o$). The market weight (P_o) used presumably should be in the optimum range where further increase in weight reduces carcass value (V_o) and hurts feed conversion $D(B_o F_{mo} + F_{po})/P_o$ more than it reduces initial parent female costs per unit of weight marketed ($1/P_o$). Thus, faster growth also will mean heavier optimum market weight (P_o) and larger mean metabolic body size (B_o) shifting the benefit of faster gain largely to reduced female costs per unit of weight marketed. However, to the extent that faster growth also means larger adult female metabolic body size (B_d), female maintenance feed per offspring will increase and optimum market weight will increase again, diminishing the net gain in efficiency even further. The desired objectives (Figure 2) are more efficient growth to market weight (P_o), and higher meat values (V_o), accompanied by earlier sexual maturity to reduce replacement costs (A) and lengthen productive life (Y) and minimum increase in size of breeding females (B_d).

The adverse effect of strong positive association between growth rate and size of breeding females is much more serious when reproductive rate (N) is low, as in cattle, because initial breeding female costs are such a large part of total cost per market animal. This limitation on the net gain in efficiency from increasing growth rate can be partially avoided in commercial production by mating males of rapidly growing, large breeds with females of smaller maternal breeds, the limitation being increased dystocia, calf mortality and delayed rebreeding. Another possible approach is "bending" the growth curve to achieve more efficient growth to market weight and earlier sexual maturity with minimum increase in size of breeding females. The earlier sexual maturity and first reproduction is necessary in order to avoid increasing total feed requirements to first reproduction. How much net increase in efficiency can be accomplished, including reduced fatness (i.e., improved meat value, V_o) of market animals?

Others (e.g., 2,3) have recently reviewed evidence from laboratory mammals concerning genetic variation in rate and composition of growth and associations with age at puberty and mature size. Some relevant conclusions are:

1. Age increase in total cell number (DNA) is essentially linear

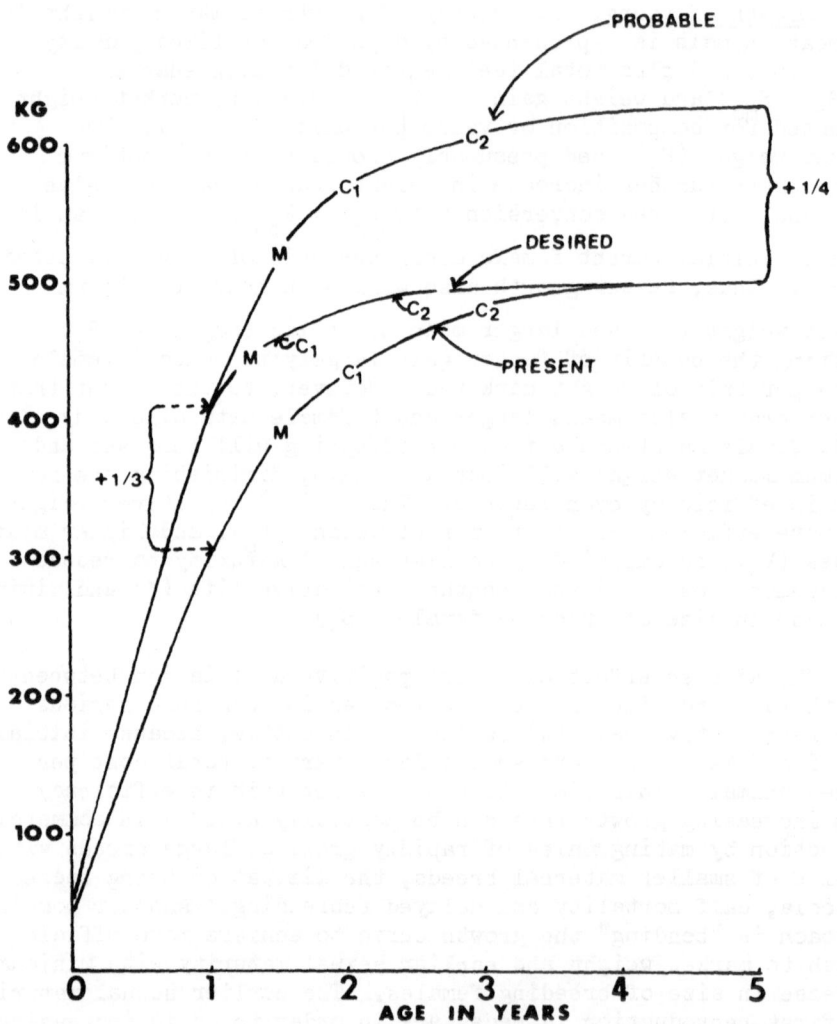

Figure 2.   Desired changes in growth curve of cattle affecting
optimum market weight (M), age at 1st calving (C_1) and mature
size relative to probable change from 1/3 increase in yearling
weight.

during last 1/3 of gestation, then asymptotes towards plateau after puberty (4).

2. Mean cell size increases at an ever decreasing rate, beginning 2/3 through gestation and asymptoting toward near maximum size at puberty (4).

3. Fat replaces water in body tissues with advancing age, especially after puberty (2).

4. Genetic fraction of variance in body weight increases somewhat during late preweaning period (2 to 3 weeks of age) but nearly doubles from weaning to pre-puberty (3 to 7 weeks of age). (See 5).

5. Maternal environment is the major source of variance in body weight before weaning (2/3) but its influence declines towards a negligible fraction of total variance at and after puberty (1/10 to 1/30). (See 5).

6. After puberty, an increasing fraction of weight gain and of genetic variance in gain is in fat deposition (2,6).

7. Selection for increased rate of pre-puberal gain (7,8,9) in body weight is effective but leads to a proportional increase in mature size and weight at puberty, to little change in age at puberty, to less increase in relative growth or gross food conversion to a constant age than to a constant weight and to little change in the intrinsic efficiency of tissue synthesis. Correlated increases in fat content of gain are greater when growth is measured to post-puberal ages (6).

These results from laboratory mammals indicate that selection intended to increase rate of lean tissue growth is likely to be most effective when rate of growth is measured to late pre-puberal age or weight. Increased pre-puberal rate of lean growth of course involves increased food intake or appetite, which would mean increased fat deposition after puberty if the elevated appetite continues under ad lib. feeding after puberty. In meat-producing animals, food intake of breeding females is usually restricted after puberty or earlier, and the market animals are slaughtered when desired degree of fatness is reached. The major limitation of selection for rate of pre-puberal growth then is its correlated effect on metabolic size of adult breeding females (B_d), which limits the net gain in life-cycle efficiency of meat production, especially in cattle and sheep in which maternal costs are such a large part of total costs per meat animal marketed.

A constantly recurring question is whether feed consumption

records are necessary in selecting for efficiency of growth in
meat animals. Possible answers are visualized more easily in
terms of growth during a fixed interval of <u>weight</u> rather than
age so that the effect of more rapid growth on feed per unit of
gain or on relative growth is not diminished by the corresponding
increase in mean maintenance body weight (B_o) to a given <u>age</u>
endpoint. For a constant weight interval (^oP_o), e.g., post-
weaning to near optimum market weight, individual costs (C_o) per
unit of <u>individual</u> <u>gain</u> in meat value ($\Delta P_o \cdot V_o$) are represented
in Formula 2 as:

$$E_o = C_o / \Delta P_o \cdot V_o = \frac{D(I_o + B_o \cdot F_{mo}) + \Delta P_o \quad \ell \cdot b_L + (1-\ell) \cdot b_F}{\Delta P_o \cdot V_o} \qquad (2)$$

Where:

ℓ = Proportion of non-fat tissue in live weight gain.

$(1-\ell)$ = Proportion of fat tissue in live weight gain.

b_L = Feed cost over maintenance per unit weight of non-fat gain.

b_F = Feed cost over maintenance per unit weight of fat gain.

Other symbols are as in Formula 1.

      Primary variables are days fed (D) and proportion of non-fat
gain (ℓ) except as both metabolic weight (B_o) and value per unit
of gain (V_o) are similarly affected by ℓ. Other quantities are
either fixed (ΔP_o and I_o) or assumed to be nearly constant (b_L
and b_F). All costs are for feed intake except those for fixed
non-feed costs per animal-day (I_o). If this model is nearly
correct, rate of growth ($\Delta P_o/D$) and proportion of non-fat gain
(ℓ) would account for most of the variation in efficiency of post-
weaning growth, primarily in feed for maintenance ($D \cdot B_o \cdot F_{mo}$),
but secondarily from variation in lean content of gain to the
extent that feed cost for synthesis of fat tissue exceeds that of
lean, i.e., $\Delta P_o \{\ell \cdot b_L + (1-1)b_F\} = \Delta P \{b_F - 1(b_F - b_L)\}$.
Measurement of feed consumption would be justified only to the
extent that appreciable genetic differences do occur in basal
metabolism and activity (F_{mo}) or in efficiency of fat and lean
synthesis (b_F and b_L). If increased growth rate ($\Delta P_o/D$) and
lean content of gain (ℓ) are considered adequate and are used in
selecting for efficiency of postweaning growth to market weight,
the emphasis on higher lean content of gain (ℓ) would be expected
to increase mature size (B_a) and delay puberty of females even
more than selecting for rate of gain alone.

If genetic variation in F_{mo}, b_F and b_L are important enough
to justify selecting directly for lower total feed consumption
($D . F_o$) and fewer days ($D . I_o$) during a specified body weight
interval, response in rate of growth and lean content of gain
could be quite different from selecting solely for rate of
gain and lean content — more efficient meat producers without
much change in body size, perhaps? Except in pigs, most direct
selection for better food conversion has used a constant age,
rather than weight, interval, a procedure which is self-defeating
because of the confounding of faster gains with heavier mean
weights on test (i.e., with higher maintenance feed required).

   Production rate of females. Higher inherited rate of female
production ($P_d.V_d$), e.g., of milk or wool, directly reduces
costs per unit of product for female replacements (A/Y), fixed
"per-head" items (I_d) and also for body maintenance feed (F_{md}),
except as metabolic size (B_d) is changed. The ratio of feed costs
above maintenance to production $F_{pd}/P_d.V_d$) is less likely to
change in composition of product alters value per unit (V_d) or
F_{pd}/P_d. Unless increasing production will also increase female
fixed and feed costs ($I_d + F_d$) more than value of product, i.e.,
(I_d+F_d) $> P_d.V_d$, a valid economic objective is increasing total
product value per female with minimum increase in body size and
fixed non-feed costs. In suckling herd management of meat
animals, increased milk production has its value in reducing
post-weaning feed costs for offspring, $N . D . F_o$ without
incurring the fixed female costs (I_d) required in market milk
production.

Application

   Development of specific criteria for selecting among or
within breeds of each species to improve performance of general
purpose or specialized maternal or paternal crossing stocks under
differing management-marketing situations is a major undertaking
and beyond the scope of this paper. However, some of the principles outlined above will be briefly illustrated for beef
(suckler management) cattle populations being selected to improve
purebred or rotation-cross performance.

   Selection within breeds. The relative importance of only
two traits will be considered: frequency (%) of twin births (T)
and yearling weight of calves (Y). Using economic values from an
earlier analysis (10) and assuming 80% fertility and 20% annual
replacement of cows, relative importance (R) of Y and T for genetic

improvement of net returns per cow-year would be roughly as follows for several levels of twinning:

Trait (P_i)	σ_{P_i} x	$g_{P_i}^2$ x	b_{EP_i}	=	R
Y (kg)	36	0.4	$0.23		$3.31
T (P=2%)	14	0.2/2	1.8(0.88)(0.8)/2		1.12
(P=4%)	20	0.2/2	0.64		1.28
(P=6%)	24	0.2/2	0.64		1.54

The R for Y is (.80-.20) x $5.50 = $3.31, where $5.50 represents net gain considering correlated effects on birth weight, calf losses and cow size (10). The R for T assumes g_T^2 = .20, divided by 2 because selection is for performance of the selected animals' dam. Also, it was assumed that an increase of 1% in twin births will increase net returns per cow-year by 1% of 180 kg of calf weaned worth $0.88/kg for 80% of all cows maintained but that about one-half of this increase will be offset by higher costs for cow and calves to reach 180 kg weight at weaning and possible delayed rebreeding. This rough approach suggests that selection for twinning deserves consideration, especially if the initial frequency of twinning is 4 to 6% or higher.

In maternal breeds or strains intended primarily for production of crossbred heifer replacements, the emphasis on yearling weight (Y) could justifiably be reduced by about one-half relative to twinning (T). In paternal or terminal-sire breeds, little attention to twinning would be justified.

Selection among breeds or crosses. If mean performance of breeds or crosses is characterized rather accurately, heritability of means ($g_{P_i}^2$) will be much higher than for individuals within breeds. Also, the range (ΔP_i) among breeds compared may be substituted for σ_{P_i} in evaluating relative importance of traits in selection. For example, the range among cattle breeds in yearling weight may be very large relative to the range in mean twinning %:

Trait	ΔP_i x	g_p^2 x	b_{EP_i}	=	R
Y (kg)	150	0.9	$0.23		$31.05
T (%)	2	0.8	0.64		1.02
	4	0.8	0.64		2.04
	6	0.8	0.64		3.07

Again, emphasis on Y would be less in maternal breeds and more in paternal breeds, relative to T. However, the small range in mean twinning rate among existing breeds would make twinning rate a definitely secondary consideration in selecting among breeds or crosses, unless the net economic advantage of rapid growth to yearling age is much less than estimated (10).

The examples given above ignore possible phenotypic and genetic association among traits and omit consideration of other potentially important components of efficiency. Comprehensive consideration of selection objectives would include such associations for all traits affecting efficiency and relative importance would need to be measured by the respective <u>standard</u> partial regressions ($\beta_{G_E \cdot P_i}$) of breeding value for efficiency on phenotypes for component traits. Under dairy management, of course, higher milk production is a major objective.

<u>Species differences</u>. Sheep differ from beef cattle primarily in the much greater existing variability in frequency of multiple births both within and among breeds (e.g. 11), which justifies much greater attention to reproductive rate in attempting to improve efficiency of meat production in sheep. Earlier sexual maturity, longer breeding season, carcass composition, rapid lean growth to heavier weights and wool yield also merit consideration (12).

In swine and meat birds, mean reproduction rate is so much higher than in cattle or sheep that' economic gain from a further increase is smaller relative to that from increasing efficiency of individual lean growth (13,14), but still merits attention in maternal crossing stocks.

SUMMARY

Selection objectives in meat producing animals can be chosen to maximize expected improvement in the net efficiency of meat production. Net efficiency is defined in terms of cost per unit of product, as affected by efficiency in reproduction, in female production and in growth of young market animals. The criterion for relative emphasis in selection is the <u>standard</u> partial regression of breeding value for net efficiency on each potential phenotypic component of performance. For independent components, relative emphasis on each trait reduces to the product of variability x heritability x partial regression of net efficiency on phenotype. Relative emphasis on components may differ considerably with management, marketing and breeding systems. Major biological objectives for reducing production costs per unit of animal product are (1) more efficient lean growth to market weight and earlier sexual maturity, with minimum increase in

birth weight or mature size, especially in cattle, (2) higher
rate of reproduction, especially in cattle, (2) higher rate of
reproduction, especially in cattle and sheep and (3) higher rate
of female production (e.g., milk, wool, eggs) relative to
metabolic body size.

REFERENCES

1. Dickerson, Gordon. (1970).   J. Anim. Sci. 30, 849.

2. Eisen, E. J. (1974). Unpubl. Amer. Soc. Anim. Sci. Growth
    Curve Symposium, Univ. Md., College Park.

3. Robertson, Alan. (1974). "Growth rate, appetite, body
    composition and efficiency." Unpublished mss.—personal
    communication.

4. Winick, M. and Noble, A. (1965). Devel. Biol. 12, 451.

5. El Oksh, H. A., Sutherland, T. M. and Williams, J. S. (1967).
    Genetics 57, 79.

6. Bakker, H. (1974). Meded. Landbouwhogeschool 74, 8.
    Wageningen, Netherlands.

7. Lang, B. J. and Legates, J. E. (1969). Theor. Appl. Genetics
    39, 306.

8. Wilson, S. P. (1973). J. Anim. Sci. 37, 1098.

9. Hanrahan, J. P., Hooper, A. C. and McCarthy, J. C. (1973).
    Anim. Prod. 16, 7.

10. Dickerson, G. E., Kunzi, N., Cundiff, L. F., Koch, R. M.,
    Arthaud, V. H. and Gregory, K. E. (1974).
    J. Anim. Sci. 39, 659.

11. Bradford, G. E. (1972). J. Reprod. Fert., Suppl. 15, 23-41.

12. Bowman, J. C. (1966). Anim. Breed. Abstr. 34, 293.

13. Smith, C. (1964). Anim. Prod. 6, 337.

14. Clayton, G. A. (1974). Worlds Poul. Sci. J. 30, 290.

USING THE WORLD'S GENETIC RESOURCES

I. L. Mason

Food and Agriculture Organization of the
    United Nations
Rome

The work of FAO is concerned largely with the developing
countries. For many people in many of these countries the
consumption of animal protein is below the desirable physiological
minimum. The consumption figures in Fig. 1 tell their own story.
The FAO recommended minima are based on an average daily intake of
12.5 MJ and on the premise that protein should supply at least 10
percent of the energy intake and 30 percent of the protein should
be of animal origin. "Other animal protein", of course, includes
fish and dairy products. In studying this diagram remember that
Mauritius stands for most of the other countries of Africa, India
for all the countries of southern Asia, and Mexico for most of the
other countries of Latin America (with the notable exceptions of
Argentina and Peru). Bear in mind also that these figures are
averages so that some people are eating much less and some much
more than indicated here.

This diagram also shows clearly that many people (especially
in Europe, North America and the Southwest Pacific), eat much more
meat than is physiologically necessary. I am frankly not interested
in using the world's genetic resources to increase a consumption
which is already excessive. Therefore this paper will concentrate
on the situation in the developing countries. Indeed I would go
so far as to say that the consumption in the rich countries should
be reduced if that can help to increase consumption in the deficient
countries.

I notice that Mr. Peart, the U.K. Minister of Agriculture,
does not agree with me; he wants people to eat more meat in
Britain. This is excellent if it is the undernourished (rather
than the already over-fed) who are enabled to do so. In Norway,

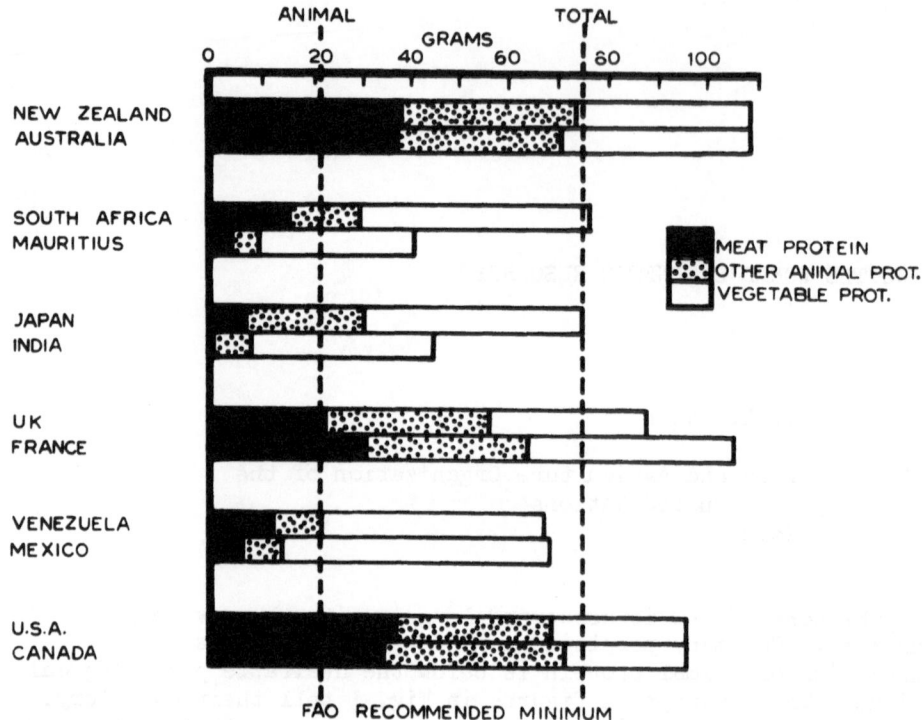

Fig. 1.   Protein contents of the food supplies available per
caput per day in selected countries (1,2).

on the other hand, it is now government policy to reduce excessive
meat consumption, not merely for idealistic reasons, but in terms
of enlightened self-interest - in order to reduce the load of
diseases such as obesity, heart disease and atherosclerosis.

Possibly the "mountain of beef" which accumulated in Europe in
1974 indicates that consumers are beginning to realize that they
have been pressurized into over-consumption and, now that the
price is so high, they are realizing that they have been keeping
up with the Jones' rather than meeting a physiological need or
achieving a "unique gustatory experience". Perhaps consumers in
the over-privileged countries may now be led to realize that the
exciting cuisines of China, Italy, Britain, and even France, were
designed to make a little meat go a long way.

ALTERNATIVE LIVESTOCK

This conference has naturally discussed chiefly cattle and
sheep, pigs and poultry. I would like to draw attention to other

animals which are important sources of meat in some countries. The figures in Table 1 show that the very low levels of protein intake, and particularly meat protein, apply not only to the countries illustrated in Fig. 1, but to very large areas of the globe. They also show that up to 24% of the meat consumed in some areas comes from animals not included in the "big four". If it were possible to obtain figures separately for town and country, the importance of "other animals" would be even more striking since it is the subsistence farmer or the farmer producing beef for sale who consumes most of the game and rodents, as well as the poultry and goats.

There are also great differences between countries. For France, Italy, Argentina, Brazil and Mongolia the "other animals" are largely horses. In Somalia and Sudan camels figure predominantly. In Peru 4% of meat comes from lamas and alpacas and 5% from guinea pigs. In the Amazon countries large rodents are exploited such as the capybara.

Table 1

Relative importance of different kinds of meat in the supply of meat protein. Other meat includes camel, horse, rabbit, game, reindeer, edible offal (3).

| Region | Meat protein per caput g/day | % supplied by | | | | |
		Beef and veal	Sheep and goat	Pig meat	Poultry meat	Other meat
Far East	3.3	22	7	45	15	11
Africa	5.5	50	16	3	7	24
Near East	5.7	36	42	0	4	18
Latin America	12.9	64	4	13	6	13
Europe	18.3	39	6	34	9	12
North America	36.4	50	2	22	20	6
World		42	5	29	13	11

In West Africa and parts of Southern Africa the most important
"other" animal is game. The list in Table 2 shows the incredible
list of species which has been recorded from markets in Ghana. In
some parts of the country as much as 78% of local meat may come
from wild animals, particularly cane rats or grass cutters. Likewise, in the northern Ivory Coast people eat more bush meat than
meat from domestic livestock (4).

This is a genetic resource which is urgently in need of
conservation. Indeed all the large game animals have already been
exterminated in West Africa. The variety of the list is due partly
to taste (like ours for pheasant or venison) but also to economic
necessity. While human population control is the only way to
alleviate this pressure on animals who have just as much right to
live on the earth as we have, our immediate job is to try to
increase the local production of meat from small domestic animals.

In Eastern and Southern Africa game control is more effective
so that some big game remains and the traditional hunting of the
local people must be categorized as "poaching". In some countries
in Southern Africa, game is an extremely important item - in
Botswana and Zaire, for instance, it accounts for about two-thirds
of the total meat consumed (see Table 3). Elsewhere the current
tendency is to replace hunting by the controlled cropping of game
in natural reserves or on ranches. On many ranches antelope are
now kept under control and some are on the way to domestication.
Certainly they have many advantages over cattle in semi-arid or
tsetse areas. They are more resistant to water shortage and to
trypanosomiasis and other diseases and they eat a much wider range
of plants so that overgrazing does not become a problem so quickly.
Furthermore, their carcasses have a lower fat content. Antelope
can have a higher reproductive rate and a higher dressing out
percentage than the local cattle, and a higher growth rate than
domestic ruminants of comparable adult size. In Southern Africa
eland, springbok, blesbok, kudu and impala are now being exploited

Table 2

Some wild animals eaten in Ghana (4)

Rodents - hares, rats, porcupines, mice, squirrels
Antelopes - duikers, bushbuck, royal antelope
Bats - fruit bats
Anteaters - pangolins and aardvark
Carnivores - civet and domestic cats, mongoose
Primates - all monkeys and chimpanzees, bush baby
Other mammals - hyrax, bush pig
Birds - including birds of prey, sunbirds, herons and egrets
Reptiles - tortoises, turtles, lizards, snakes
Invertebrates - ants, beetle larvae, giant snails

for meat. These species number over 1 million head on farms,
national parks and nature reserves in South Africa and Namibia,
and their numbers are increasing (6).

Clearly this is a resource which must be conserved and
exploited not only in Africa but also throughout the world. For
instance, in Scotland the productivity of red deer is being compared
with that of sheep. Even the Australians after years of sacri-
ficing their interesting feral animals for the benefit of the
Merino and the Shorthorn, are now beginning to explore and exploit
them rationally. The buffalo and the banteng of the Northern
Territory are being retamed and used for meat. As for kangaroos,
if it is necessary to continue to kill them at all, I hope their
carcasses also can be used as food instead of being left to rot
by the roadside.

However, care must be taken in any domestication programme.
Control will lead to the possibility of artificial selection. If
this is applied solely to growth rate and carcass quality under
improved conditions of feeding and management, the wild animals
may well lose the advantages in hardiness and disease resistance
which now make them interesting. The need for fencing, housing,
feeding, and disease control will eventually render them much less
economic than our present cattle and sheep.

Goats

The minor species of livestock are neglected because they lack
prestige – they have no religious or ritual value and they do

Table 3

Rural and urban meat consumption in six southern
African countries and the importance of game meat (5)

| Country | Consumption (kg /head/year) of | | Game animals |
| | Domesticated mammals | | |
	Urban	Rural	
Botswana	19.5	8.0	13.8
Malawi	9.3	3.9	–
Swaziland	44.4	19.7	–
Tanzania	23.0	8.9	3.4
Zaire	7.7	1.8	8.7
Zambia	21.2	5.2	2.2

not get into the western textbooks. I shall discuss only two
species — goat and buffalo. The importance of goat meat in
particular is concealed by the statistics which lump it with
mutton (of Table 1). In fact it is often much more important
especially in North and West Africa and in southern Asia (see
Table 4).

Goat meat is disparaged because of the smell of entire males
and because it lacks fat. This latter characteristic should be
an advantage. Indeed French (11) writes: "Provided young kids
are well fed, the meat produced can be delicious, palatable,
tender and attractive". Certainly goat meat is highly prized in
Sudan, Pakistan, Malaysia, Fiji and Ceylon to mention only a few
countries. In none of these areas has there been a systematic
selection for improvement in meat production (12).

Such programmes are overdue. However, they must not sacrifice
the features of the goat which make its meat the only one within
economic reach of so many consumers, namely its hardiness,
catholicity of taste in food, prolificacy and independence. Of
particular importance are its resistance to drought and its
ability to thrive on sparse vegetation.

Even in its unimproved state the goat is important economically,
French (11) writes: "Recent studies in the Peleponnesus hillside
villages have demonstrated that the proportionate monetary return
per year from goats, sheep and cattle respectively gave the
following ratios: 122:100:72. Similarly, data from Pakistan and

Table 4

The contribution of goats to meat consumption
in some tropical and subtropical countries

Country	Meat from goats (%)	Reference
Libya	36	7,8
India	35	7,9
Niger	35	10
Nigeria	22	3
Morocco	16	7,8
Turkey	16	7,8
Cyprus	14	7,8
Iraq	12	7,8
Indonesia	11	7,8

and Lebanon showed the revenue from goats to be higher than that
from sheep. In Venezuela, the policy of goat elimination, introduced approximately 20 years ago, has had to be revised and relaxed
and the Government is now studying goat productivity under more
controlled conditions." Of course these goats would be producing
milk as well as meat.

I am pleased to say that FAO has active projects for stimulating goat production in Kenya, Fiji, Argentina and Brazil;
projects for Fiji and West Africa are in the planning stage. In
addition, we are advising several countries (currently, for
instance, Thailand and Malaysia) on planning goat improvement and
production programmes.

Buffalo

By buffalo, I mean, of course, the water buffalo of southern
Asia where they number nearly half as many as the cattle. Indeed,
in several countries of southeast Asia, as well as in Egypt,
buffaloes outnumber cattle. However, just as goat meat is con-
founded with mutton so the statistics conceal buffalo meat under
the heading "beef". Nor is it easy for the consumer to distinguish
between the two. Nevertheless, in India he can be fairly sure he
is eating buffalo beef; and a visit to the docks and slaughterhouses
in Hong Kong immediately indicates their importance there.

Table 5

Comparison of young male buffaloes and local (Jenubi)
cattle in Iraq; the former were about 12 and the
latter about 15 months old. They were fed for
4 months on alfalfa and wheat straw ad libitum plus
34% of the feed intake as concentrate fed
according to body weight.

	Cattle	Buffalo
Number	10	10
Initial weight (kg)	122	200
Daily gain (g)	889	1163
Feed utilization (kg TDN/kg gain)	4.60	4.32
Carcass weight (kg)	117	162
Dressing percentage	51.2	48.4
Hide as % live weight	7.8	10.8
% lean meat in 3-rib sample	61.9	52.9
% fat in " " "	17.2	24.5

The buffalo is ideally suited for meat production in hot humid climates whether as an adjunct to milk production (e.g. in India) or to traction (e.g. in south-east Asia). Some figures comparing their productivity with that of cattle are given in Tables 5-7. They are not very satisfactory since they are based on small numbers: also they appear to be contradictory - the results depend on whether the buffaloes were compared with a larger or a smaller cattle breed. However, I think one could safely conclude that buffaloes are not inferior in weight gain, food utilization, or carcass quality to the local cattle living in a similar climate.

The extent of this unexploited resource is indicated by the estimate that in Bombay alone some 10,000 buffalo calves die of starvation each year. This figure is taken from an encyclopaedic book (16) on "The Health and Husbandry of the Domestic Buffalo" just published by FAO. The chapter on meat production concludes "An expansion of buffalo meat production and consumption would make a notable contribution to human welfare and the internal economy of many countries, including India, Indonesia, Pakistan, the Philippines and, indeed, wherever water buffaloes are part of the agricultural scene. When they are reared for meat production and the carcass is properly handled and dressed, the resulting product is palatable, nutritious and highly acceptable. It compares favourably with the meat of other domestic animals, whether fresh, chilled, frozen or as a constituent of manufactured products."

CATTLE

I am not trying to pretend that all our meat requirements can be met by the alternative livestock species discussed above. What I am stressing is their vital importance in many developing countries where their full exploitation has been neglected - particularly by advice and assistance coming from countries which

Table 6

Weight gains of buffaloes and cattle in Trinidad (14)

	n	Initial weight (kg)	Daily weight gain (g)	
			1st 10 weeks (poor pasture)	2nd 10 weeks moderate pasture
Brahman	6	140	0	295
Jamaica Red	6	190	0	477
Buffalo	6	184	213	617

Table 7

Carcass composition of 18-22 month old Swamp buffalo and
and grade Brahman bulls dissected in Australia,
but not reared together (15)

	Buffalo	Brahman
Number	3	3
Carcass weight (kg)	129	181
Dressing percentage	52.3	52.7
% lean meat	71.3	68.5
% fat	4.9	8.1
% muscle in "expensive" muscle groups	56.4	56.1

lack these species or, at least, lack knowledge of their importance
and of their possibilities.

Cattle are still the most numerous species of farm mammal
and produce the majority of the world's meat. For the wisest and
most profitable exploitation of this resource I want to emphasize
two points: 1. Choice of the most appropriate breed, and 2. The
place of crossbreeding.

Choice of Breed

A just balance must be kept between on the one hand the
conservatism which advises clinging to the traditional, locally
adapted breed and on the other so-called progress which is always
looking for something different and, hopefully, better. Caution
and the pressure of vested interest confined British beef breeding
to the local breeds until the introduction of the Charolais in 1961.
Good salesmanship based on their beautiful appearance and fat
carcasses had enabled the Hereford, Shorthorn and Aberdeen-Angus to
nearly eliminate the other beef breeds in Britain. Now there is a
flood of large breeds from the continent of Europe to Britain and
America and they are being carefully evaluated. Fig. 2 shows
recent results from the Meat and Livestock Commission's recording
of growth rate in crossbreds. The reason for the rising popularity
of the Continental breeds is clear. The reason for the earlier
rise of the Hereford and Angus (compared with other British breeds)
is not. Certainly in a pure breed other characters besides growth
rate are important (e.g. fertility) but this would not apply to
crossbreds. Food utilization and carcass quality are also
important and Table 8 shows that the Continental breeds are as good

as, if not better than the British breeds in these respects. The Limousin in particular demonstrates, as in previous trials, a high lean/bone ratio and a large eye-muscle area.

While these and other trials have demonstrated the large size and growth rate of the Charolais and Simmental, the Chianina and Maine-Anjou have the reputation of being even better in these respects. Preliminary results from USA are given in Table 9. They should be compared with previous results using Charolais bulls, viz. calving difficulty of 24.1 percent and 200 day weight ratio of 106.3. The Chianina seems to deserve its reputation. Results from small numbers of Chianina steers at Minnesota Agricultural Experiment Station also show a higher growth rate than the Charolais and no inferiority in carcass quality (17).

As for the Maine-Anjou, the U.S. trial indicates that it is no bigger than the Charolais. On the other hand a small-scale trial in Canada indicates that it may exhibit superiority on a high-energy ration (See Table 10).

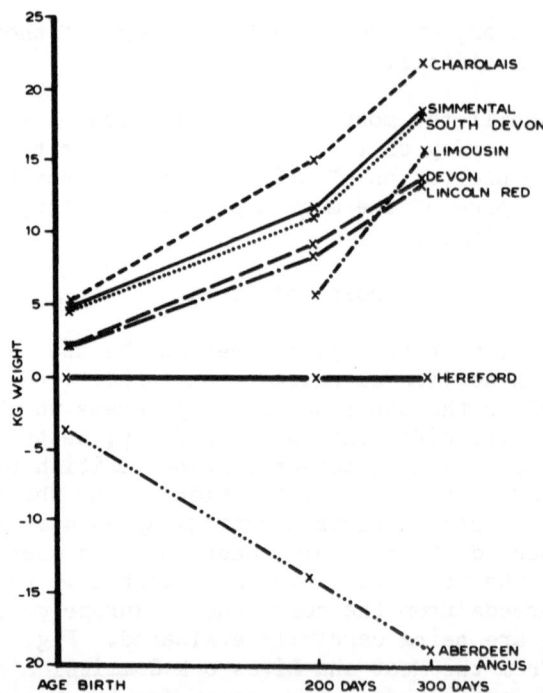

Fig. 2.   Weights of crossbred calves by various breeds of bull compared with the weights of Hereford cross calves in the same herds at the same time, recorded by the Meat and Livestock Commission during 1972-74. A total of 5704 calves were weighed at birth, 13528 at 200±49 days and 10158 at 300±49 days. (Based on figures in Deeble (18)).

Table 8

Comparison of carcasses of Friesian steers with crossbred by beef bulls, born in 1972, fed ad libitum on a complete diet at four Experimental Husbandry Farms in England and at the Norfolk Agricultural Station, and slaughtered serially at four ages between 375 and 487 days. The German and Swiss Simmentals were very similar and their results have been combined (19).

	Friesian	Hereford cross	Limousin	Simmental
Number	26	31	36	49
Overall breed averages:				
Growth rate (kg/day)	0.39	0.90	0.89	0.95
% high-priced cuts	45.7	45.9	46.8	45.8
Lean/bone ratio	3.7	3.8	4.3	3.9
Breed values adjusted to 20% fat in carcass:				
Cold carcass weight (kg)	237	300	233	245
Dressing %	52.8	51.2	55.0	52.7
% lean in carcass	61.9	62.3	64.0	62.5
% bone " "	16.6	16.0	14.6	16.0
Eye muscle area (cm^2)	58.8	52.6	69.8	64.6
Food consumption per unit weight lean	20.8	17.6	18.0	19.6
Weight of lean (kg)	147	124	149	154

These breed comparisons, provided they are based on all the important economic characters, and not only on growth rate, are essential in order to choose the most profitable breed to use in temperate areas with developed farming systems. However, the same breed may not be appropriate for use on poor pasture in the topics. Breeds must be tested under the conditions in which they are going to be used. This is well demonstrated by results from Zambia where bulls of local zebu breeds and Herefords were tested under uniform conditions which alternated between feedlot and pasture. On pasture the zebu breeds grew more quickly; on feedlot the Hereford (see Fig. 3).

A similar example of breed x environment interaction is provided by the FAO feedlot project in Kenya. The crossbreds between European breeds and the local zebu (Boran) were no better than improved Borans on the low concentrate ration but they grew

Table 9

Calving difficulty and 200-day weight of crossbred calves born in
1973 from Hereford and Angus 4-8 year old cows at U.S. Meat
Animal Research Center, Nebraska (20)

Breed of sire	No. of calves weaned	Calving difficulty (%)	200-day weight ratio
Hereford and Angus	92	8.1	100.0
Red Poll	88	3.5	100.5
Brown Swiss	95	9.2	105.6
German Yellow	95	11.8	107.5
Maine-Anjou	80	21.7	106.1
Chianina	80	14.3	110.0

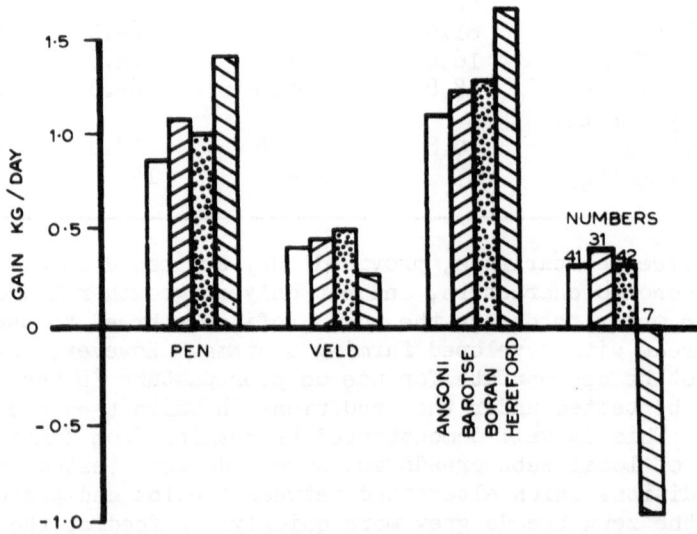

Fig. 3.   Liveweight gains of the same bulls of four breeds
during performance tests alternating between feedlot (pen) and
pasture (veld) (each lasting 105-120 days) during 1967-69 at the
Central Research Station, Mazabuka, Zambia (22).

Table 10

Comparison of crossbred bull calves out of Hereford cows at a Canadian feedlot. Group 1 were heavier than group 2 and were initially fed a higher energy ration; a similar difference applied to groups 2 and 3 (21).

Feedlot group	Average daily gain (kg)		
	1	2	3
Breed of sire			
South Devon	1.16	1.24	1.13
Simmental	1.39	1.31	1.08
Maine-Anjou	1.71	1.36	1.19

10% faster on the high concentrate ration (see Table 11). In a later test the unimproved and local Borans were similar and again the crossbreds by large beef breeds had no advantage on the low energy ration but they grew 30 percent more quickly than the

Table 11

Breed x ration interaction in the Kenya feed lot 1971 trial. The high concentrate ration contained 33% roughage and the low concentrate ration 67%. The feed cost item is defined as "Daily added value minus daily feed and non feed costs including interest and mortality)" (23).

	Unimproved Boran	Improved Boran	Friesian cross	Hereford cross
High concentrate ration				
Average daily gain (kg)	0.99	1.26	1.39	1.38
kg feed per kg gain	8.7	7.4	7.1	7.2
Feed cost (US cents/kg gain	34	28	26	30
Low concentrate ration:				
Average daily gain (kg)	0.88	1.10	1.06	1.05
kg feed per kg gain	9.5	7.9	7.9	9.1
Feed cost (US cents/kg) gain	31	26	26	30

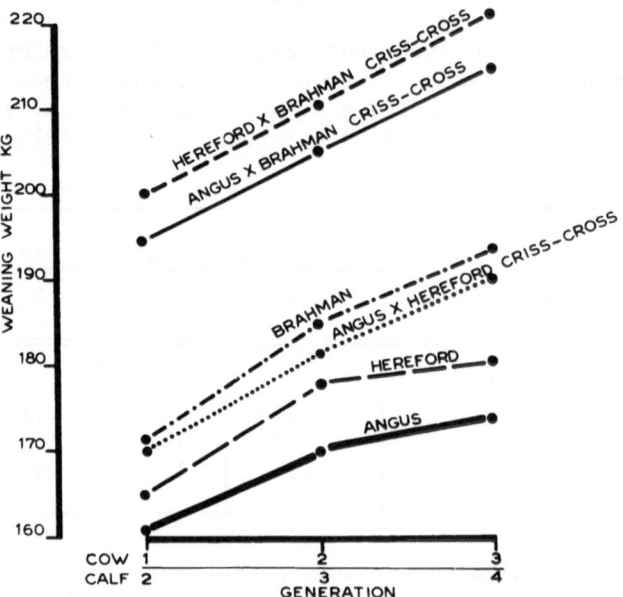

Fig. 4.   Crisscrossing compared with pure breeding at the
Agricultural Research and Education Centre, Belle Glade, Florida.
(After (25)).

Borans on the high concentrate ration which caused the Borans to
develop laminitis (24).

Crossbreeding

The last table leads naturally to a discussion of crossbreeding. This has long been the traditional method of producing

Table 12

Breed	no.	Weight at 18 months kg	Weight relative to Polled Sinu (%)
Polled Sinu	241	226	100
Brahman	72	270	119
Sinu ♂ x Brahman ♀	36	312	138
Brahman ♂ x Sinu ♀	84	307	136

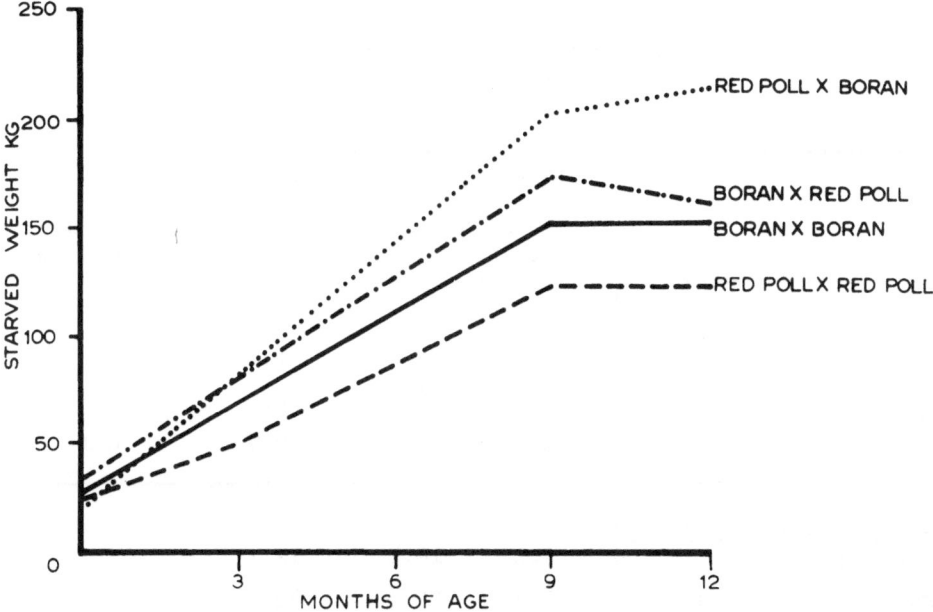

Fig. 5.   Body weights of Boran (zebu) and Red Poll (European) calves and their reciprocal crosses on unsupplemented pasture at Ruhengere Field Station in southwest Uganda during 1967/68 (27).

beef in Britain. The cows not needed for breeding replacements for dairy herds or for herds on the hills and marginal land are mated to beef bulls. Often a second cross is used, e.g. the Blue Grey cows from the Whitebred Shorthorn x Galloway cross are mated to bulls of the large beef breeds.

Presumably hybrid vigour is involved in the Blue Grey cross but I know of no experiment designed to prove it. Such hybrid vigour has been demonstrated in crosses between British beef breeds in U.S.A. and is sufficient to justify crossbreeding for this reason alone. Although the amount of hybrid vigour shown by individual traits may be small, e.g. about 5% for calf crop, weaning weight and post-weaning gain, these items are cumulative and give a total economic advantage of crossbreds over purebreds of the order of 20%.

This diagram refers to one character, one area and one cross-breeding system. However, the results are typical. Table 12 and

Table 13

Superiority of various crossbreds over purebreds
for beef production in America (28)

	% advantage of crossbreds over purebreds			
	Weaning rate	Weaning weight	Production per cow	Daily gain in feedlot
European breed crosses				
F_1 progeny	1.7	3.3	1.7	4.6
Progeny of F_1 dams	4.4	6.2	12.5	2.3
Rotation crossbreds	3.3	8.4	16.9	6.1
European x zebu crosses				
F_1 progeny	5.1	10.8	12.1	13.3
Progeny of F_1 dams	19.4	22.6	46.3	2.2
Rotation crossbreds	9.6	25.1	30.9	0.7
F_2 (= crossbred x crossbred)	11.7	13.4	26.6	-1.8

Fig. 5 give some results from Colombia and Uganda. Table 13
summarizes the results of all the experiments reported at the
Twentieth Annual Beef Cattle Short course at the University of
Florida, 1971. The greatest amount of heterosis is obtained by
breeding from F_1 dams, especially if a third breed of bull is
used. However, this system of breeding needs a somewhat sophisticated organization. It involves keeping a purebred herd to breed
F_1 replacement cows or else buying in replacements from outside
the herd; also bulls of three breeds are required. The crisscrossing has the advantage that the relacement cows arise from the
same matings that produce the calves for slaughter. It can, of
course, be made more complicated by using a terminal crossing sire
of a different breed on the criss-cross cows. Or it may be turned
into a rotational crossing system by using three or four breeds in
succession.

CONCLUSION

I have confined this paper to ruminants and I have concentrated on those which have the reputation of living on poor
pastures and high roughage fodders. Ruminants were the first

animals to be domesticated and they were probably chosen because they did not compete with man for food. Domesticated non-ruminants started as scavengers. I think that in the future we shall be forced more and more to exploit this invaluable characteristic. I am a pessimist. I see the Sahel drought not as a passing phenomenon but as a southward spread of the Sahara. For me the fuel crisis and the fertilizer shortage are signs that we are over-exploiting the mineral resources of the globe. The lesson is that we must more and more rely on renewable resources and so use our animal heritage that its exploitation needs the minimum of grain feeding and, therefore, the minimum of oil, of fertilizer and of water.

REFERENCES

1. Rao, M. Narayana (1973). Wld Anim. Rev. No. 5, p.38.

2. FAO (1971). "Production Yearbook",Vol. 25. (FAO, Rome).

3. FAO (1974). "Livestock development for milk and meat production." Working paper for Committee on Agriculture, Second Session, April 1974. (Mimeograph).

4. Asibey, E.S.A. (1972). "Wildlife as a source of protein in Africa." FAO working paper for fourth session of Ad Hoc Working Party on Wildlife Management, African Forestry Commission. (Mimeograph).

5. UNECA/FAO/OAU (1973). "African Livestock Development Study." Part One. Southern and Central Africa. 2 vols. (UN Economic Commission for Africa, Addis Ababa).

6. Skinner, J. D. (1973). Z. Tierzücht. ZüchtBiol. 90, 263.

7. Devendra, C. (1974). Z. Tierzücht. ZüchtBiol. 91, 246.

8. MacKenzie, D. (1967). "Goat Husbandry". 2nd ed. (Faber and Faber, London). (Cited by Devendra (7)).

9. Singh, S. N. (1968). "Investigations on milk and meat production in Indian goats." Abstr. Pap. Am. Dairy Goat. Ass., Bethesda, Md 64th ann. Meet. (Cited by Devendra (7))

10. Robinet, A. H. (1973). In:"1le. Conference Internationale de l'Elevage Caprin, Tours, France 1971" p.127. (Institut Technique de l'Elevage Ovin et Caprin, Paris).

11. French, M. H. (1970). "Observations on the Goat." FAO Agricultural Studies No. 80. (FAO, Rome).

12. Devendra, C. and Burns, Marca. (1970). "Goat Production in
     the Tropics." (Commonwealth Agricultural Bureaux, Farnham
     Royal, Bucks.)

13. Kassir, S. M., McFetridge, D. B. and Hansen, N. G. (1968).
     "Studies on the growth, feed costs and carcass composition of
     young male cattle and buffalo fed under comparable conditions."
     Technical Report No. 21. UNDP/FAO Animal Husbandry Research
     and Training Project, Baghdad. (FAO, Rome).

14. Shute, D. J. (1966). Dip. Trop. Agric. Thesis, University of
     the West Indies. (Cited by Ognjanovic (20)).

15. Charles, D. D., Johnson, E. R. and Butterfield, R. M. (1970)
     Proc. Austr. Soc. Anim. Prod. 8, 95.

16. Cockrill, W. Ross (Ed.) (1974). "The Husbandry and Health of
     the Domestic Buffalo." (FAO, Rome).

17. Meiske, J. C. and Goodrich, R. D. (1973). In:"Minnesota
     Cattle Feeders' Report", p. 19 (Department of Animal Science,
     University of Minnesota).

18. Deeble, F. K. (1974). "The evaluation of the Limousin and
     Simmental breeds as sires of suckled calves." Report No. 7,
     Limousin and Simmental Tests Steering Committee. Reading,
     11 November 1974. (Mimeograph).

19. Deeble, F. K. (1974). "The food conversion efficiency and
     carcass quality of steers by Friesian, Hereford, Limousin,
     German Simmental and Swiss Simmental bulls out of Friesian
     cows." Report No. 6, Limousin and Simmental Tests Steering
     Committee. Reading, 9 September 1974 (Mimeograph).

20. US Meat Animal Research Center (1974). "Germ Plasm Evaluation
     Programme." Progress Report No. 1.

21. Rao, M. Narayana (1973). Wld Anim. Rev. No. 5, p.38.

22. Maule, J. P. (1973). S. Afr. J. Sci. 3, 111.

23. Creek, M. J. (1972). Wld Anim. Rev. No. 3, p. 23.

24. Squires, H. (1974). Paper for conference on "Beef Cattle
     Production in Developing Countries", Edinburgh, September,
     1974.

25. Franke, D. E. and Crockett, J. R. (1974). Span 17, 64.

26. Salazar, J. J. (1973). In:"Crossbreeding Beef Cattle. Series 2 "(Ed M. Koger, T. J. Cunha and A. C. Warnick) p. 402. (University of Florida Press, Gainesville).

27. Sacker, G. D., Trail, J. C. M. and Fisher, I. L. (1971). Anim. Prod. <u>13</u>, 181.

28. Koger, M. (1973). In:"Crossbreeding Beef Cattle Series 2" (Ed. M. Koger, T. J. Cunha, and A. C. Warnick) p. 434 (University of Florida Press, Gainesville).

DISCUSSION

In reply to Mr. Mason on the leanness of red deer, *Dr. Fuller* mentioned the comparison of lambs and deer made by Miss Ann Pollock at the Rowett Institute. Given the same food intake, deer retained less energy, but with the same rate of energy retention more of their retained energy was protein. *Dr. Webster* added that this difference was similar to that between Pietrain and Large White pigs discussed by Dr. Lister.

Dr. Frisch felt that part of the genetic component of appetite lay in the control of milk yield since, in her view, early food intake set the pattern of appetite in later life. *Dr. Turner* thought that the partition of retained energy between protein and fat might be controlled by the same hormonal system which regulated appetite.

The Challenge of New Food

VEGETABLE PROTEIN AS A HUMAN FOOD -

BACKGROUND AND PRESENT SITUATION

Tokuji Watanabe

National Food Research Institute

Ministry of Agriculture and Forestry, Tokyo, Japan

INTRODUCTION

In recent years there has been growing concern that production has been unable to keep pace with the increase in world population. In particular, crop production and fish catching have been locally very poor because of the sporadic incidence of unusually dry or cold weather which has led to acute and dramatic shortages of food. But it is evident that the consumption of feed crops including wheat, corn, milo and soybean is rapidly increasing in many countries to provide for the increased requirement of animal foods.

Human foods of animal origin require animals to be fed 7-10 times more calories than their own value, in the form of crops and grass. In other words, in countries where animal foods are more popular crops are used less efficiently for food than where animal foods are rarely consumed. Fig. 1 shows that the total consumption (i.e. both direct and indirect) of crops in the U.S. per capita per year is 10 times greater than in Brazil, although the direct consumption of crops as food in the U.S. per capita per year is less than that in Brazil. Fig. 2 compares the calorie intakes of several countries using the notion of 'original' calories, which are calories derived from crops plus the calorie cost of producing the animal, i.e. animal calories x 8. The original calorie intake by the U.S. people is 3 times their usual calorie intake, whereas that of the people of India is only 20% greater. Thus chronically low yields and poor harvests anywhere in the world influence prices for meat and meat products even in such countries as the U.S. and Japan.

As demand for animal foods in the world will continue to

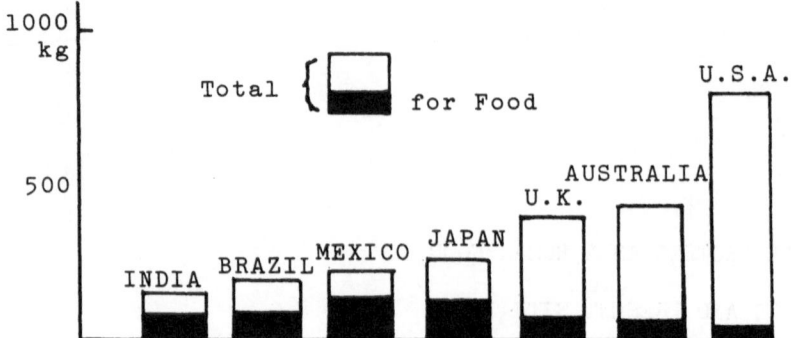

Fig. 1.   Consumption of total and food crops per capita per year
in several countries.

increase as indicated in Table 1, it is urgently necessary to
increase production from crops for animal use, and to search for
and develop new feed resources.   Single-cell and leaf protein
and oil cakes are now being discussed in this connection.   At
the same time the possibility should be considered of using soybean meal, wheat flour and peanut for human food.   Soybean and
wheat have been used traditionally as foods in the Orient as
recorded by Watanabe (1) and contribute to the protein nutrition
of the people there as shown in Table 2.   The following are

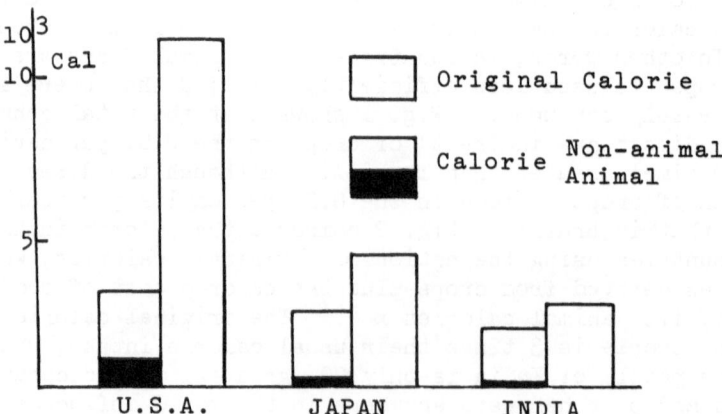

Fig. 2.   Calorie supply per capita per day.   (For detailed
explanation see text.)

Table 1

Supply of animal (incl. fish) protein (g)
per capita per day in several countries

	1960/62	1963/65	1966/68	1969/70
United States	65.1	66.4	68.7	71.5
Australia	59.7	61.3	65.4	68.2
United Kingdom	56.0	56.3	56.3	53.4*
France	55.8	58.7	62.7	64.0
Spain	24.8	28.6	32.4	34.6
Japan	22.6	25.5	28.4	30.8
India	6.1	6.0	5.5	5.6
Argentine	52.4	50.7	60.0	62.3*

* 1970/71

suggested as the reasons why these foods were and are still so
popular in the Orient.   One is that the people have known for
many years technology about the use of microorganisms as enzyme
sources to convert starch, protein and fat to the low molecular
compounds which may play a role in the development of the flavour
and digestibility of products, such as miso, soysauce and tempeh.
Another is that they developed methods of modifying the proteins

Table 2

Protein supply (g) from various food items
in Japan per capita per day

	1960	1968	1970	1972
Grains	28.8	26.7	25.6	25.3
Soybeans, miso, soysauce	10.8	9.9	9.9	10.2
Other legumes	2.8	2.7	2.6	2.5
Fruits and vegetables	4.3	5.6	5.1	5.4
Animal foods	6.6	13.2	15.7	17.1
Fish	14.6	16.1	15.8	16.6
Others	1.6	1.3	1.5	1.3
Total	69.5	75.5	76.2	78.4

of soybean and wheat to make them more acceptable in texture and
eating characteristics generally.   Soybean milk is a liquid containing oil and protein and is similar to cow's milk.   Tofu is
calcium gel prepared from heat-denatured protein.   Aburage is
deep-fried tofu swollen to a sponge-like condition during heating.
Dried fu is prepared by baking a mixture of fresh gluten and rice
or wheat flour.

    These foods provide examples of the different ways that the
chemical or physical properties of vegetable protein can be modified to fit consumer preferences, just as margarine can be prepared
as a substitute for butter.   Vegetable proteins have also been
used as substitutes for or extenders of meat or chicken products
in China and Japan.   There is, for instance, a traditional Chinese
dish which is made from wet gluten seasoned by soy or other sauces
which has a very similar texture to meat.   Another dish is made
by pressing out films of soybean protein, sticking them to each
other and adding seasoning.   The films used are those formed on
the surface of soybean milk during heating.   In Japan, tofu, the
calcium gel of soybean protein, can be ground, moulded and deep
fried to make gammodoki, the texture of which resembles chicken.
The name literally means wild goose meat analogue.

    Several foods have been prepared in the U.S. from soybean
and wheat protein products; these include soybean milk, soybean
cheese and gluten meat.   They are mainly consumed as vegetarian
foods and in diets for allergic patients.   In the last ten years
these foods have also been evaluated as dietic foods in the prevention of cardiovascular disease.   Other uses of vegetable
protein as food are as ingredients of traditional foods such as
bread, sausage and cake.   Vegetable proteins can be used to retard
the deterioration of starch in bread, promote the emulsification of
fat in sausage and prevent oil penetration in doughnuts.   The
amount used is usually less than 3% in the final products.

                NEW FOODS FROM VEGETABLE PROTEIN PRODUCTS

    The significance of vegetable proteins in the world food
supply has greatly increased in recent times.   To use these
protein sources more effectively, soybean and wheat protein products
are now industrially processed into such food materials as textured
vegetable protein, spun protein and gel protein.   These particular
materials are easily prepared on account of the inherent characteristics of soy and wheat protein.   They are mainly used as extender
of meat and meat products but some of them are processed to meat
analogue.

    I will now describe briefly the chemical and physical
characteristics of vegetable protein and the present situation

regarding their industrial processing for food.   Research and
development in these fields will be described in the next chapter
by Saio.

Soybean and its Derived Products

Soybean contains 30-40% of protein and 20% of oil.   It is
one of the most important vegetable oil resources in the world.
After the removal of oil by hexan-extraction, the resulting soybean meal contains 45-50% of protein and is mainly used as feed.
90% of protein in the meal can be extracted with water, if the
meal is produced without heating.   Protein becomes more or less
insoluble, possibly through denaturation, when the meal is heated
during or after defatting.   This might be for heat denaturation
of protein.   Protein extracted from the meal with water can be
precipitated with acid at its isolectric point, pH 4.5 and the
remaining low molecular nitrogen compounds, sugar and other minor
components are retained in the whey.   The precipitated protein is
separated, washed and dried to make so-called isolated soy protein,
the yield of which is about 30-35% of the meal.   It is now
industrially manufactured in various countries such as the U.S.,
Japan, European countries and perhaps in China.   Acid-precipitated
protein is neutralized, if necessary, and spray-dried.

If the meal is treated with alcohol or dilute acid, low
molecular nitrogen compounds, sugar and other minor components are
eluted as whey; the main protein components remaining in the meal.
The meal produced contains 70% protein on a dry basis and is called
70% soybean meal or soybean protein concentrates.   It is rich in
protein and has a less 'beany' flavour.   Protein concentrates can
also be made by heating meal to make the protein insoluble and
repeatedly washing with water.

Other products derived form the meal are powdered soybean
milk which is a spray-dried product of cold or hot water extract of
meal or whole soybean.   Although a beany flavour is still retained
in this product, it is more nutritious because of its high digestibility, which results when the insoluble, fibrous components are
removed by filtration.

Residues in the extraction water used in the production of
isolated soy protein have no economical use other than for feeds.
Almost all whey is now discarded and is responsible for much of
the pollution of rivers and the sea.   Concentration by ultra-
filtration or reverse osmosis may provide alternative applications.
It has also been used in growth media for microorganisms such as
yeast.

There is some confusion concerning the use of the term

'vegetable protein' or 'soybean protein'.   Strictly speaking,
vegetable protein is isolated protein, such as acid-precipitated
protein of soybean or gluten from wheat flour.   But conventionally
soybean meal, soybean protein concentrates and dried soy milk are
all called soybean protein, because they contain 50% or more protein
on a dry basis.   The soybean protein products considered in this
paper include the various products which contain more than 50%
protein including textured vegetable protein and spun protein
mentioned below.

Soybean and Wheat Protein Products as
Meat Extender or Meat Substitute

Before vegetable protein products are used as extenders or
meat or meat product substitutes, it is necessary to give them the
desirable texture for manufacture and eating.   Fig. 3 shows
briefly a flowsheet of production of various soybean protein products from defatted soybean meal.   These processes have been
reviewed by Smith and Circle (2).

Isolated soy protein.   Neutralized isolated soybean protein,
after mixing with the same amount or more of water, is converted
to elastic hard gel by heating.   It is used as an extender in
sausage, meat loaf, fish sausage and fish paste products.   This
property can be employed to some extent in flour, concentrates and
soybean milk powder.   Circle et al. (3) reported gel-forming
property of isolated soy protein.   The isolated protein is the

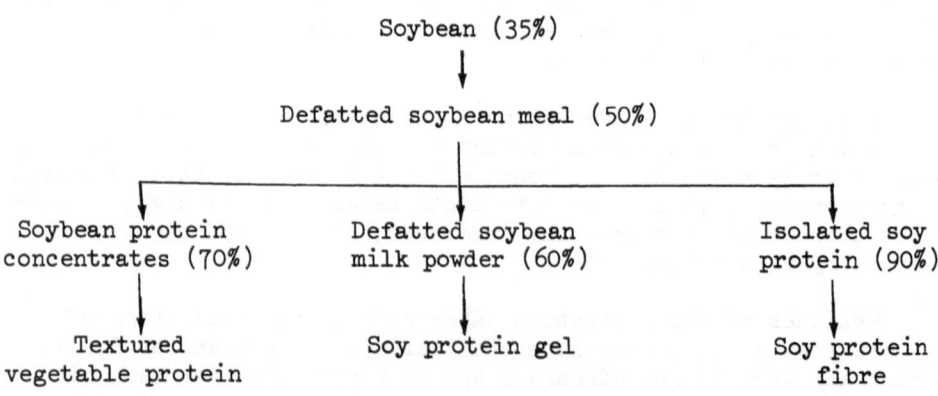

The protein content on a dry basis is shown as (-%)

Fig. 3.   Flowsheet of production of soybean protein products.

material used for spun protein mentioned later.   The price of
isolated soy protein is about 90¢/kg, compared with 16¢/kg for soybean meal or flour.

   Textured vegetable protein (TVP) from soybean meal.   Textured
vegetable protein from soybean meal is perhaps the most popular
material in use as meat extender or substitute.   It is made from
soybean meal or soybean protein concentrates and sometimes mixed
with isolated soy protein.   They are mixed with water and ingredients, if necessary, and treated with cooking extruder to apply
when the thermoplastic extrusion technique is to be used (4).
In the course of this somewhat drastic, physical treatment, the
protein denatures which results in its stretching and twisting.
The reduction in solubility of protein plus these changes make the
denatured proteins capable of reorientation into the desired
'muscle' structure.   The size and shape of the products can be
controlled by modification of the design of the dies used and
speed of the cutting knife.   As extruded materials are usually
moist, they are dried and cooled before packing.   In the case of
products made from soybean meal, the beany flavour can be removed
by washing.   TVP is superior as meat extender not only for its
eating character but also for its ability to retain the natural
juices and flavours that ordinarily would be lost during cooking.
Furthermore, TVP can reduce the shrink loss which is observed
during the cooking of meat.   The use of TVP as meat extender
requires its hydration at the rate of two parts water to one part
product by weight and adding this blend to the total product mix
at the rate of 20-30%.   Hydration with more than two parts water
is not desirable from the point of acceptability, nutrition and
reduction of cooking shrink.   The price of TVP is about 75¢/kg
or more dry, and 30¢/kg when hydrated.

   According to Cumming et al. (5), the physical properties of
TVP such as its shear characteristics (Kramer Shear Press) are
markedly affected by process temperature and indicate that maximum
texturization may occur under their experimental conditions between
175-192°C.   Hashida et al. (6) measured the texture change of TVP
induced by heating for canning with texturometer.   They found
decrease in hardness and chewiness after heating and differences
in the texture of the reconstituted uncooked and cooked materials
which depended on the temperature and pH of the water used.   They
also showed that the heat induced decrease in hardness and chewiness
could be prevented by reconstituting with dilute $CaCl_2$ solution.

   Yoshioka et al. (7) examined the texture of TVP using
Tensilon and Texturometer, as it was affected by the pH of water
used for reconstitution, which influences water holding capacity
and hardness, and by precooking.

   There are different types of TVP other than that produced by

the thermoplastic extrusion technique.   Structural protein is a
TVP made from isolated protein by a patented method.

   Spun protein from soybean.   Spun protein is made from isolated
soy protein by Boyer's method.   Proteins are dissolved in alkaline
at pH 11-13 as a spinning dope.   This is extruded through a fine
spinneret into an acetic acid bath containing NaCl.   Fibre protein
molecules are stretch oriented by rolling.   The tenderness and
diameter of the fibres can be adjusted by concentration of the
protein and alkalinity of the dope, diameter of spinnerets and also
by stretching (4).   When the fibres are used for producing meat
substitute, they are treated with flavouring solution, colouring
and seasoning substances, mixed with coagulating agent such as egg
white and heated in a bath for moulding.   They are then cut,
dried or frozen for marketing as ground beef, beef bits, analogues
or other commodities.   Spun proteins are sometimes used as
extender in meat products, being prepared in a chosen length and
strength.   The price of spun protein is about 130-140¢/kg of
dried product, which is about twice that of TVP.   It is said,
however, that spun protein can be used in meat products at a higher
rate than TVP and thus it may not necessarily be uneconomic.

   Using the Instron Universal Testing Machine, Stanley (8)
examined the physical properties of fibre protein and meat in
terms, for instance, of its breaking strength and break elongation
and found distinct differences between the two.

   Wheat gluten.   Wheat flour contains 10-15% protein.   Flour
from hard wheat contains more protein than that from soft wheat
and low grade flour contains more protein than high grade flour.

   Gluten is obtained by kneading flour with water, washing out
soluble substances and starch repeatedly until only an elastic soft
mass remains.   This material, so-called wet gluten, which contains
80% protein on dry basis, can be coagulated by heating to a soft
gel which is chewy.   This is why wet gluten is used as an extender
of fish sausage and some fish paste products in Japan.   Wet gluten
can be air-dried after cutting into pieces or spray-dried after
dissolving in acetic acid.   TVP from gluten is now produced on a
commercial scale in Japan by a 'shearing' process which gives the
product a fibrous texture.   This texture can be formed also by
coagualting dissolved gluten under special conditions.   Gluten
based TVP is mainly used as an extender of meat and meat products.

               Nutritive Value of Vegetable Protein Products

   Although the sulphur amino acid content of soybean protein is
less than that of animal protein, its nutritive value is high.
The Protein Efficiency Ratio (PER) of soybean protein is estimated

as 2.5, or about 80% that of milk casein.   But TVP and spun protein may have lower nutritive value as a consequence of the
destruction of some amino acids, if the processing conditions are
not satisfactory.   The PER of TVP is 1.8-2.0, or 20-25% lower than
the original soybean protein.

Constance et al. (9) discussed the nutritional value of TVP
and methionine-fortified TVP.   Both meet the requirements of
adult men when fed at the 8 g nitrogen intake level.   At the 4 g
nitrogen intake level, beef was found to be superior to a TVP
product as a source of protein, but this superiority could partially
be overcome with 1% fortification of the TVP product with DL-
methionine.

According to Shirata et al. (10) rats fed a 1:1 mixture of
meat and TVP showed no differences in body weight gain, feed
efficiency and other characteristics from rats fed meat alone.

Fujita (11) found some destruction of the basic amino acids
in fibre protein which was possibly attributable to the alkaline
treatment of the dope.   This may result in a decrease of nutritive
value of this product relative to the original isolated soy protein.
Animals and children fed spun protein (12) accepted it readily and
the results of the tests suggested that the protein was equivalent
to 80% cow's milk.

Wheat gluten is distinctly low in lysine, but if it is used
as meat extender, this is not serious.   Combinations of wheat
gluten with soy protein are preferable because lysine and methionine
supplement each other.   Lysine fortification should also be considered.

Soybean meal or flour causes flatulence when fed in large
amounts.   It is thought that oligosaccharides such as raffinose
and stachyose are the constituents responsible.   Protein concentrate and isolate do not cause such troubles, because they
contain little oligosaccharides.   TVP produced from protein concentrates is even more desirable from this viewpoint.   Rachic (13)
has reviewed the literature pertaining to these problems.

The Flavouring of Vegetable Protein Products
Used for Meat Substitutes

The beany flavour is undesirable when soybean is used as a
basis for meat extender and meat substitute.   It is removed from
soybean by making protein concentrate or isolated protein.   Careful and repeated treatments such as steaming are necessary if they
are to be effective.

If meat substitutes derived from vegetable protein are to taste and look like meat or meat products, flavouring and colouring must be added.

Flavouring substances need to be absorbed on the surface of the protein of the cell structure.   They need not be too strongly bound, because they are not then easily eluted in the process of chewing.

Beef steak or roast beef flavours are created by the inter- action of amino acids and reducing sugars which exist in the meat as precursors.   Cystine or cysteine and fat are important for developing such flavours.   The essential compounds are commercially available as flavouring substances.   Soaking the protein products in a mixture of protein hydrolyzate, salt, sugar and seasonings at $60^{\circ}C$ is effective in conferring flavour on the product (14).   In the case of TVP, flavouring substances may be dissolved in water together with the heat coagulant agent.   They are fixed in the structure of TVP by heating.

LABELLING AND SPECIFICATION OF NEW FOODS
FROM VEGETABLE PROTEINS PRODUCTS

The amount of vegetable protein used as meat extender and meat substitute is increasing in Japan and in the U.S.A. especially since the U.S. Government authorized the use of TVP in the schools lunch programmes (15).

TVP, fortified by minerals and vitamins, can be used in com- bination with meat.   The ratio of hydrated protein to cooked meat, poultry or fish should not exceed 30 parts to 70 parts on the basis of weight.   The USDA lower limit in the PER of TVP is 1.8 and that of combined mixture of TVP and meat, 2.5.   TVP must meet the composition requirements shown in Table 3.   All values are expressed on dry basis and are applicable to dry or hydrated forms. The moisture content of the hydrated form shall not exceed 65% nor be less than 60%.

In Japan, the Government approval of processed foods is indicated by the 'JAS' mark on the products, and various processed meat products such as pressed ham and mixed ham are allowed to include vegetable protein products to the extent of 3 or 5% including starch in the total product (Table 4).   As this system is not compulsory, there are meat products available which do not bear the 'JAS' mark, in which vegetable protein is added in excess of 3-5%.   There are now no restrictions controlling the quality and quantity of vegetable protein used in meat products.   The only requirement is for the material used in the processed food to be indicated.

Table 3

Composition requirement for TVP

		Minimum	Maximum
Protein (Nx 6.25)	%	50.0	–
Fat	%	–	30.0
Magnesium	mg%	70.0	–
Iron	mg%	10.0	–
Thiamin	mg%	0.30	–
Riboflavin	mg%	0.60	–
Niacin	mg%	16.0	–
Vitamin B_6	mg%	1.4	–
Vitamin B_{12}	mg%	5.7	–
Pantothenic acid	mg%	2.0	–

In many European countries, food laws prohibit the use of vegetable protein in meat and meat products on the basis of definition of the commodity. It is natural that those traditional foods should be made from their original material, but several countries are now discussing the possible use of vegetable protein in meat products as a consequence of recent trends in the world food situation. However, until the technology to produce independent foods exclusively from vegetable protein is developed, vegetable protein will be used to decrease or overcome some of the inherent disadvantages or physiological problems associated with meat and meat products.

Detection of Vegetable Protein in Meat Products

Sometimes it is necessary to check for the presence and amounts of vegetable protein in meat products. The methods considered for this purpose are 1) comparison of the amino acid content of the product relative to pure meat products, and 2) immuno-chemical methods for identifying the pure protein components. The former method is not applicable when the amount of vegetable protein used is small, because it is difficult to differentiate between the constituent acids. The latter needs some experience and special expertise and training before reproducible results can be obtained. Recent reports show the effectiveness of acrylamide gel-electrophoresis, which differentiate vegetable protein as independent peaks from those of meat protein components (16,17). In these methods, extracts of meat products are made with urea

Table 4

JAS specification on the use of soybean protein
in ham, fish ham, fish sausage and kamaboko

Product		Limitation	Binder
Pressed ham	Special	below 3%	
	High class	below 5%	starch, wheat, flour,
	Standard	below 5%*	corn meal, vegetable
			protein or skim milk
Mixed ham		below 5%*	
Fish ham and fish sausage			starch, egg white, vegetable protein, casein, gluten
		no limitation	
Casing kamaboko			egg white, starch, vegetable protein

* Below 3% in the case of starch, wheat flour or corn meal

solution at room temperature or 100°C before subjecting them to
gel electrophoresis.

PRESENT SITUATION AND FUTURE PROSPECT

According to studies at Cornell University, the use of meat
extenders and analogues, from vegetable protein sources, may reach
the equivalent of 10% of all U.S. meat consumption by 1985 and
certainly by the year 2000.   This growth would carry the use of
protein ingredients in the meat industry from their present level
of 65,000 tons to approximately 1,100,000 tons in 15 years, which
represents an annual growth rate of 19.3%.

In Japan, the production of vegetable proteins for meat
extender is increasing (Table 5).   TVP from soybean meal and
wheat gluten is most popular and is used in hamburger, sausage and
Chinese foods which are mainly distributed as frozen foods.   As
soybean and wheat gluten have been popular as traditional foods in
Japan, those materials are easily accepted by the Japanese people.
Unlike Europe, there is no tradition in Japan for eating animal
foods, and it is likely that vegetable proteins will be used
primarily as meat extenders or meat substitutes.

Table 5

Production (tonnes) of vegetable protein foods
in Japan

	Total	Textured and spun (frozen)	Powder	From Soybean	From Wheat
1971	27.309	9.591 (8.200)	17.718	15.895	11.414
1972	28.016	10.637 (9.016)	17.379	14.485	13.531
1973	35.593	18.668 (13.819)	17.925	12.878	22.810

CONCLUSION

The world situation of the supply and demand for human food and animal feeds is very serious. The future increase in the consumption of animal foods with increases in national incomes in many countries will make the situation even worse.

With the need to identify and develop new feed sources, the promotion of vegetable protein sources, which are now mainly used as animal feed, for human use is urgently necessary from the viewpoint of the economy of resources.

ACKNOWLEDGEMENT

I am very grateful for the kind assistance of the American Soybean Association which provided me with the latest information on the use of soy protein for food in the U.S.

REFERENCES

1.   Watanabe, T. (1969). Industrial production of soybean foods in Japan. In: Report of UNIDO's Soybean Expert Meeting .

2.   Smith, A. K. and Circle, S. J. (1972). Soybean: Chemistry and technology, Vol. 1 (Proteins) (AVI Publishing Inc.)

3.   Frank, S. S. and Circle, S. J. (1959). The use of isolated
      soybean protein for non-meat simulated sausage products.

4.   Horan, F. (1973). Soy protein products and their production.
      In: Proceedings of World Soy Protein Conference , p. 67A

5.   Cumming, D. B., Stanley, D. W. and de Man, J. M. (1972).
      J. Inst. Can. Sci. Technol. Aliment. 8, 124

6.   Hashida, W., Mori, T. and Baba, A. (1971). Texture of vege-
      table protein foods. In: Proceedings of 20th Annual Meeting
      of Japan Canned Foods Association

7.   Yoshioka, K. and Koda, Y. (1973). J. Fd Sci. Technol.(Japan),
      20, 95

8.   Stanley, D. W., Cumming, D. B. and de Man, J. M. (1972).
      J. Inst. Can. Sci. Technol. Aliment. 5, 118

9.   Constance, K. and Hazel, M. F. (1971). J. Fd Sci. 36, 841

10.   Shirata, K., Anzai, K., Saito, Y. and Suzuki, A. (1970).
      Nutrition and Foods (Japan) 73, 71

11.   Fujita, T. (1969). Soy protein for new protein foods. In:
       Proceedings of Japanese Society of Food Science and Tech-
      nology , p. 57

12.   Bressani, R., Viteri, F., Elias, L. G., de Zaghi, S.,
      Alvasado, J. and Odell, A. D. (1967). J. Nutr. 93, 349

13.   Rochis, J. J. (1973). Biological and physiological factors
      in soybean. In: Proceedings of World Soy Protein Conference ,
      p. 161A

14.   Hashida, W. (1974). Flavor potentiation in meat analogue.
      In: Food Trade Review , January, p. 21

15.   McCloud, J. T. (1973). Soy protein in school feeding
      program. In: Proceedings of World Soy Protein Conference ,
      p. 141A

16.   Parsons, A. L. and Lawrie, R. A. (1972). J. Fd Technol. 7,
      455

17.   Guy, R. C. E., Jayaram, R. and Willcox, C. J. (1973). J.
      Sci. Fd Agric. 24, 1551.

VEGETABLE PROTEIN AS A HUMAN FOOD -

RESEARCH AND DEVELOPMENT IN THE NATIONAL FOOD RESEARCH INSTITUTE

Kyoko Saio

National Food Research Institute

Ministry of Agriculture and Forestry, Tokyo, Japan

INTRODUCTION

Beside the traditional use of soybean, a wide variety of soybean proteins have emerged as new food materials in Japan. The change of eating habits of the Japanese people towards increased meat consumption has particularly encouraged industrialists to produce meat-like materials from vegetable proteins.

Our research activities at the National Food Research Institute (Tokyo, Japan) are concerned with developing new protein food technology based upon our research results on the traditional soybean foods, which became a major research topic after World War II.

Here, we will introduce briefly our recent researches on the utilization of soybean protein, and we will outline the fundamental data which let us arrive at these experimental products.

RESEARCHES

Sponge-like Protein Materials from Soybean

In the processing of Kori-tofu (frozen, dried Tofu), the aging process is important in order to dry the fresh Tofu without case-hardening it and to obtain the characteristic sponge-like texture.

The sponge-like protein materials recently developed by our Institute are products made by following the basic procedure for

Kori-tofu.   The material is prepared from isolated soybean protein solution by freezing, for instance, at -5°C for 24 hrs. with
a small amount of calcium chloride and then aging for several days.
After thawing, almost all the protein forms a tight network which
can be easily squeezed out.   The density of the network depends
on the temperature of freezing and the rate of freezing.   The
binding force of this network is highly related to the formation
of S-S linkages during concentration of protein solution by
freezing and successive aging.   The resulting materials are
successfully used as ingredients to supplement meat or fish sausage and other products.

Practical Separation of 7S and 11S Components
of Soybean Protein

In the course of our investigation on Tofu-making, it has
been recognized that the ratio of 7S to 11S among Japanese soybean varieties differed significantly and the ratio was related
to the physical properties of final products.   7S and 11S form
the main components of glycinin, the soybean reserve protein.
Using partly purified 7S and 11S components, it has been found
that 11S forms a harder and more elastic calcium-gel or heat-
induced gel than the 7S component.

We proposed, therefore, a method to fractionate 7S and 11S
components, based on their different precipitation behaviour with
calcium salt as shown in Fig. 1.   Separated fractions were
designated 7SPRF (the 7S Protein Rich Fraction) and 11SPRF (the
11S Protein Rich Fraction).

Differences in the physical properties between the two can
be seen in calcium-gel, heat-induced gel, giving cheese- and jelly-
like products respectively.   In the detailed investigations on
heat-induced gels prepared by kneading 7SPRF or 11SPRF with water
and heating at 60°C to 100°C, the gel containing 11SPRF showed
higher tensile strain, tensile stress and shear strength than that
of 7SPRF and SPI (Soybean Protein Isolate).   Water retention was
remarkably higher in 11SPRF than in 7SPRF.

These results suggest that a trial to find effective ways of
utilizing them for food use might be profitable.

High Temperature Expansion Gel

In the course of our researches on the production of Aburage
(deep-fried Tofu), it has been recognized that the expansion ratio
and textural changes caused by moisture flashing were influenced
by various conditions of preparation.   The expansion characteris-

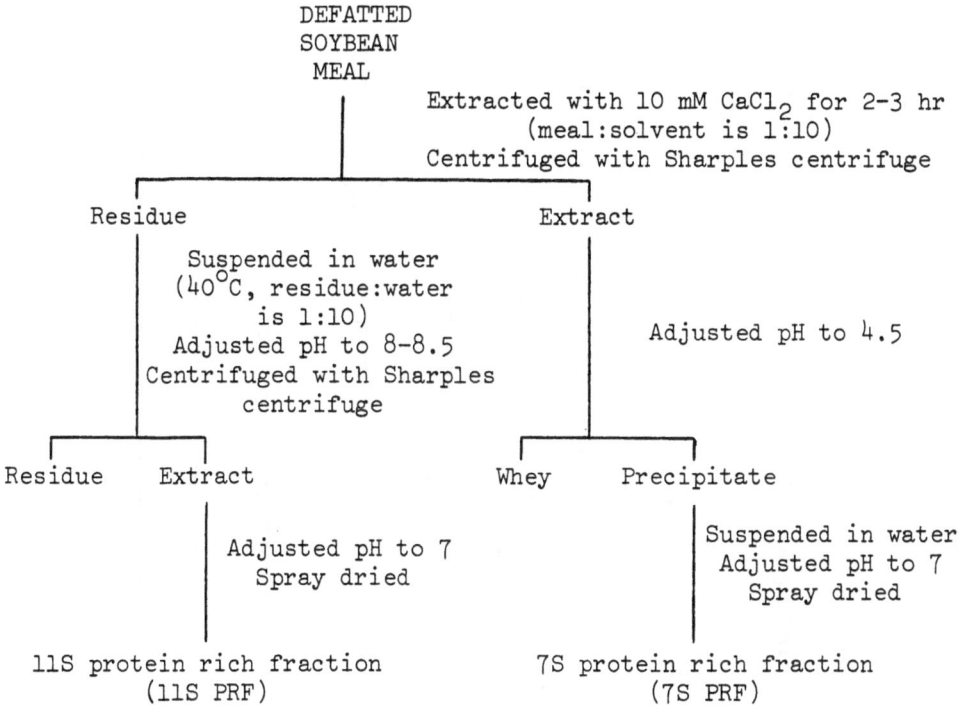

Fig. 1.   Flow sheet of fractional extraction of 7S and 11S protein rich fractions.

tics in Aburage, in our opinion, might be essentially similar to those shown in Textured Vegetable Protein extruded at high temperature.

High temperature expanded gel is a product made by following the procedure for Aburage.   The material is prepared from calcium-gel of soybean protein by autoclaving in dilute alkaline buffer solutions, the most acceptable of which were ammonium citrate or phosphate buffer.

The expansion characteristics of the porous forms of the gels could be introduced by autoclaving without deepfrying in oil. The relationships between the temperature of autoclaving and changes in characteristics of the protein are summarised in Table I.

Table I.   Relation between temperatures of autoclaving and changes in characteristics of protein.

Temperature of autoclaving (°C)	100	105	110	120	130	140	150	160	170
gross-structure of subunits		intact			little degradated			degradated	
solubility	rapid decrease		slow increase					rapid increase	
binding force (degree of aggregates)	rapid increase		slow decrease					rapid decrease	
expansion property		increase					rapid decrease		
texture	hard fragile		soft elastic					like sol	

REFERENCES

1. Hashizume, K., Kakinchi, K., Koyama, E. and Watanabe, T.
    (1971). Agr. Biol. Chem. $\underline{35}$, 449.

2. Hashizume, K., Kosaka, K., Koyama, E. and Watanabe, T. (1974).
    Nippon Shokuhin Koggo Gakkaishi $\underline{21}$, 136.

3. Hashizume, K., Nakamura, N. and Watanabe, T. (1974). Nippon
    Shokuhin Koggo Gakkaishi $\underline{21}$, 141.

4. Saio, K., Kaji, M. and Watanabe, T. (1973). J. Food Sci. $\underline{38}$,
    1139.

5. Saio, K., Sato, I. and Watanabe, T. (1974). J. Food Sci. $\underline{39}$,
    777.

6. Saio, K., Sato, I. and Watanabe, T. (1974). Nippon Shokuhin
    Koggo Gakkaishi $\underline{21}$, 234.

7. Saio, K., Terashima, M. and Watanabe, T. (1975). J. Food
    Sci. - in the press.

8. Saio, K., Terashima, M. and Watanabe, T. (1975). J. Food
    Sci. - in the press.

SINGLE CELL PROTEIN AS A FEEDSTUFF

T. Walker

BP Proteins Limited, London

INTRODUCTION

The products which form the basis of this paper have, in recent years, come to be referred to as Single-cell Protein (SCP), a term coined at a Symposium held in Massachusetts (1). The term embraces such micro-organisms as bacteria, fungi (including yeasts) and algae, and whilst the term is not strictly accurate, in all cases its meaning is now generally understood in the context of protein production. The so-called SCP are not new in themselves and their novelty lies in the projected large scale production and in their use as major contributors of protein in the diets of animals.

The production of algae appears to hold very little immediate promise in the context of feed protein supplies and it will not be considered in this paper. Much more imminent is the impact that will be made by the processes currently being developed for the production of yeast and bacteria. These microbes will grow on a wide variety of substrates and there seem to be very few organic substrates which will not support the growth of one type or other. Developments in this field have been rapid and in this paper some examples of SCP processes which appear to hold potential as suppliers of feed protein in the next decade or so are discussed.

EXAMPLES OF SCP PRODUCTION PROCESSES

Carbohydrate Substrates

Carbohydrates for microbial culture are available from a variety of sources. The oldest established process is that for the

production of baker's yeast (<u>Saccharomyces cerevisiae</u>) which is
grown on molasses with the addition of ammonium hydroxide or sulphate as the nitrogen source. Other commercial processes are the
production of food yeast (<u>Candida</u> (<u>Torula</u>) <u>utilis</u>) and <u>Saccharomyces
fragilis</u>. Food yeast has the advantage over baker's yeast in that
it is able to utilise pentose sugars and can utilise a wider range
of substrates. Waste sulphite liquor (from paper pulp manufacture)
is used in Russia and Czechoslovakia; molasses in Taiwan, South
Africa, Cuba and the Philippines. A few small units produce
<u>S. fragilis</u>, which can utilise lactose, using cheese whey as the
substrate.

Despite the apparent variety and availability of substrates,
SCP production from carbohydrates appears to be restricted to a few
thousand tonnes per annum (2). Carbohydrates are frequently present
in waste products from such activities as cheese and wood pulp production, brewing, distilling, and the industrial preparation of
vegetables and fruits. More stringent attitudes towards the disposal of such industrial and farm wastes have recently stimulated
developments in the utilisation of these materials as SCP substrates. Possibly the most advanced of the new processes in this
field is the Pekilo process which has been developed by a number of
companies at the Finnish Pulp and Paper Research Institute (3). In
this process sulphite spent-liquor, available in substantial
quantities from the Finnish wood-pulping industry, is the substrate.
The process involves the continuous culture of filamentous micro-
fungi of the classes Ascomycetes and Fungi Imperfecti. Mono and
polysaccharides, pentose sugars and acetic acid are removed from
the liquor. Control of river pollution is acknowledged as an
important factor in this development. The Pekilo process has an
important "innovation" compared with the conventional SCP processes
based on carbohydrates. The microbe used is filamentous and simple
filtration replaces centrifugation as the method of separating the
product from the spent culture medium; this reduces the cost of the
process. This type of process appears most likely to be applied in
countries with substantial wood pulping industries. The construction of a Pekilo unit of 10,000 t.p.a. capacity has recently
been announced in Finland, and there have been reports of similar
developments in Norway. The lack of any published estimate of the
world wide availability of substrates of this type makes an assessment of the potential of these processes impossible.

Hydrocarbon Substrates

<u>The Preferred Hydrocarbons</u>. Knowledge of hydrocarbon micro-
biology has improved substantially in the past decade and it is now
apparent that yeasts or bacteria could be found to utilise almost
any hydrocarbon. In practice those utilising methane (CH_4) and
C_{10} to C_{20} n-paraffins show the greatest potential for industrial

development. According to Evans (4) n-paraffins of C_2 to C_{10} are less interesting on account of their being less liable to microbial attack and also because substrate specificity is marked in this range; n-paraffins of C_{20} and higher are acceptable from a purely microbiological aspect but since they are solids at normal fermentation temperatures they cannot be used directly. Microbial growth on aromatic, alicyclic and branched chained hydrocarbons appears to be much less common than on the aliphatic n-paraffins.

The Supply of Hydrocarbons. Estimates of the supply and reserves of hydrocarbon substrates vary, but all indicate very large quantities of apparently suitable materials. Laine (5) has estimated that, on average, crude oil contains about 2% wt./wt. of the appropriate n-paraffins. With world production of crude oil exceeding 2,300 million tonnes in 1970 the current annual availability of n-paraffin substrate appears to be about 50 million tonnes. Assuming a conversion of n-paraffin to protein of 50 per cent (4) the potential production of protein from n-paraffins alone exceeds 20 million tonnes per annum. Moreover, with proved crude oil reserves at the end of 1970 exceeding 84,000 million tonnes the supply of n-paraffin substrates seems assured for several decades.

The supply of methane is more difficult to define. The methane content of natural gas varies from less than 50 per cent (Kapuri field, New Zealand) to more than 95 per cent (West Sole field, North Sea). Vast, unmeasured quantities of natural gas are currently wasted in locations where they could not be utilised economically in the past. Total utilisation of natural gas in Western Europe, North America, Japan and Australasia amounted to over 600 million tonnes of crude oil equivalent in 1970. World reserves of natural gas were estimated to be equivalent to about 28,000 million tonnes of crude oil in 1970. Assuming a conversion of methane to protein of 50 per cent and a methane content of 75 per cent in natural gas, the potential production of protein from methane appears to be many times greater than that from the liquid n-paraffins.

It is the enormous potential supply of hydrocarbon substrates which has attracted so much attention and which attaches special significance to the development of hydrocarbon-based processes.

a) Processes Using Methane. The principal attractions of methane as a substrate are that it is plentiful and in some locations comparatively inexpensive. All the methane utilising microbes identified so far have been bacteria (6) but, until recently, a barrier to the industrial development of methane-based SCP processes has been the poor productivity of methane cultures. However, there is evidence of recent progress in this respect; Sheehan and Johnson (7) reported a mixed, methane-utilising, bacterial culture of high productivity and the chemical composition of their mixture showed considerable improvement over that of the early methane utilisers. Wilkinson and co-workers in

Table 1

Some chemical characteristics of three samples
of methane-utilising bacteria compared with soya bean meal

Constituent	Soya bean meal	Methane-utilising bacteria		
		Ribbons (9)	Sheehan and Johnson (7)	D'Mello (8)
Crude Protein (N x 6.25) %	45	52	76	62
Selected amino acids g/100g dry matter				
Lysine	3.0	2.6	3.2	3.7
Methionine	0.7	0.8	2.3	1.5
Cystine	0.7	not stated	0.2	0.5
Arginine	3.5	2.6	3.8	3.5
Threonine	1.8	2.2	3.7	3.7
Tryptophane	0.7	not stated	2.0	3.8

the University of Edinburgh have isolated many methane-utilising
bacteria and the best look promising from a nutritional point of
view (8). Table 1 compares the early (9) and more recent methane-
utilising bacteria with soya bean meal. Those strains reported
recently contain substantially more lysine and "S" amino acids and
it seems likely that further improvements will be made.

The Shell Petroleum Company (10) have very recently announced
the construction of a pilot plant facility and the industrial
development of processes using methane seems likely during the
next decade.

b) Processes Using Liquid n-paraffins. A group from the
Societe Francaise des Petroles BP were the first to describe an
industrial process for cultivating a micro-organism on n-paraffins.
Since then the development by British Petroleum (BP) and its
associated French company, of two processes, both producing yeasts,
has been widely disclosed and the reports by Champagnat, Vernet,
Laine and Filosa (11) and Evans (4) describe the processes in
detail.

Fig. 1, taken from the paper by Evans, shows the principal

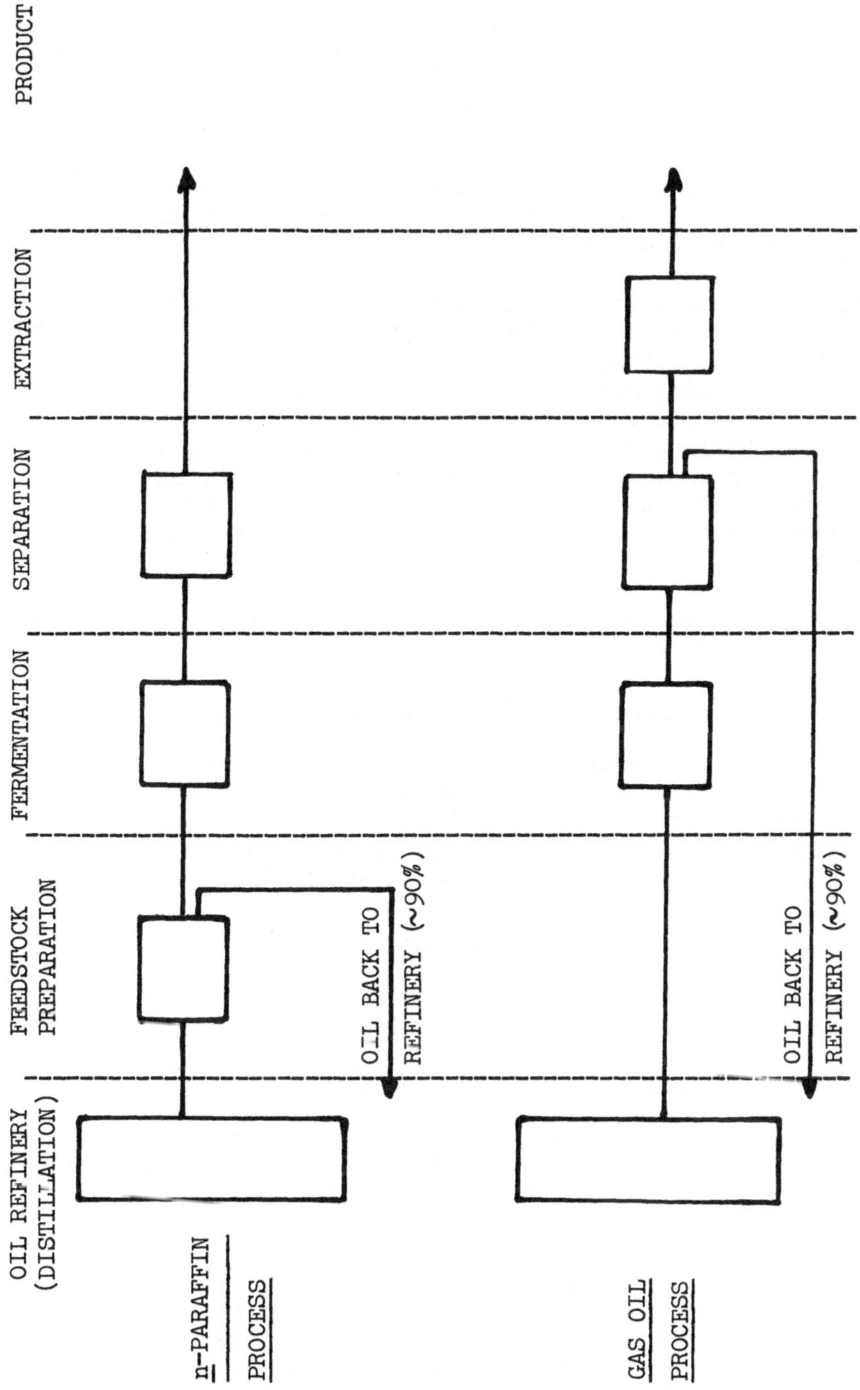

PRODUCT

EXTRACTION

SEPARATION

FERMENTATION

FEEDSTOCK
PREPARATION

OIL REFINERY
(DISTILLATION)

OIL BACK TO
REFINERY (~90%)

OIL BACK TO
REFINERY (~90%)

n-PARAFFIN
PROCESS

GAS OIL
PROCESS

Fig. 1.   COMPARATIVE PROCESS - FLOW DIAGRAMS

stages in the two processes. The first process uses a substrate of
high-purity n-paraffins and these are almost completely consumed
during fermentation. The second process uses standard heavy gas-
oil (i.e. C. 300-380°C TBP) from which the n-paraffins are preferentially consumed. In this case only about 10% of the substrate
is utilised and the remainder is recovered for re-use as a component of a range of petroleum products. The gas-oil process involves
a solvent extraction step which removes all traces of gas-oil
trapped in the yeast cells.

Production units for each of the two BP processes have been
built and are in operation. The earliest unit using pure n-paraffins
is at Grangemouth, Scotland, and has a capacity of 4,000 t.p.a.
The second unit using gas-oil as substrate is at Lavera, near
Marseilles, France, and that was built to produce 16,000 t.p.a. of
dried yeast. Currently BP and an Italian Company A.N.I.C. S.p.A.
are building a plant, using a n-paraffins, to produce 100,000 t.p.a.
of dried yeast, and this plant will be commissioned late in 1975.
Further investment in the two BP processes can be anticipated in
the near future.

Other companies known to have developed processes, all using
liquid n-paraffins, include:-

Kanegafuchi Chemical Industry Company Limited of Japan

This process has been licensed to Liquichimica Biosintesi
S.p.A. who are currently constructing a plant to produce 100,000
t.p.a. of dried yeast at Saline di Montebello, Italy. This plant
will also probably commission in 1975. Some features of this
process have been described in the publications 12,13, and 14.

Gulf Oil Company (15, 16) of the U.S.A.

This company have constructed a large pilot plant at Wasco,
California, but no plans for commercialisation have been announced.

Dainippon Ink and Chemical Company of Japan.

This company is involved in a venture in Rumania for the
construction of a plant to produce 60,000 t.p.a. of yeast. The
plant is scheduled for completion by the end of 1976.

Groupement Francaise des Proteines (17)

The Institute Française du Petrole (IFP) has been conducting
research on SCP for many years. Recently, in an association with
the state owned oil companies, Cie. Française de Raffinage and
Erap, known as the Groupement Française des Proteines, the development of an SCP process using pure n-paraffins has been disclosed.

A novel feature of this process appears to be that cultivation is at $40°C$, which confers certain advantages over the other processes in this category which appear to operate at $30-35°C$.

c) <u>Processes Using Hydrocarbon Derivatives</u>. In view of the early difficulties with methane fermentation it is not surprising that there have been reports of possible SCP processes using simple petro-chemicals produced from methane or other hydrocarbons. Acetic acid, methanol and ethanol have all been suggested as substrates and the process using methanol being developed by Imperial Chemical Industries (ICI) in this country is probably as advanced as any of this type. Details of the ICI process have been reported by Maclennan, Gow and Stringer (18) and by Stringer and Litchfield (19). After an extensive screening programme this group selected a bacterium, in fact a strain of <u>Pseudomonas.</u>

Other groups working with methanol include Kanegafuchi (yeast), Northern Illinois Gas Company (bacterium) and the Mitsubishi Gas and Chemical Company (yeast and bacterium).

Processes using ethanol have also been disclosed but these are believed to be inherently more expensive than those using either methanol or simple hydrocarbons and generally aimed at human food applications rather than animal feed.

DISTINCTIVE FEATURES OF THE SCP PROCESSES

The essential difference between hydrocarbon and carbohydrate substrates is that the latter furnish some of the elements essential to cell growth, carbon (C), hydrogen (H) and oxygen (O) in a soluble form, whereas a hydrocarbon supplies only carbon and hydrogen in a form which is virtually insoluble in water. Oxygen must then be supplied in greatly increased quantities, and in practice this is likely to be from air blown into the fermentor. Other nutrients essential to both substrates are the cations:-

$$NH_4^+, K^+, Mg^{2+}, Fe^{2+}, Zn^{2+} \text{ and the anions } SO_4^{2-} \text{ and } PO_4^{3-}.$$

Specific growth factors may also be needed to achieve maximum performance. The overall conversion of hydrocarbons to yeast cells may be expressed as follows (in kg. moles):-

$$2nCH_2 + 2nO_2 + 0.19n\ NH_4^+ + \text{other essential elements}$$

$$(P, K, S, \text{etc.}) \qquad\qquad n(C\ H_{1.7}\ O_{0.5}\ N_{0.19}Ash)$$

$$+ nCO_2 + 1.5n\ H_2O + 200,000n\ kcals.$$

This may be compared with a similar equation for a carbohydrate substrate using the same empirical formula for the cells:-

$$1.8nCH_2O + 0.8nO_2 + 0.19nNH_4^+ + \text{other essential elements}$$

(P, K, S, etc.) $n(C\ H_{1.7}\ O_{0.5}\ N_{0.19}\ \text{Ash}) + 0.8nCO_2$

$$+ 1.3n\ H_2O + 80,000n\ \text{kcals.}$$

These heat releases correspond to about 7.600 and 3,000 kcals. respectively per kg. cell dry weight.

Thus the use of n-paraffins instead of carbohydrates involves the supply of about 2.5 times as much oxygen and the removal of about 2.5 times as much heat. Also, because the two liquid reaction phases are virtually immiscible, sufficient agitation must be provided to dispense the smaller volume hydrocarbon phase thoroughly into the larger volume aqueous phase in order to ensure efficient hydrocarbon mass transfer. The transfer mechanism of the hydrocarbon to the cell has been the subject of much study and speculation. Evans[4] indicated, from a knowledge of the productivity of liquid hydrocarbon fermentation systems, that a combination of intermediate solution of the hydrocarbon in water, and of direct contact between the droplet of hydrocarbon and the cell was responsible. In practice, according to Bennett and Knights [20], oxygen transfer is more likely to be the first limiting step in productivity.

The use of methanol requires less oxygen than does growth on n-paraffins and the process is less exothermic; MacLennan et al. [18] produced the following equation to describe the ICI methanol process.

$$1.72nCH_3OH + 0.23n\ NH_4^+ + 1.51nO_2 + \text{other essential elements}$$

(P, K, S, etc.) $n(C\ H_{1.68}\ O_{0.36}\ N_{0.23}\ \text{Ash})$

$$+ 0.72nCO_2 + 2.94n\ H_2O + 185,000n\ \text{kcals.}$$

THE TOXICOLOGY OF SCP

When the toxicological programme for the BP yeasts was started in 1964 there was no precedent for it. Since that time the SCP Working Group of the FAO/WHO/UNICEF Protein Advisory Group (PAG) has published a Guideline for the toxicological testing of novel protein sources intended for human food [20] and, more recently, one for the testing of novel proteins for animal feeding [21].

The PAG Guidelines do not, of course, have the force of law. However, in the absence of specific legislation, it is anticipated that many Governments will adopt them as the basis of their individual requirements for demonstrating the safety of SCP.

The methods of evaluation now recommended are based on those developed for the testing of drugs and other additives. It has been necessary to adapt the classical toxicological approach to take account of the practical situation, which is that the SCP materials will be used as major components of animal feeds rather than as additives included at a level of a few - or even a few hundred - parts per million. Basically this involves the measurement of the effects of short-, medium- and long-term administration of graded doses of the product under examination to a variety of animals and the comparison of these effects with those produced by materials regarded as safe and acceptable.

The programme adopted for the BP yeasts has been described in detail by Shacklady (22) and Engel (23) and the various experiments are listed as an example of what will be required of each SCP as a demonstration of safety in use.

Initially six separate tests were applied to yeast grown on both gas-oil and n-paraffins:-

1) Acute toxicity tests of six weeks' duration using 40% yeast in diets of rats.

2) Sub-chronic toxicity tests of 90 days' duration using 10, 20 and 30% yeast in diets of rats.

3) Chronic toxicity tests of 2 years' duration using 10, 20 and 30% yeast in diets of rats.

4) Carcinogenicity test of 2 years' duration using 10, 20 and 30% yeast in diets of rats.

5) Carcinogenicity test of 18 months' duration using 10, 20 and 30% yeast in diets of mice.

6) Reproduction test up to the F3 generation of rats using 10, 20 and 30% yeast in the diets.

This series of tests was completed by 1970 and detailed results have been published in the scientific literature by Engel (23) and by de Groot, Til and Feron (24, 25 and 26). Multiple generation tests with rats and quail have continued and, so far, 15 successive generations of rats and more than 20 generations of quail have been produced on diets containing up to 30% of yeast with no evidence of deleterious effects. Specific tests for

Table 2

Some chemical components of SCP described in the literature

Constituent	Palmer & Smith (41) Series				Sheehan & Johnson (7) MB	Pekilo product FF	Conventional fodder yeast Y
	Y_2	Y_5	B_6	B_7			
Nitrogen % of dry matter	10.5	8.7	14.2	13.6	12.1	8.8 – 10.0	8.0
Nucleic acid nitrogen g per 100g total N	9.0	15.0	18.4	7.5	23.7	NK	NK
Total lysine g per 16 g N	7.6	7.5	5.9	6.0	4.3	6.5	6.8
g per 100g dry matter	5.0	4.1	5.2	5.1	3.2	3.8	3.4
Total 'sulphur' amino acids g per 16 g N	2.8	2.4	2.8	2.8	3.4	2.4	2.0
g per 100g dry matter	1.8	1.3	2.5	2.4	2.5	1.4	1.0

Y = Yeast
B = Bacteria
MB = Mixed bacterial culture
FF = Filamentous fungi
NK = Not known

teratogenicity and mutagenicity have also been completed, again
with no adverse effects from the yeasts.

The programme of toxicological examination of the two BP
yeasts was carried out at the Central Institute for Nutrition and
Food Research (C.I.V.O.) at Zeist in Holland.

THE NUTRITIVE VALUE OF SCP

Chemical Composition

There are substantial differences between the various SCP in
their chemical composition, and in Table 2 I have attempted to
demonstrate both the variety and the principal differences between
yeasts and bacteria by selecting examples from the literature and
from the products already mentioned in this paper. From the data
in the table it appears that nitrogen concentrations are higher in
bacteria than in yeasts. Non-protein nitrogen in the form of
nucleic acids also appears to be higher in bacteria, although
nucleic acid levels vary with culture conditions and comparisons of
cultures of unknown origin are not necessarily valid. The Palmer
and Smith collection was of SCP grown on hydrocarbons or hydrocarbon derivatives and the lysine levels in both the yeast and the
bacteria of that series are somewhat higher than those of either
the Pekilo product or the fodder yeast. The exceptions seem to be
the sample of Sheehan and Johnson and the other methane utilisers
shown in Table 1 which had lower lysine contents. Both yeasts and
bacteria are generally poor in the sulphur amino acids, with
bacteria having an advantage by virtue of their higher nitrogen
contents. Deficiencies of methionine are less serious than those
of lysine because chemically synthesised DL-methionine is widely
available and comparatively inexpensive.

Animal Feeding Experiments

There are now numerous publications describing feeding trials
in which SCP has replaced conventional proteins in the diets of
farm animals. The majority of the experiments have involved the
two BP yeasts and these are again taken as the example in this
section.
  <u>Breeding Animals</u>. In 1970 Shacklady (22) reported the early
results of long-term experiments in which BP yeast was given to
breeding pigs and poultry. These experiments have continued and
more recent information is shown in Table 3 (pigs) and Table 4
(poultry). The experiments are being continued and with detailed
systematic examination of the tissues and products they represent
an important additional demonstration of the safety of the yeasts.
  <u>Growing and Fattening Animals a)Pigs</u>. Clausen (27) and
Nielsen, Sriwaranard, Danielsen and Eggum (28) have published the
results of experiments in which the two BP yeasts were used as the

Table 3

The effect of feeding diets containing BP yeast on the
reproductive and subsequent performance of pigs
over three generations

	Control diet (Fishmeal + soya)	Experimental diet (10% yeast grown on gas-oil)
Number of litters	133	119
Mean number of live piglets per litter	10.2	10.2
Mean live piglet birth weight (g)	1315	1244
	Control creep feed (Fishmeal + soya)	Experimental creep feed (15% yeast grown on gas-oil)
Daily liveweight gain of piglets during suckling (g)	370	361

sole supplementary source of protein in the diets of early-weaned
piglets from 3 to 10 weeks of age. The performance of the pigs
receiving the diets containing yeast was indistinguishable from
that of the controls. These experiments included diets containing
up to 29% of yeast grown on pure n-paraffins and even at this high
level there appeared to be no palatability problems. The authors
concluded that the yeast could supply virtually all the supplementary protein in the diets of early weaned pigs. Henry, Pion
and Rerat (29) also considered BP yeast grown on gas-oil to be
suitable for piglets as a substitute for milk and fish proteins.

There are many published reports of experiments in which SCP
has successfully replaced soya bean meal and/or fishmeal in the
diets of growing pigs. In experiments with several yeasts and the
ICI bacterium Oslage and Schulz (30) found mixtures of barley and

Table 4

Egg production over 6 generations on a diet based
on fishmeal and soya and a diet containing 10% BP yeast

Generation	Control	10% yeast	
		Actual	% of control
P	59.9	61.7	103
F1	60.1	56.7	94
F2	57.5	59.8	104
F3	56.4	55.9	99
F4	57.3	59.6	104
F5	58.8	61.0	104
Mean	58.3	59.1	101

SCP gave the same results as barley plus fishmeal. Bergonzini and
Fabbri (31) replaced soya bean méal with BP yeast, grown on pure
n-paraffins, in semi-synthetic diets. There were no statistically-
significant differences in performance but the feed conversion was
better and the carcasses less fat when yeast was used. Russo,
Catalano, Mariani and Delmonte (32) reported satisfactory performance with yeast produced by the Liquichimica process. One of
the most interesting experiments so far published was that by
Barber, Braude, Mitchell and Myres (33): they replaced white fish-
meal with BP yeast, grown on n-paraffins, at two protein levels in
the diets of growing pigs. The results are summarised in Table 6.
At both protein levels the yeast marginally improved both the rate
and the economy of liveweight gain. Similar results with yeast
grown on gas-oil were reported by van der Wal, Shacklady and van
Weerden (34).

   b) Poultry. Stringer and Litchfield (19) reported good performance in broiler chickens given diets containing 5 or 10% of the
ICI bacterium. Waldroup and Payne (35) found 5 or 10% of another
bacterium grown on methanol to be satisfactory in pelleted diets
for broiler chickens, but higher levels gave poor results, and in
diets fed as mash even 5% of the bacterium depressed performance.

Table 5

A summary of an experiment with early-weaned pigs given
diets containing a mixture of conventional proteins, or
BP yeast grown on n-paraffin (from Nielsen et al (28))

Treatment	1	2	3	4	5	6
Cereals	1	1	1	0	0	0
Tapioca	0	0	0	1	1	1
Conventional protein	1	0.5	0	1	0.5	0
BP yeast	0	0.5	1	0	0.5	1
Mean daily liveweight gain (g)	359	348	326	358	376	360
Feed conversion kg. feed/ kg. gain	2.4	2.3	2.5	2.3	2.3	2.2

Note: There was a significant difference in feed
conversion between cereals and tapioca ($P < 0.01$).

Shannon and McNab (36) replaced herring meal and a mixture of
herring meal and soya bean meal with BP yeast, grown on n-paraffins,
in the diets of broiler chickens. These diets were also fed as
mash and the results are summarised in Table 7. More than 10%
yeast depressed performance on the starter feeds but up to 20% was
satisfactory in the finisher feeds. Similarly van Weerden,
Shacklady and van der Wal (37) detected some depression in chick
growth with 15% of BP yeast grown on gas-oil, again in diets fed
as mash.

Thus some reports indicate that there could be factors
associated with SCP which limit their inclusion in the diets of
young chickens. In the case of the two BP yeasts an examination of
the many experiments made with young chickens has indicated that
sub-optimal intakes of selenium and arginine could be implicated,
as could reduced feed intake on diets containing high levels of
yeast and fed as mash.

Table 6

Performance of pigs given diets supplemented with either white fish meal (WFM) or n-paraffin grown yeast (Y) at a 'standard' and a 'low' level of protein supplementation (from 33)

	'Standard' level		'Low' level		Significance of differences	
	WFM	Y	WFM	Y	WFM v Y	Standard v Low
Daily liveweight gain (g)	703	730	621	633	**	***
Feed consumed (kg/kg gain)	2.87	2.80	3.21	3.18	*	***

'Standard' Protein level = 15% up to 60 kg. liveweight 13% thereafter.

'Low' Protein level = 13% up to 60 kg. liveweight 12% thereafter.

10 pigs per treatment on experiment from 20 to 90 kg. liveweight.

*P 0.05; **P 0.01; ***P 0.001.

In recent experiments with either cereal-based or semi-synthetic diets, the yeasts have been used as the sole source of supplementary or of dietary protein. In pelleted diets containing adequate selenium, and equal concentrations of lysine, methionine plus cystine, and arginine, performance on yeast diets was indistinguishable from that with a mixture of soya-bean meal and fish-meal (38). Feed intake tended to be reduced in mash feeds containing yeast, but this appears to be of no practical significance because almost all commercially-produced feed for broiler chickens is pelleted.

Similarly, the two BP yeasts have been used as the sole supplementary protein source in the diets of laying hens (39). The complete replacement of a combination of soya bean meal, fish-meal and meat meal had no effect on egg numbers or egg weight.

No data were found in the literature describing the use of bacteria in the diets of laying hens.

c) Other Species. Though less intensive than the work already mentioned, the use of the two BP yeasts has been investigated in the diets of pre-ruminant calves and lambs, rabbits, dogs, cats, mink and various species of fish. Similar studies with

Table 7

The performance of broiler chickens given diets
containing BP yeast grown on pure n-paraffins

	% yeast			
	0	5	10	20
Liveweight gain (g)				
0-4 weeks	719	740	723	594
4-8 weeks	1043	1079	1043	1081
Feed conversion efficiency*				
0-4 weeks	0.64	0.64	0.62	0.55
4-8 weeks	0.37	0.37	0.39	0.41

*g liveweight gain per g feed consumed.

other SCP are known to be in progress in laboratories throughout
the world.

CONCLUSIONS

In this paper I have attempted to indicate the very considerable efforts being made to develop processes for the production of
single cell proteins which are both nutritious and safe in use.

Known current world production of yeasts from carbohydrate
substrates is probably about 330,000 tonnes per annum, although
probably more than half is fed directly to human beings (40). The
development of new processes such as the Finnish Pekilo process
may make this route more attractive and public concern about the
methods of disposal of industrial wastes containing carbohydrates
may also influence investment in such processes.

The use of hydrocarbon or petro-chemical substrates is
currently on a much more limited scale. The only two facilities
known to be operational use liquid n-paraffins and have a combined
production capacity of 20,000 tonnes per annum. Additional manufacturing units with a combined capacity of 200,000 tonnes per
annum are due for completion in Italy within one year and a unit
of 60,000 tonnes per annum is apparently under construction in
Rumania. Considerable extension of such facilities is generally
anticipated and it seems likely that processes using methane and
methanol will be seriously considered within the decade which I
originally set as the time scale for this paper. Thus capital
expenditure sufficient to produce an additional 260,000 tonnes
per annum is already committed. The ultimate extent of investment
in SCP production is quite impossible to predict. The supply of
hydrocarbon substrates does not appear to be a limiting factor.
It is certain, however, that investment on a hitherto unprecedented
scale will be necessary if SCP is to become significant source of
feed protein.

Those products which have been subjected to the most prolonged
toxicological testing, namely the two BP yeasts, appear to be per-
fectly safe and also to be useful components of animal feeds. It
seems likely that given time other products will similarly prove
themselves. Exploitation will ultimately depend on whether investment in these processes will yield a satisfactory return. Few
people would deny that additional feed protein will be required
and it seems probable that substantial exploitation of several
routes to SCP will eventually occur.

ACKNOWLEDGEMENT

Permission to publish this paper has been given by The
British Petroleum Company Limited.

REFERENCES

1. Mateles, R. I. and Tannenbaum, S. R. (1968). "Single-cell
    Protein". (M.I.T. Press, Cambridge, U.S.A.).

2. Heydeman, M. T. (1973). In: "The Biological Efficiency of
    Protein Production". (Ed. J. G. W. Jones) (University Press,
    Cambridge).

3. Forss, K. (1972). Symposium on new developments in the pro-
    vision of amino acids in the diets of pigs and poultry, Geneva.

4. Evans, G. H. (1969). 4th Petroleum Symposium. (U.N. Economic
    Commission for Asia and the Far East).

5. Laine, B. M. (1973). In: Proc. UNIDO Conference, Vienna.

6. Whittenbury, R., Phillips, K. C. and Wilkinson, J. F. (1970).
    J. Gen. Microbiol., $\underline{61}$, 205.

7. Sheehan, B. T. and Johnson, M. J. (1971). Appl. Microbiol.,
    $\underline{21}$, 511.

8. D'Mello, J. P. F. (1973). Br. Poult. Sci., $\underline{14}$, 291.

9. Ribbons, D. W. (1968). In: Proc. of Symposium on Microbiology.
    (Institute of Petroleum, London).

10. Shell Petroleum Company (1974). Press Release, November 1974.

11. Champagnat, A., Vernet, C., Laine, B. M. and Filsoa, J.
    (1963). Nature, $\underline{187}$, 13.

12. Kanegafuchi Chemical Industry Company Limited. Kanepron
    Product Bulletin.

13. Giacobbe, F. (1973). Notizario, $\underline{14}$, 41.

14. Giacobbe, F. (1973). In: Proc. UNIDO Conference, Vienna.

15. Silver, R. S. and Cooper, P. G. (1972). In: Proc. 72nd
    National Meeting, Am. Inst. Chem. Eng., St. Louis, Missouri.

16. Cooper, P. G., Silver, R. S. and Boyle, J. P. (1973). In:
    "International Conference on SCP". (M.I.T., Cambridge, Mass.).

17. European Chemical News (1973), $\underline{21}$, 8.

18. MacLennan, D. G., Gow, J. S. and Stringer, D. A. (1973).
    Proc. Biochem., $\underline{8}$ (6), 22.

19. Stringer, D. A. and Litchfield, M. H. (1973). FEBS Meeting, Dublin.

20. PAG Guideline No. 6. March 1970, Revised June 1972.

21. PAG Guideline No. 15. September 1974.

22. Shacklady, C. A. (1970). In: Proc. 3rd International Congress of Food Science and Technology, Washington, D.C.

23. Engel, C. (1972). In: "Proteins from Hydrocarbons". (Ed. H. Gounelle), p. 53. (Academic Press, London and New York).

24. de Groot, A. P., Til, H. P. and Feron, V. J. (1970). Fd Cosmet. Toxicol., 8, 267.

25. de Groot, A. P., Til, H. P. and Feron, V. J. (1970). Fd Cosmet. Toxicol., 8, 499.

26. de Groot, A. P., Til, H. P. and Feron, V. J. (1971). Fd Cosmet. Toxicol., 9, 787.

27. Clausen, H. (1971). Danish Agricultural Institute Year Book (Copenhagen).

28. Nielsen, H. E., Sriwaranard, P., Danielsen, V. and Eggum, B. O. (1973). Z. Tierphysiol., Tierernähr. u. Futtermittelk., 33, 151.

29. Henry, Y., Pion, R. and Rérat, A. (1974). In: Proc. EAAP, 25th Annual Meeting, Copenhagen.

30. Oslage, H. J. and Schulz, E. (1974). ibid.

31. Bergonzini, E. and Fabbri, R. (1974). ibid.

32. Russo, V., Catalano, A., Mariani, P. and Delmonte, P. (1974). ibid.

33. Barber, R. S., Braude, R., Mitchell, K. G. and Myres, A. W. (1971). Br. J. Nutr., 25 (2), 285.

34. van der Wal, P., Shacklady, C. A. and van Weerden, E. J. (1969). In: Proc. 8th Int. Cong. Nutr., p. V-3. (Prague).

35. Waldroup, P. W. and Payne, J. R. (1974). Br. Poult. Sci., 53, 1039.

36. Shannon, D. W. F. and McNab, J. M. (1972). Br. Poult. Sci., 13, 267.

37. van Weerden, E. J., Shacklady, C. A. and van der Wal, P.
     (1970). Br. Poult. Sci., $\underline{11}$, 189.

38. Shacklady, C. A. and Walker, T. (1974). Unpublished observa-
     tions.

39. Shacklady, C. A. (1974). In: Proc. Venezuelan Congress of
     Microbiology, Caracas.

40. Engel, C. (1972). "Proteins from Hydrocarbons". (Ed.
     H. Gounelle), p. 53. (Academic Press, London and New York).

41. Palmer, R. and Smith, R. H. (1971). Proc. Nutr. Soc. $\underline{30}$, 60A.

DISCUSSION

In reply to questions, *Dr. Walker* said that the nitrogen
source used for fermentation was the cheapest available, which was
generally ammonium sulphate. The n-paraffins were normally available in larger quantities than were required for other uses.
Applying to a question by Dr. Widdowson on the possible contamin-
ation of the product, *Dr. Walker* replied that the known toxic
compounds of crude oil were eliminated in the preparation of the
feedstuff. Replying to Dr. van Es, *Dr. Walker* said that the
harvesting costs of bacteria and yeast were not greatly different.
Prof. Ingram asked whether Dr. Walker's optimistic view for the
prospects for hydrocarbon grown protein took account of the rising
costs of oil. *Dr. Walker* replied that the economics of a
particular investment are specific to that investment; economic
viability depends, not only on commodity prices, but on related
political decisions. *Dr. Fuller* asked whether lignified residues
available in the tropics, such as bagasse and forestry by-products,
could be used as substrates for industrial fermentations.
Dr. Walker thought that this was possible. He did not know the
details of any such process, but foresaw that separation of the
microorganism from the residual substrate might present difficulties. *Dr. Turner* commented that there were not only negative,
especially toxicological considerations in the use of microbial
protein, but also possible positive qualities, such as hypochol-
esterolaemic factors. *Dr. Rhodes* referred to his comparison of
the organoleptic properties of meat from pigs given diets containing single-cell protein or fish meal. The only adverse
comments related to the pigs given fish meal.

Final Perspectives

FINAL PERSPECTIVES

Summary of the Conclusions of the Final Joint Working Session

The need to make more effective use of resources may require that we modify our view of what constitutes an acceptable carcass.

The proposition, derived mainly from work with small animals, that nutrition in utero and in early life affects subsequent appetite, growth and mature size, if applicable to large species, has important implications for animal production and should be investigated.

More information is needed on the time course of absorption of food and the uptake by tissues of the products of digestion and their interactions.

At present, growing animals typically retain only about 30% of their dietary protein. A greater understanding of the mechanisms controlling amino acid incorporation is necessary if this efficiency is to be improved.

There is dissatisfaction with systems of feed evaluation which take no account of the contribution of protein synthesis to the energy cost of growth. It appears that such a refinement can be applied in the case of information collected for the pig, but not for ruminants.

Better methods are needed for the assessment of the nutritive value of high-fibre forages and by-products both alone and in mixed diets. The amounts that can be consumed need to be known if the greater utilisation of such materials is to be achieved.

There are examples of changes occurring in the roles of hormones as animals grow. We need to know more about the changes in neuro-humoral function which appear to be involved in growth and development.

For genetic selection, better models are needed of the inter-relationships between appetite and the utilisation of nutrients for protein and fat production. These models would be simplified if it could be shown that, for practical purposes, the unitary energy costs of maintenance and of fat and protein accretion could be regarded as constant.

Immediate improvements in the efficiency of meat production by pigs are likely to be achieved by the increased use of boars for meat, by early weaning and by the elimination of specific pathogens.

In ruminants, the major emphasis should be on improving the reproductive rate, especially by multiple pregnancy.

Genetic differences in the rate of fat deposition may be related to differences in the efficiency of food conversion, in sensitivity to stress and, hence, the commercial quality of meat.

A minimum level of fatness may be necessary for the initiation of reproduction; more information on such a necessity should be sought.

Alternative non-genetic means should be sought of improving the efficiency of food utilisation for lean tissue production by reducing variable fat deposition and, perhaps, protein turnover.

It is necessary to develop a capability to adapt systems of animal production so that they can accommodate acute and chronic variations in the production of forage and feed grains associated with changes of weather and climate.

We have a clear picture of the factors affecting the efficiency of the various components of meat production but more work is needed on their integration into complete models which identify the critical resources and their optimal use.

LIST OF PARTICIPANTS

Alexander, Prof. R. McN.
    Department of Pure and Applied Zoology, University of Leeds

Armstrong, Prof. D. G.
    Department of Agricultural Biochemistry, University of
    Newcastle upon Tyne

Bichard, Dr. M.
    Pig Improvement Company, Fyfield Wick, near Abingdon,
    Berkshire

Braude, Dr. R.
    National Institute for Research in Dairying, Shinfield,
    Reading

Burch, Prof. P. R. J.
    Department of Medical Physics, University of Leeds

Burleigh, Dr. I. G.
    Meat Research Institute, Langford, Bristol

Buttery, Dr. P. J.
    School of Agriculture, University of Nottingham

Cahill, Prof. G. F., Jr.
    Joslin Research Laboratory, Department of Medicine, Harvard
    Medical School

Dickerson, Dr. G. E.
    U.S. Meat Animal Research Center, Agricultural Research
    Service, U.S. Department of Agriculture, Nebraska

Duckworth, Dr. J. E.
    Meat and Livestock Commission, Bletchley

Elsley, Prof. F. W. H.
    Department of Agriculture, University of Edinburgh

Enser, Dr. M. B.
    Meat Research Institute, Langford, Bristol

Es, Dr. A. J. H. van
    Institute for Animal Feeding and Nutrition Research, Hoorn,
    and Department of Animal Physiology, Wageningen

Fowler, Dr. V. R.
    Rowett Research Institute, Aberdeen

Frisch, Dr. R. E.
    Centre for Population Studies, Harvard University

Fuller, Dr. M. F.
    Rowett Research Institute, Aberdeen

Garton, Dr. G. A.
    Rowett Research Institute, Aberdeen

Goss, Prof. R. J.
    Division of Biological and Medical Sciences, Brown University,
    Providence, Rhode Island

Hardwick, Dr. D. C.
    Ministry of Agriculture, Fisheries and Food, Great Westminster
    House, Horseferry Road, London SW1

Ingram, Prof. M.
    Meat Research Institute, Langford, Bristol

Kielanowski, Prof. J.
    Institute of Animal Nutrition, Jabłonna, Warsaw

King, Dr. J. W. B.
    Animal Breeding Research Organization, Edinburgh

Lamming, Prof. G. E.
    School of Agriculture, University of Nottingham

Large, R. V.
    Grassland Research Institute, Hurley, Maidenhead

Leat, Dr. W. M. F.
    Institute of Animal Physiology, Babraham, Cambridge

Lewis, Prof. D.
    School of Agriculture, University of Nottingham

Lister, Dr. D.
    Meat Research Institute, Langford, Bristol

Lucas, Prof. I. A. M.
    Department of Agriculture, University College of North Wales

Mason, I. L.
    Animal Production and Health Division, Food and Agriculture
    Organization, Rome

Monteiro, Dr. L. S.
    Animal Breeding Research Organization, Edinburgh

Norris, Prof. J. R.
    Meat Research Institute, Langford, Bristol

Perry, Dr. B. N.
    Meat Research Institute, Langford, Bristol

Pomeroy, Dr. R. W.
    Meat Research Institute, Langford, Bristol

Rerat, Dr. A.
    Laboratoire de Physiologie de la Nutrition, C.N.R.Z., Jouy-
    en-Josas

Rhodes, Dr. D. N.
    Meat Research Institute, Langford, Bristol

Robertson, Prof. A.
    Institute of Genetics, Edinburgh

Rook, Prof. J. A. F.
    Hannah Research Institute, Ayr

Saio, Dr. Kyoko
    National Food Research Institute, Ministry of Agriculture
    and Forestry, Tokyo

Spedding, Prof. C. R. W.
    Department of Agriculture, University of Reading

Steinhauf, Prof. D.
    Institute for Animal Production, Technical University of
    Berlin

Turner, Dr. M. R.
    Department of Physiology and Biochemistry, University of
    Southampton

Walker, Dr. T.
    B.P. Proteins Ltd., London

Watanabe, Prof. T.
   National Food Research Institute, Ministry of Agriculture
   and Forestry, Tokyo

Webster, Dr. A. J. F.
   Rowett Research Institute, Aberdeen

Widdowson, Dr. Elsie M.
   Department of Investigative Medicine, University of Cambridge

Wilson, Dr. P. N.
   B.O.C.M.-Silcock, Basingstoke

Wood, Dr. J. D.
   Meat Research Institute, Langford, Bristol